FOURIER METHODS
for Mathematicians, Scientists and Engineers

MATHEMATICS AND ITS APPLICATIONS
Series Editor: G. M. BELL, Professor of Mathematics,
King's College London, University of London

STATISTICS, OPERATIONAL RESEARCH AND COMPUTATIONAL MATHEMATICS
Editor: B. W. CONOLLY, Emeritus Professor of Mathematics (Operational Research), Queen Mary College, University of London

Mathematics and its applications are now awe-inspiring in their scope, variety and depth. Not only is there rapid growth in pure mathematics and its applications to the traditional fields of the physical sciences, engineering and statistics, but new fields of application are emerging in biology, ecology and social organization. The user of mathematics must assimilate subtle new techniques and also learn to handle the great power of the computer efficiently and economically.

The need for clear, concise and authoritative texts is thus greater than ever and our series endeavours to supply this need. It aims to be comprehensive and yet flexible. Works surveying recent research will introduce new areas and up-to-date mathematical methods. Undergraduate texts on established topics will stimulate student interest by including applications relevant at the present day. The series will also include selected volumes of lecture notes which will enable certain important topics to be presented earlier than would otherwise be possible.

In all these ways it is hoped to render a valuable service to those who learn, teach, develop and use mathematics.

Mathematics and its Applications
Series Editor: G. M. BELL, Professor of Mathematics, King's College London, University of London

Anderson, I.	Combinatorial Designs: Construction Methods
Artmann, B.	Concept of Number: From Quaternions to Monads and Topological Fields
Arczewski, K. & Pietrucha, J.	Mathematical Modelling in Discrete Mechanical Systems
Arczewski, K. and Pietrucha, J.	Mathematical Modelling in Continuous Mechanical Systems
Bainov, D.D. & Konstantinov, M.	The Averaging Method and its Applications
Baker, A.C. & Porteous, H.L.	Linear Algebra and Differential Equations
Balcerzyk, S. & Jösefiak, T.	Commutative Rings
Balcerzyk, S. & Jösefiak, T.	Commutative Noetherian and Krull Rings
Baldock, G.R. & Bridgeman, T.	Mathematical Theory of Wave Motion
Ball, M.A.	Mathematics in the Social and Life Sciences: Theories, Models and Methods
de Barra, G.	Measure Theory and Integration
Bartak, J., Herrmann, L., Lovicar, V. & Vejvoda, D.	Partial Differential Equations of Evolution
Bell, G.M. and Lavis, D.A.	Statistical Mechanics of Lattice Models, Vols. 1 & 2
Berry, J.S., Burghes, D.N., Huntley, I.D., James, D.J.G. & Moscardini, A.O.	Mathematical Modelling Courses
Berry, J.S., Burghes, D.N., Huntley, I.D., James, D.J.G. & Moscardini, A.O.	Mathematical Modelling Methodology, Models and Micros
Berry, J.S., Burghes, D.N., Huntley, I.D., James, D.J.G. & Moscardini, A.O.	Teaching and Applying Mathematical Modelling
Blum, W.	Applications and Modelling in Learning and Teaching Mathematics
Brown, R.	Topology: A Geometric Account of General Topology, Homotopy Types and the Fundamental Groupoid
Burghes, D.N. & Borrie, M.	Modelling with Differential Equations
Burghes, D.N. & Downs, A.M.	Modern Introduction to Classical Mechanics and Control
Burghes, D.N. & Graham, A.	Introduction to Control Theory, including Optimal Control
Burghes, D.N., Huntley, I. & McDonald, J.	Applying Mathematics
Burghes, D.N. & Wood, A.D.	Mathematical Models in the Social, Management and Life Sciences
Butkovskiy, A.G.	Green's Functions and Transfer Functions Handbook
Cartwright, M.	Fourier Methods: for Mathematicians, Scientists and Engineers
Cerny, I.	Complex Domain Analysis
Chorlton, F.	Textbook of Dynamics, 2nd Edition
Chorlton, F.	Vector and Tensor Methods
Cohen, D.E.	Computability and Logic
Cordier, J.-M. & Porter, T.	Shape Theory: Categorical Methods of Approximation
Crapper, G.D.	Introduction to Water Waves
Cross, M. & Moscardini, A.O.	Learning the Art of Mathematical Modelling
Cullen, M.R.	Linear Models in Biology
Dunning-Davies, J.	Mathematical Methods for Mathematicians, Physical Scientists and Engineers
Eason, G., Coles, C.W. & Gettinby, G.	Mathematics and Statistics for the Biosciences
El Jai, A. & Pritchard, A.J.	Sensors and Controls in the Analysis of Distributed Systems
Exton, H.	Multiple Hypergeometric Functions and Applications
Exton, H.	Handbook of Hypergeometric Integrals
Exton, H.	q-Hypergeometric Functions and Applications
Faux, I.D. & Pratt, M.J.	Computational Geometry for Design and Manufacture
Firby, P.A. & Gardiner, C.F.	Surface Topology
Gardiner, C.F.	Modern Algebra

Series continued at back of book

FOURIER METHODS
for Mathematicians, Scientists and Engineers

MARK CARTWRIGHT M.A., D.Phil.
Department of Mathematics
University of Nottingham

ELLIS HORWOOD
NEW YORK LONDON TORONTO SYDNEY TOKYO SINGAPORE

First published in 1990 by
ELLIS HORWOOD LIMITED
Market Cross House, Cooper Street,
Chichester, West Sussex, PO19 1EB, England
A division of
Simon & Schuster International Group

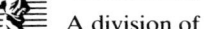

© Ellis Horwood Limited, 1990

All rights reserved. No part of this publication may be
reproduced, stored in a retrieval system, or transmitted,
in any form, or by any means, electronic, mechanical,
photocopying, recording or otherwise, without the prior
permission in writing, from the publisher

Typeset in Times by Ellis Horwood
Printed and bound in Great Britain
by Hartnolls, Bodmin

British Library Cataloguing in Publication Data

Cartwright, Mark
Fourier methods.
1. Fourier analysis
I. Title
515.2433
ISBN 0-13-327016-5 (Library Edn.)
ISBN 0-13-327008-4 (Student Pbk. Edn.)

Library of Congress Cataloging-in-Publication Data

Cartwright, Mark, 1961–
Fourier methods for mathematicians, scientists and engineers/
Mark Cartwright
p. cm. — (Mathematics and its applications)
ISBN 0-13-327016-5 (Library Edn.)
ISBN 0-13-327008-4 (Student Pbk. Edn.)
1. Fourier analysis. I. Title. II. Series: Mathematics and its
applications (Chichester, England: 1988)
QA403.5.C37 1990
515'.2433—dc20   89-71633
                                                      CIP

Table of contents

Preface 11
Table of symbols used in the text 13

PART I — FOURIER THEORY

1 HARMONIC MOTION 19
    1.1   The harmonic equation 19
    1.2   The extension oscillator 20
    1.3   Harmonic approximation 22
    1.4   A two-mode system 24
    1.5   Linear systems 27
    1.6   Damping 28
    1.7   Forced oscillation 32
    1.8   The vibrating string 35
    Exercises 41

2 FOURIER SERIES 44
    A. Using Fourier series
    2.1   Calculating Fourier coefficients 44
    2.2   Guitar strings 46
    2.3   Periodic functions 49
    2.4   Amplitude and phase 52
    B. Some useful functions
    2.5   Odd and even functions 53
    2.6   The triangle wave 53
    2.7   The square wave 54
    2.8   Impulsive 'functions' 55
    C. Properties of Fourier series
    2.9   Arithmetic properties 59

	2.10	Differentiation and integration 61
	2.11	Parseval's equation 63
	2.12	Decay of coefficients at large n 64
D.	**Theory of Fourier series**	
	2.13	Formulation using complex numbers 65
	2.14	Approximation of functions 67
	2.15	Discontinuities and the Gibbs phenomenon. 68
	2.16	Convergence of Fourier series 70
	Exercises 73

3 FOURIER TRANSFORMS 77

A. Definition of Fourier transforms

3.1 The Fourier transform operator 77
3.2 Inversion 78
3.3 Spectra 79
3.4 Power spectral density functions 80

B. Some useful functions

3.5 The bell curve 80
3.6 Step functions 81
3.7 Limiting arguments and the function $\sin x/x$ 82
3.8 More on impulsive 'functions' 85
3.9 Periodic functions 85

C. Properties of Fourier transforms

3.10 Arithmetic properties 89
3.11 Differentiation and integration 91
3.12 Other properties 93

D. Using Fourier transforms

3.13 Heat conduction 94
3.14 Interpreting spectral peaks 96

E. Fourier transform theory

3.15 Angular frequency versus frequency 98
3.16 Applicability of the inversion theorem 99
3.17 Fourier transforms and Fourier series 99
Exercises 103

4 TWO AND MORE DIMENSIONS 106

A. Formulation and properties

4.1 Fourier transforms 106
4.2 Fourier series 109
4.3 Vector formulation 113
4.4 Properties of vector transforms 115

B. Some useful functions

4.5 The n-dimensional bell curve 117
4.6 Impulsive 'functions' in n dimensions 118
4.7 Periodic functions 122
4.8 Lattices 123
Exercises 126

5 CONVOLUTION 128
A. Definition and properties
5.1 Echoes 128
5.2 Convolution 129
5.3 Convolution and Fourier transformation 131
B. Uses of convolution
5.4 Smoothing 133
5.5 Truncation 135
5.6 Amplitude modulation 137
5.7 Frequency modulation 138
C. Convolution and Fourier theory
5.8 The power theorem 140
5.9 The uncertainty theorem 141
5.10 The transfer function 144
5.11 Deconvolution 145
D. Functions of several variables
5.12 Vector convolution 145
5.13 Frosted glass 146
Exercises 148

6 FILTERING 150
A. Noise
6.1 Types of noise 150
6.2 Signal-to-noise ratio 152
6.3 Signal averaging in the x-domain 153
6.4 Signal averaging in the ω-domain 155
B. Filtering techniques
6.5 Filtering for convenience 157
6.6 Mask filters 158
6.7 Amplitude demodulation 162
6.8 Analogue filters 163
Exercises 166

7 CORRELATION 168
7.1 The correlation coefficient 168
7.2 Correlation of two functions 169
7.3 The cross-correlation function 171
7.4 The autocorrelation function 175
7.5 Correlation and Fourier transformation 177
7.6 Coherence 179
Exercises 180

8 OTHER SERIES AND TRANSFORMS 182
A. Series
8.1 Orthogonal functions 182
8.2 Bessel functions 184
8.3 Fourier–Bessel series 186
8.4 Circularly symmetric waves 187

B. Transforms
- 8.5 The Laplace transform 191
- 8.6 The Mellin transform 193
- 8.7 Self-inversion and the Hankel transforms 194
- 8.8 The Radon transform 197
- Exercises 198

9 DISCRETE FOURIER TRANSFORMS 201
A. Computing Fourier transforms
- 9.1 The discrete Fourier transform 201
- 9.2 Properties of the DFT 202
- 9.3 Sampling: resolution and aliasing 205
- 9.4 Transform of a sinusoid 206
- 9.5 Fourier transform algorithms 208
B. Using discrete transforms
- 9.6 Discrete convolutions 210
- 9.7 Windowing 212
- 9.8 Coherence of sampled functions 214
- 9.9 Digital filters 214
- 9.10 Phase response of filters 218
- Exercises 219

PART II — FOURIER APPLICATIONS

10 MATHEMATICS 223
- 10.1 Sums of series 223
- 10.2 The isoperimetric problem 225
- 10.3 Differential equations 229
- 10.4 Statistics 231
- 10.5 The prime number theorem 235

11 PHYSICS 247
- 11.1 Quantum theory 247
- 11.2 Lasers 250
- 11.3 Scattering 252
- 11.4 Diffraction 256
- 11.5 Telescopes 259
- 11.6 Interferometry 261
- 11.7 Impedance 264

12 CHEMISTRY 270
- 12.1 Mass spectrometry 270
- 12.2 NMR spectroscopy 273
- 12.3 Infra-red spectroscopy 275
- 12.4 Visible light spectroscopy 276
- 12.5 Crystallography 277

13 LIFE SCIENCE 280
- 13.1 Sight 280

13.2	Hearing	282
13.3	Speech analysis	286
13.4	Morphogenesis	289
13.5	Medical diagnostics	291

14 MISCELLANEOUS 293
- 14.1 Water waves 293
- 14.2 Turbulence in fluids 297
- 14.3 Meteorology 299
- 14.4 Glacier beds and 'roughness' 301
- 14.5 Seismology 305
- 14.6 Vibration analysis 307
- 14.7 Economics 308

Solutions to selected exercises 312

Bibliography 317

Index 321

*To Leslie,
with all my love*

Preface

Few areas of mathematics have found a range of applications so widespread and so pervasive as Fourier analysis. But most teaching of Fourier techniques, for obvious reasons, concentrates on their application to a particular discipline — chemical spectroscopy, say, or electric circuit analysis — and ignores their wider usage. This is a shame, for it is instructive to see the theory in action elsewhere. It is, therefore, the aim of this book to provide an introduction to Fourier methods which is as comprehensive as possible, as readable as possible, and suitable for as wide a variety of people as possible.

Most of the book (Part I) is 'common ground', taking the reader through all the basic devices and stratagems of Fourier theory. This needs little comment. The student should be familiar with complex numbers and simple differential equations; other than that, there are no stringent requirements. There is, inevitably, much missing that a more specialized reader might wish to have: theory of integration for mathematicians, properties of materials for engineers, digital filter design for software writers, and so on. For such people, I have compiled what I hope is a useful Bibliography.

The remainder (Part II) is a collection of 'anecdotes' and these need some justification. In them, I have tried to collect a wide variety of real applications of Fourier methods. Each is (perforce) very sketchy. These anecdotes, too, require on occasion somewhat more background than Part I. I have tried to explain the technicalities as clearly as possible, and I hope that these chapters will be read by a good proportion of readers.

For Part II, I have had to scour a range of subjects which are not my own. In doing so I may have made errors of interpretation (or even of fact). Of course I regret this possibility; if any kind specialist spots such an error in the text, I would appreciate it greatly if he would let me know. More generally, the topics I have chosen to discuss are a personal selection; if the kind specialist notices what he regards as a serious *omission* from the list of applications, please let him tell me. I will endeavour to understand his story; and (who knows?), if there is ever a new edition of this book, its contents will be that much more representative of Fourier methods.

In order to get as far as I have with the applications, I have had to strain the friendship of a number of people, including Handel Davies, Campbell Gemmell, Leslie Knoop, Pam Nickell, John Parker, Peter Ratoff, Andrew Turberfield, and Richard Wayne (who started the whole thing off). To all these: many thanks for your patience and willingness to explain.

Table of symbols used in the text

Symbol	Meaning	Page
Constants and parameters		
A_n	Fourier series: cosine coefficient	44
B_n	Fourier series: sine coefficient	44
c	wave equation: wave speed	38
C_n	Fourier series: amplitude	52
K	Fourier series: constant term	52
Q	damped harmonic motion: selectivity	31
Z	impedance	265
Z_k	characteristic impedance	267
Z_n	Fourier series: complex coefficient	65
σ	signal-to-noise ratio	152
φ_n	Fourier series: phase coefficient	52
ω	angular frequency	19
$\boldsymbol{\omega}$	vector angular frequency	113
ω_0	natural angular frequency	22
Functions		
$\text{cov}(X,Y)$	covariance	168
$D_N(x)$	Dirichlet kernel	68
$F(\|\mathbf{q}\|^2)$	form factor of a nucleus	254
$F_{a,c}(x)$	confluent hypergeometric function	253
$f_N(x)$	Fourier approximation	67
$H(x)$	Heaviside function	84
$H_n(x)$	Hermite polynomial	184
$J_0(x)$	Bessel function of order 0	122
$J_v(x)$	Bessel function of order v	184
$N(x)$	bell curve	80
$N_n(\mathbf{x})$	n-dimensional bell curve	117
$p_n(x)$	Legendre polynomial	182

Table of symbols used in the text

Symbol	Meaning	Page
$r(X,Y)$	correlation coefficient	169
$\text{Saw}(x)$	sawtooth function	73
$\text{sinc}(x)$	$\sin\pi x/\pi x$	84
$\text{Sq}(x)$	square wave	54
$\text{Tr}(x)$	triangle wave	53
$\text{var}(X)$	variance	168
$\delta(x)$	(Dirac) delta 'function'	55
$\delta_D(x)$	discrete delta 'function'	316
$\delta_L(\mathbf{x})$	lattice function of L	123
$\delta_n(\mathbf{x})$	n-dimensional delta function	118
$\delta_r(x,y)$	$(1/2\pi)\delta(\sqrt{(x^2+y^2)}-1)$	120
$\delta_x(x,y)$	$\delta(x)1(y)$	120
$\zeta(s)$	Riemann zeta function	238
$\Lambda(x)$	lambda (triangle) function	105
$\Pi(x)$	pi (rectangle) function	105
$\pi(n)$	prime counting function	235
$\sigma(\omega)$	signal-to-noise ratio	153
$\tau(n)$	prime density function	235
$\chi_I(x)$	characteristic function of interval I	81
$\text{III}(x)$	sampling function	57
$\text{III}_n(\mathbf{x})$	n-dimensional sampling function	120
$\text{III}_x(x,y)$	$\text{III}(x)1(y)$	120
$1(x)$	unity function	53
$1_n(\mathbf{x})$	n-dimensional unity function	119
$\mathbf{a\cdot b}$	dot (scalar) product	113
$\|\mathbf{a}\|$	norm (length)	114
A^*	$(\overline{A}')^{-1}$	116

Operators

Symbol	Meaning	Page
A	amplitude spectrum	80
C	(Fourier) cosine transform	77
$_1C$	one-side cosine transform	195
$D_{n,\tau}$	discrete Fourier transform	202
F	(complex) Fourier transform	77
F^{-1}	inverse Fourier transform	79
F_n	n-dimensional Fourier transform	108
$\hat{F}$	Fourier frequency transform	98
$_\nu H$	Hankel transform of order ν	195
Ha	Hartley transform	195
K	consensus cross-spectral power	179
L	Laplace transform	192
$_2L$	two-sided Laplace transform	192
M	Mellin transform	193
P	power spectrum	80
p	power spectral density	80
R	Radon transform	197

Table of symbols used in the text

Symbol	Meaning	Page
S	(Fourier) sine transform	77
$_1S$	one-sided sine transform	195
$Z_{n,\tau}$	z-transform	201
γ	(complex) coherence	179, 214
Π	consensus cross-spectral power	156
Φ	phase spectrum	80
∇	vector derivative	109
$*$	convolution	129
$*_D$	discrete convolution	211
$*_n$	n-dimensional convolution	145
$R[f,g](x)$	cross-correlation function	172
$S[f](x)$	autocorrelation function	175
$T[f,g](x)$	covariance function	174
$U[f](x)$	variance function	175
$u[v_i](x)$	variance density function	298
$\rho(f,g)$	correlation coefficient	169

Part I
Fourier theory

1
Harmonic motion

1.1 The harmonic equation
The *harmonic equation* is a differential equation which appears very frequently in the analysis of physical phenomena. It is:

$$\frac{d^2y}{dx^2} = -\omega^2 y \ . \tag{1.1}$$

To solve this equation, we put $v = dy/dx$; then

$$\frac{d}{dx}(v^2) = 2v\frac{dv}{dx} = 2v\frac{d^2y}{dx^2} \ ,$$

while

$$\frac{d}{dx}(y^2) = 2y\frac{dy}{dx} = 2yv \ .$$

It follows from (1.1) that

$$\frac{d}{dx}(v^2) = 2v\frac{d^2y}{dx^2}$$

$$= 2v(-\omega^2 y)$$

$$= -\omega^2 \frac{d}{dx}(y^2) \ ;$$

so, upon integration,

$$v^2 = -\omega^2 y^2 + k \tag{1.2}$$

for some constant k. Since v, y, and ω are all real, $v^2 + \omega^2 y^2 \geq 0$, so $k \geq 0$, and $a = \sqrt{k}$ is real. Then (1.2) can be rewritten as

$$v = \frac{dy}{dx} = \sqrt{(a^2 - \omega^2 y^2)} \ ,$$

so that

$$\frac{dx}{dy} = \frac{1}{\sqrt{(a^2 - \omega^2 y^2)}} \; .$$

Integrating again,

$$x = \int \frac{1}{\sqrt{(a^2 - \omega^2 y^2)}} \, dy \; .$$

This is a standard integration, and the integral is

$$x = (1/\omega) \cos^{-1}(\omega y/a) + c \; .$$

Rearranging gives

$$y = (a/\omega)\cos(\omega(x - c)) \; .$$

For convenience, we replace the two arbitrary constants a and c by $C = a/\omega$ and $\varphi = -\omega c$. The complete solution of equation (1.1) is then

$$y = C\cos(\omega x + \varphi) \; . \tag{1.3}$$

Since $-C\cos\alpha = C\cos(\alpha + \pi)$, we can always take C to be *positive*, replacing φ by $\varphi + \pi$ if necessary. Moreover, since $\cos(\alpha + 2\pi) = \cos\alpha$, we can always take φ to be in the range $0 \leq \varphi < 2\pi$ (or, indeed, in any given range of length 2π).

Using the double-angle formula $\cos(\alpha + \beta) = \cos\alpha\cos\beta - \sin\alpha\sin\beta$, we can rewrite (1.3) as

$$y = A\sin(\omega x) + B\cos(\omega x) \; . \tag{1.4}$$

where $A = -C\sin\varphi$ and $B = C\cos\varphi$. In what follows, we shall use either (1.3) or (1.4) as expressions of the solution of the harmonic equation, according to which is the more convenient in each case.

1.2 The extension oscillator

Let us now analyse a simple engineering structure: the extension oscillator. Fig. 1.1 shows a schematic diagram. A rigid body of mass m is suspended from a fixed ceiling by an extensible string of spring constant k. (The string is assumed to behave according to *Hooke's law*: when stretched, it provides a restoring force proportional to its extension. The constant of proportionality is the *spring constant*.)

We make some simplifying assumptions about the way the system behaves. For a start, we shall assume that the body moves only vertically: if it moves horizontally as well, the oscillator behaves as a pendulum too. Again, we assume that the body does not 'rock', but stays in the same orientation in space. In effect, we are treating the body as a point mass. The upper end of the string is taken to be fixed: the ceiling does not 'give' when a string is pulled. And finally, we assume that the string has no mass.

Suppose that at time t, the body is suspended at height $x(t)$ above some reference mark. Let the force on the body, upwards, due to the stretching of the string, at time

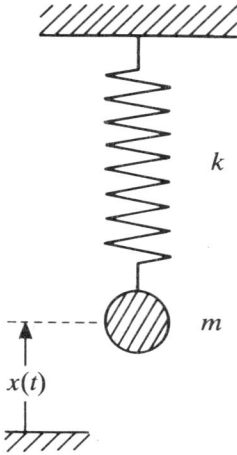

Fig. 1.1 — The extension oscillator.

t, be $F(t)$. Finally, let the string be unstressed when the bob is at height x_0. From Hooke's law,

$$F = k(x_0 - x) ,$$

where k is the spring constant of the string; while Newton's second law tells us that

$$F - mg = m \frac{d^2x}{dt^2} .$$

Combining these equations gives

$$\frac{d^2x}{dt^2} = \frac{k}{m}(x_0 - x) - g$$

$$= -\frac{k}{m}(x - c) , \qquad (1.5)$$

where $c = x_0 - mg/k$. This is the *equation of motion* of the extension oscillator.

Now change the variable to $y = x - c$, and put $\omega_0 = \sqrt{(k/m)}$. Then (1.5) becomes $d^2y/dt^2 = -\omega_0^2 y$: this is the harmonic equation. The solution, from equation (1.4), is $y = C\cos(\omega_0 t + \varphi)$, and therefore

$$x = c + C\cos(\omega_0 t + \varphi) . \qquad (1.6)$$

The bob oscillates sinusoidally about its equilibrium postition at $x = c$.

The constants C and φ have real significance. C is the *amplitude* of the motion; it is the distance between the extreme points the bob reaches and the equilibrium point. φ is the *phase*, and determines the stage of the bob's oscillation at $t = 0$. They cannot, however, be determined, given only equation (1.5). Equation (1.6) is a solution to (1.5) for *every* pair of values of C and φ. Physically, the bob can move in many different ways, each of which has the form of (1.6).

There are various pieces of additional information we can ask for about any particular bob motion that will enable us to find its amplitude and phase. Most commonly, we might be given the (vertical) *position* and *velocity* of the bob at time $t = 0$; that is, the values of $x(0)$ and $(dx/dt)(0)$. These items of information are together known as the *initial conditions*.

More important than these is the quantity ω_0, called the *angular frequency* of the system. Unlike C and φ, it is not dependent on the initial conditions: it is present in the equation of motion. In the case of the extension oscillator, ω_0 is known to be $\sqrt{(k/m)}$; in other systems involving the harmonic equation, there will usually be a corresponding expression in terms of known quantities. Its significance follows from the fact that an increase if t by $2\pi/\omega$ leaves $A\cos(\omega t + \varphi)$ unchanged; so, in terms of the motion of the bob, $x(t + 2\pi/\omega_0) = x(t)$. The bob is said to oscillate with *period* $2\pi/\omega_0$. The *frequency* of the oscillation is the reciprocal of the period, namely $\omega_0/2\pi$; it is the number of cycles the oscillator makes in unit time. Both angular frequency and frequency are measured in inverse seconds, s^{-1}; for frequency the name hertz (symbol Hz) is applied to this unit.

Notice that the angular frequency of the bob's oscillation does not depend on the gravitational term mg at all (although the equilibrium position does). Thus, an exactly similar oscillation would arise in the absence of gravity, or with the oscillator in any nonvertical orientation.

Example. A mass of 5 kg is suspended from a Hookean string of spring constant 2000 N m^{-1}. It is disturbed slightly, and allowed to vibrate vertically. What is the frequency of the oscillation?

The preceding analysis gives $\omega_0 = \sqrt{k/m)}$, so the angular frequency of the oscillations is $\sqrt{(2000/5)}$ s^{-1}, or 20 s^{-1}. The frequency is this divided by 2π, or about 3.186 Hz.

In real life, the oscillations of the bob die away eventually, while the prediction of the mathematical model, equation (1.6), is that it continues to oscillate for ever. To account for this behaviour, we have to refine the model, and we shall see how to do this later.

1.3 Harmonic approximation

A more familiar oscillating system is the pendulum (Fig. 1.2). To analyse this, we again make some simplifying assumptions. The pendulum bob is a *point* mass m, attached to a *massless* and *inextensible* rod of length l. The rod is attached to a *fixed* ceiling by a *frictionless* pivot. The pendulum, moreover, is assumed to swing *in one plane* only, rather than both side-to-side and front-to-back. Finally, the only force on the bob is a constant gravitational one, acting vertically downwards.

Let us use $\theta(t)$ to denote the angle that the pendulum makes with the vertical at time t. Then the equation of motion is, under these assumptions,

$$\frac{d^2\theta}{dt^2} = -\frac{g}{l}\sin\theta . \tag{1.7}$$

There are two things to note about equation (1.7). First, the mass, m, of the bob does not appear. This means that the motion of the pendulum *is not affected* by the

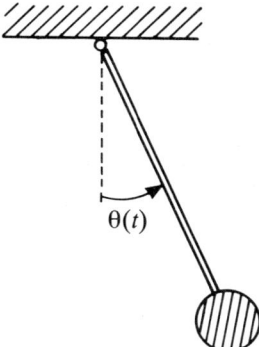

Fig. 1.2 — The pendulum.

mass of the bob. The only parameters that appear are the acceleration due to gravity g and the length of the pendulum l.

Secondly, the equation is not the harmonic equation, for the right-hand side is a multiple of $\sin\theta$ rather than of θ. This makes the solutions difficult to extract. We can, however, simplify matters considerably by making a final assumption: that the oscillations of the pendulum are 'small'. This allows us to approximate this 'difficult' term: for $\sin\theta \simeq \theta$ for small angles θ (provided θ is measured in radians). Thus, for small θ, equation (1.7) becomes

$$\frac{d^2\theta}{dt^2} \simeq -\frac{g}{l}\theta , \qquad (1.8)$$

which is now (approximately) the harmonic equation. Put $\omega_0 = \sqrt{(g/l)}$, and the solutions are

$$\theta \simeq C\cos(\omega_0 t + \varphi) . \qquad (1.9)$$

Thus, for small oscillations, the pendulum is *almost* harmonic, oscillating with angular frequency $\sqrt{(g/l)}$. (As with the extension oscillator, we cannot say anything about the amplitude C or the phase φ without knowing more about the pendulum's starting state; but we do have an expression for the angular frequency, $\omega_0 = \sqrt{(g/l)}$, which can be checked against the motion of real pendulums.)

Although the extraction of such a harmonic approximation might seem to be nothing more than a happy accident of this particular physical situation, it is not. A very large number of oscillatory systems can be reduced to the harmonic equation, simply by assuming the oscillations to be small. Suppose we have a quite general physical system, whose state is described by a single variable x. Let $V(x)$ denote the potential energy of the system when it is in state x. In a given experiment, x will vary with time as the system evolves; thus the state of the system follows the variable $x(t)$, and its potential energy is $V[x(t)]$. The *total* energy E of the system is constant; so the kinetic energy is $T(t) = E - V[x(t)]$.

The evolution of the system depends largely on the nature of the function $V(x)$ at and near the starting point $x(0) = x_0$. Denote dx/dt by v; then it is fair to assume that T is proportional to v^2, at least when v is small — say $T(t) = kv^2$. Thus we have the equation

$$kv^2 + V(x) = E \tag{1.10}$$

as the system's equation of motion. Differentiating (with respect to t) gives $kv(dv/dt) + v(dV/dx) = 0$; so where $v \neq 0$, $dv/dt = -(1/k)(dV/dx)$. But dv/dt is the system's 'acceleration': if it is nonzero at $x = x_0$, then x_0 cannot be an equilibrium point of the system. Since $dv/dt = 0$ precisely when $dV/dx = 0$, it is the turning points of V that determine the equilibria.

Now suppose x_0 is an equilibrium point. The Taylor series expansion of V about x_0 gives $V(x_0 + h) \simeq V(x_0) + h\,(dV(x_0)/dx) + \frac{1}{2}h^2\,(d^2V(x_0)/dx^2)$ for small h. But the second term vanishes. Put $\alpha = d^2V(x_0)/dx^2$; then $V(x_0 + h) \simeq V(x_0) + \frac{1}{2}\alpha h^2$. If $h(t) = x(t) - x_0$, then $dh/dt = v$, and equation (1.10) becomes

$$k\left(\frac{dh}{dt}\right)^2 + \tfrac{1}{2}\alpha h^2 = E - V(x_0) \ . \tag{1.11}$$

Differentiating now gives

$$\frac{dh}{dt}\left[2k\frac{d^2h}{dt^2} + \alpha h\right] = 0 \ .$$

Thus, wherever $dh/dt \neq 0$, we have

$$\frac{d^2h}{dt^2} + \frac{\alpha}{2k}h = 0 \ . \tag{1.12}$$

If $\alpha/2k$ is negative (corresponding to x_0 being a *maximum* for V), the solutions are exponential; since $h(0) = 0$, we have $h(t) = A\sinh[t\sqrt{(-\alpha/2k)}]$ for some A. The system *diverges* from the equilibrium point (at least until h is no longer 'small'), and the equilbrium at x_0 is *unstable*. On the other hand, if $\alpha/2k$ is positive (when x_0 is a *minimum* of V), then putting $\omega^2 = \alpha/2k$ turns (1.12) into the harmonic equation. The solution then is sinusoidal: $h(t) = A\sin[t\sqrt{(\alpha/2k)}]$, and therefore $x = x_0 + A\sin[t\sqrt{(\alpha/2k)}]$. In this case the system *oscillates* around the equilibrium x_0, which is therefore *stable*. (If $\alpha = 0$, the analysis breaks down.)

So in this very general situation, the harmonic equation arises as *the* linear approximation whenever there is a second-order minimum of V.

Now in actuality, the Taylor series is only approximate, so the differential equation (1.12) is not exact. For small t, the solution is good, but after a time inaccuracies will build up. The looser the approximation, the quicker the inaccuracies will build; in many cases the system will not execute even a single wave before being significantly nonharmonic. For 'small' oscillations, the harmonic approximation is good, but it is not always easy to predict where oscillations cease to be 'small'.

1.4 A two-mode system

The two systems we have analysed so far have both been relatively simple, for they are both systems which can be completely specified by a single variable (height of bob for the extension oscillator, angle for the pendulum). They are said to be systems *with one degree of freedom*. This means the resulting equation of motion is an equation in one dependent variable (dependent, that is, on the 'independent' variable, t).

Fig. 1.3, by contrast, shows a schematic diagram of a double oscillator, which is

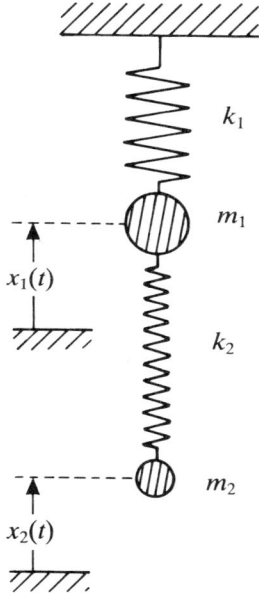

Fig. 1.3 — The double oscillator.

simply two extension oscillators connected in series. There are two masses here, one hanging below the other. The strings obey Hooke's law; the motion of the bobs, as in the simple oscillator, is assumed to be purely vertical. The question naturally arises of how, if at all, the motion of the double oscillator is related to that of the simple oscillator (Fig. 1.1).

The double oscillator clearly has *two* degrees of freedom. To specify the state of the system at time t we need two numbers: the vertical position of the upper bob $x_1(t)$, and the vertical position of the lower bob $x_2(t)$. For simplicity, we shall take the reference marks for x_1 and x_2 to be the equilibrium (rest) positions of the two bobs. The equations of motion are then

$$m_1 \frac{d^2 x_1}{dt^2} = (k_1 - k_2) x_1 + k_2 x_2$$

and

$$m_2 \frac{d^2x_1}{dt^2} = k_2(x_1 - x_2) \ . \tag{1.13}$$

Rather than attempt to solve these directly, we reason as follows. In the simple extension oscillator, we have the harmonic equation, and therefore sinusoidal motion. Since we have here a physically similar system, it is not unlikely that some, at least, of the solutions to equations (1.13) will be sinusoidal. We therefore look for solutions of the form

$$\begin{cases} x_1 = C_1\cos(\omega_1 t + \varphi_1) \\ x_2 = C_2\cos(\omega_2 t + \varphi_2) \end{cases} \ . \tag{1.14}$$

Differentiating twice,

$$\frac{d^2x_1}{dt^2} = -\omega_1^2 C_1 \cos(\omega_1 t + \varphi_1)$$

$$\frac{d^2x_2}{dt^2} = -\omega_2^2 C_2 \cos(\omega_2 + \varphi_2) \ .$$

Substituting these into (1.13) and rearranging gives

$$(-m_1\omega_1^2 - (k_1 - k_2))C_1\cos(\omega_1 t + \varphi_1) = k_2 C_2 \cos(\omega_2 t + \varphi_2)$$
$$(m_2\omega_2^2 + k_2)C_2\cos(\omega_2 t + \varphi_2) = k_2 C_1\cos(\omega_1 t + \varphi_1) \ . \tag{1.15}$$

Thus for any sinusoidal solution, the constants C_1 and C_2, ω_1 and ω_2, φ_1 and φ_2 must satisfy these equations.

Now equations (1.15) say that the functions $\cos(\omega_1 t + \varphi_1)$ and $\cos(\omega_2 t + \varphi_2)$ are multiples of each other. It follows that $\omega_1 = \omega_2$ ($=\omega$, say), and $\varphi_1 - \varphi_2$ is a multiple of π. Since we may pick $\varphi_1 \leq \varphi_2 < \varphi_1 + 2\pi$, we have two cases: either $\varphi_2 = \varphi_1$ (case I), or $\varphi_2 = \varphi_1 + \pi$ (case II). In case I, $\cos(\omega t + \varphi_2) = \cos(\omega t + \varphi_1)$, and (1.15) reduces to

$$(-m_1\omega^2 + (k_1 - k_2))C_1 = k_2C_2$$
$$(-m_2\omega^2 + k_2)C_2 = k_2C_1 \ ; \tag{1.16a}$$

in case II, $\cos(\omega t + \varphi_2) = -\cos(\omega t + \varphi_1)$, and we have

$$(-m_1\omega^2 + (k_1 - k_2))C_1 = -k_2C_2$$
$$(-m_2\omega^2 - k_2)C_2 = k_2C_1 \ . \tag{1.16b}$$

In either case, we obtain, and can solve, a quadratic equation for ω^2. We cannot solve for the amplitudes C_1 and C_2 individually, just as we could not solve for the amplitude C of the extension oscillator without the initial conditions; but we can solve for their quotient $\gamma = C_2/C_1$. The solutions are left as an exercise for the reader; let us denote them by ω_1, γ_1 (in case I), and ω_2, γ_2 (in case II). Thus there are two harmonic solutions to the double oscillator:

$$\begin{cases} x_1 = C\cos(\omega_1 t + \varphi) \\ x_2 = \gamma_1 C\cos(\omega_1 t + \varphi) \end{cases} \tag{1.17a}$$

and

$$\begin{cases} x_1 = C\cos(\omega_2 t + \varphi) \\ x_2 = \gamma_2 C\cos(\omega_2 t + \varphi + \pi) \ (= -\gamma_2 C\cos(\omega_2 t + \varphi)) \end{cases} \quad (1.17b)$$

Here, C and φ are arbitrary constants, the amplitude and phase of the upper bob.
What these solutions mean in terms of real vibrations is illustrated in Fig. 1.4.

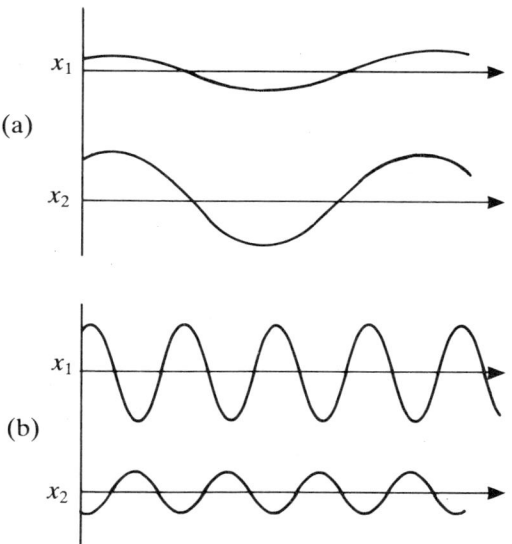

Fig. 1.4 — Two modes of harmonic motion of the double oscillator. (a) The lower-frequency mode, in which the bobs move in phase. (b) The higher-frequency mode, in which the bobs move in antiphase.

Fig. 1.4(a) depicts case I: the bobs move with the same phase φ, and are said to move *in phase*. Fig. 1.4(b) depicts case II, where the bobs are moving *in antiphase*. (It turns out that $\omega_2 > \omega_1$, i.e. that the antiphase oscillation has a higher angular frequency than the in-phase oscillation.)

These two harmonic solutions are the two *(natural) modes of vibration* of the double oscillator, which therefore has two *natural frequencies*. More complex systems can have more than two modes: normnally one for each degree of freedom.

1.5 Linear systems
This is not a complete solution of the double oscillator. We have only found the *harmonic* solutions to its equations of motion. There are others, as simple experiment will show.

To obtain the remaining solutions, we use the fact that equations (1.13) form a *linear system*: each equation consists of a sum of terms, and each term is a multiple of x_1, x_2, or one of their derivatives. There are no products like $x_1 x_2$ or $(dx_1/dt)(d^2x_1/dt^2)$, no constants, and no functions like $\sin x_1$ or $\sqrt{x_2}$. (The coefficients of the terms need not be constant: they can involve functions of the 'independent' variable, in this case t. For instance, the equation $(dy/dx) - x^2 y = 0$ is linear.)

Linearity has an important consequence. We know that, for constants α and β, $d(\alpha f + \beta g)/dx = \alpha(df/dx) + \beta(dg/dx)$. The same is therefore true of second and higher derivatives. Then, if f and g are both solutions of a linear system of equations, we see that $\alpha f + \beta g$ is also a solution. This is called the *principle of superposition*.

Here is an example: the harmonic equation, $d^2x/dt^2 = -\omega^2 x$, is linear. (In this case, we happen to know the solutions explicitly, but let us ignore that for the moment.) Suppose the two functions f and g are solutions: that is, that $d^2 f/dt^2 = -\omega^2 f$ and $d^2 g/dt^2 = -\omega^2 g$. It follows that

$$\frac{d^2(\alpha f + \beta g)}{dt^2} = \alpha \frac{d^2 f}{dt^2} + \beta \frac{d^2 g}{dt^2}$$

$$= \alpha(-\omega^2 f) + \beta(-\omega^2 g)$$

$$= -\omega^2(\alpha f + \beta g) .$$

Thus $x = \alpha f + \beta g$ is itself a solution.

Applying the principle of superposition to the double oscillator, we obtain (from (1.17)) the more general solution

$$\begin{cases} x_1 = A\cos(\omega_1 t + \varphi) + B\cos(\omega_1 t + \psi) \\ x_2 = \gamma_1 A\cos(\omega_1 t + \varphi) + \gamma_2 B\cos(\omega_1 t + \psi + \pi) . \end{cases} \quad (1.18)$$

Indeed, it can be shown that (1.18) is a *complete* solution to (1.13).

1.6 Damping

In real physical situations, as we have said, oscillations die away, unless they are maintained by continuous shaking. This is usually a result of the presence of additional forces in the system, which act to dissipate kinetic energy; or as a result of thermodynamic departures from strict linearity in physical laws like Hooke's. These latter effects, collectively called *hysteresis*, also act as if they result from energy-dissipating 'forces', even though there is no real force present.

Such energy-dissipating forces, real or conventional, are called *damping* forces. Among the more common are air resistance; the viscous forces arising from fluid components, like lubricants; and friction.

Damping forces are *velocity-dependent*. A damping force will (usually) have a magnitude dependent on the (relative) velocity of the components involved, and will act in the same *direction* as the relative velocity, but in the opposite sense. The accelerations that result thus *decrease* the relative velocity. Thus, a stationary car on a calm day experiences no wind resistance. The same car begins to feel a force ('drag') when it starts moving, which grows as the speed of the car increases; and this force acts to brake the car.

Damping forces are of widely varying properties, and no single mathematical model can account for them all with total accuracy. However, they are often — at least in oscillating systems — small. Thus we may often use a linear approximation. This leads us to the concept of the *ideal* (or *Newtonian*) *damper* (see Fig. 1.5).

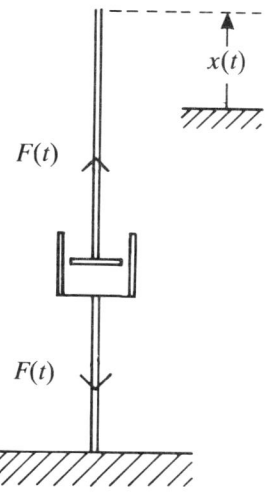

Fig. 1.5 — Representation of an ideal (Newtonian) damper. The equation governing this system is $F + \eta(dx/dt) = 0$.

The principle of the ideal damper is similar to that of the ideal spring. In the Hookean spring, there is a force provided by extension or compression. The force and the extension are proportional to each other, and the constant of proportionality is called the spring constant, k. In the Newtonian damper, there is a force provided by *rate of change of* extension. Here again, the two are made proportional; the constant of proportionality is the *damping constant*, often denoted by η. In equation form, this becomes

$$F = \eta \frac{dx}{dt} . \tag{1.19}$$

The ideal damper is a significantly poorer model, in many cases, than the ideal spring. Many damping forces do not show this linear behaviour. Sliding friction, for example, is a force of constant magnitude, and only its direction is affected by the velocities involved. On the other hand, air resistance is often proportional to the *square* of velocity. Viscous forces are often linear, but hysteresis is not.

As an example of the use of the ideal damper in modelling, consider again the extension oscillator; but this time, in order to make the spring more realistic, let us include a small component of Newtonian damping. Since the spring and damping forces are independently determined by the position and velocity of the bob, the two components of the model must be in parallel (Fig. 1.6).

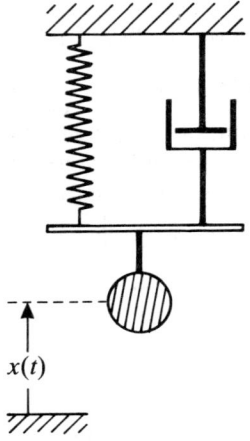

Fig. 1.6 — The damped oscillator.

The total upward force on the bob at time t is

$$F = -kx - \eta \frac{dx}{dt}$$

where the position x of the bob is measured relative to its rest position. It follows that the equation of motion of the damped oscillator is

$$m \frac{d^2x}{dt^2} = -kx - \eta \frac{dx}{dt}$$

or, rearranging,

$$m \frac{d^2x}{dt^2} + \eta \frac{dx}{dt} + kx = 0 \qquad (1.20)$$

We solve this equation by trying *complex-valued* functions of the form Ce^{zt}, for complex constants C and z. Substituting $x = Ce^{zt}$ in (1.20) gives $C(mz^2 + \eta z + k)e^{zt} = 0$; so C is arbitrary, and z is a root of $mz^2 + \eta z + k = 0$:

$$z = -\frac{\eta}{2m} \pm \frac{\sqrt{(\eta^2 - 4km)}}{2m} . \tag{1.21}$$

When the damping constant η is zero, then (1.20) is the harmonic equation, and (1.21) reduces to $z = \pm\sqrt{(-4km)}/2m = \pm i\sqrt{(k/m)}$. The general solution of the oscillator is then $x = C_1\exp(it\sqrt{(k/m)}) + C_2\exp(-it\sqrt{(k/m)})$; when this is real, it is the same as the harmonic solution we obtained before, equation (1.3).

More generally, suppose η^2 is small compared to km. In this case, (1.21) reduces to $z \simeq (-\eta/2m) \pm i\sqrt{(k/m)}$. The real solutions of (1.20) are then those of the form

$$x \simeq Ce^{-(\eta/2m)t}\cos(\sqrt{(k/m)}t + \varphi) . \tag{1.22}$$

Thus, the effect of a small amount of damping on the extension oscillator is that the bob oscillates at almost the same angular frequency as when undamped; but its amplitude gradually diminishes to zero, according to a negative exponential. The new solution is illustrated in Fig. 1.7.

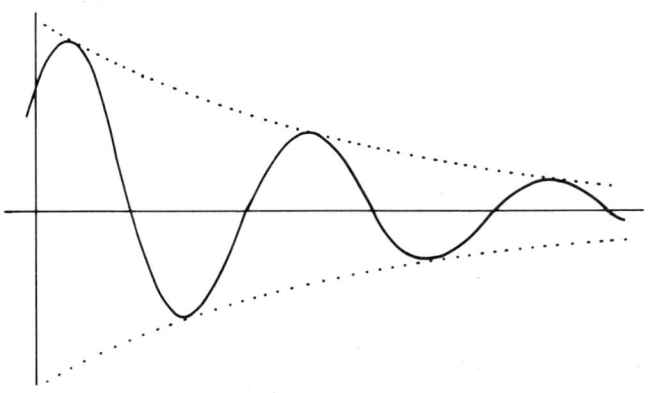

Fig. 1.7 — Motion of the damped oscillator.

A useful parameter of the damped system is the *selectivity*, or *Q-value*, defined as $Q = \sqrt{(km)}/\eta$. It is related to the proportional decrease in amplitude of the oscillation over a single cycle, and it therefore indicates how strongly damped the system is. The period of the oscillation is $2\pi\sqrt{(m/k)}$, and in this time the amplitude decays by a factor of $\exp(-(\eta/2m)2\pi\sqrt{(m/k)})$, or about $1-(\pi/Q)$. A high Q means a very lightly damped system; a low Q, a strongly damped system.

When η is not small, (1.22) no longer holds. Indeed when $\eta \geq 2\sqrt{(km)}$, then z has no imaginary component; so in a system with $Q \leq \frac{1}{2}$, the bob no longer oscillates at all. A system with $Q = \frac{1}{2}$ is called *critically damped*. (See Question 10 at the end of the chapter.)

Equation (1.20) can be simplified, by putting $s = \omega_0 t$, where $\omega_0 = \sqrt{(k/m)}$. We then have $dx/dt = (dx/ds)(ds/dt) = \omega_0 (dx/ds)$, and $d^2x/dt^2 = \omega_0^2 (dx/ds^2)$. Substituting these into (1.20) and dividing by k gives

$$\frac{m\omega_0^2}{k}\frac{d^2x}{ds^2} + \frac{\eta\omega_0}{k}\frac{dx}{ds} + x = 0 .$$

But $m\omega_0^2/k = 1$, and $\eta\omega_0/k = 1/Q$, so this is

$$\frac{d^2x}{ds^2} + \frac{1}{Q}\frac{dx}{ds} + x = 0 . \tag{1.23}$$

By virtue of this simplification, we can see that the solution to the damped oscillator depends only on the natural angular frequency ω_0 and on the selectivity Q.

1.7 Forced oscillation

So far we have looked at naturally oscillating systems, and found that sinusoidal motion, or an approximation to it, appears frequently. This is not the only way in which a system can oscillate: it can be shaken. The oscillations that arise are called *forced*.

Consider, once again, the extension oscillator — a mass m hung from a ceiling by a string of spring constant k. This system, we have seen, has a natural frequency of vertical oscillation of $\sqrt{(k/m)}/2\pi$. *If the oscillator is left to itself*, then all it can do is vibrate at this frequency, or not at all. Moreover, if there is any damping at all, any such vibration will eventually die away.

But now suppose that the ceiling is itself moving up and down sinusoidally. Let $y(t)$ be the vertical position of the ceiling at time t (Fig. 1.8); then

$$y = y_0 + A\cos(\omega t) . \tag{1.24}$$

(We ignore any phase angle here by starting our time variable t when the ceiling reaches the appropriate point of its oscillation.)

In this situation, ω can take any value. In real life, such vibration could arise from a nearby oscillating system vibrating with natural angular frequency ω; thus we are effectively investigating how one vibrating system affects another.

First, we need to change the equation of motion to take account of the shaking ceiling. The length of the string at time t is $y - x$; if its equilibrium length is l, then the upward force on the bob is $F = k(y - x - l)$. As before, the bob's weight is a force mg acting vertically downwards. Newton's second law then gives us the equation:

$$m\frac{d^2x}{dt^2} = k(y - x - l) - mg . \tag{1.25}$$

Since the input vibration is sinusoidal at angular frequency ω, and the equation of

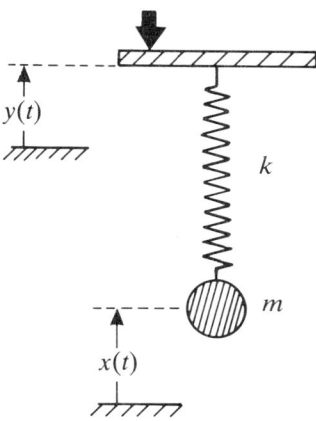

Fig. 1.8 — The forced oscillator.

motion of the extension oscillator is linear, we guess that the resulting motion of the bob will also be sinusoidal of angular frequency ω. We therefore look for solutions of the form

$$x = x_0 + B\cos(\omega t + \varphi) \ . \tag{1.26}$$

Now we have an expression for y, and a trial expression for x, in equations (1.24) and (1.26) respectively. Putting them in equation (1.25) gives

$$-m\omega^2 B\cos(\omega t + \varphi) = k(y_0 + A\cos(\omega t) - x_0 - B\cos(\omega t + \varphi) - l) - mg \ . \tag{1.27}$$

Matching the constant terms gives $x_0 = y_0 - l - mg/k$. That leaves

$$-m\omega^2 B\cos(\omega t + \varphi) = k(A\cos(\omega t) - B\cos(\omega t + \varphi)) \ ,$$

which we expand with the double angle formula as

$$-m\omega^2 B(\cos(\omega t)\cos\varphi - \sin(\omega t)\sin\varphi) = k(A\cos(\omega t) - B(\cos(\omega t)\cos\varphi - \sin(\omega t)\sin\varphi)) \ .$$

If this is true for all time, it is certainly true for $t = 0$ (at which point $\sin(\omega t) = 0$ and $\cos(\omega t) = 1$) and for $t = 2\omega/\pi$ (at which point $\sin(\omega t) = 1$ and $\cos(\omega t) = 0$). So this yields the two equations

$$-m\omega^2 B\cos\varphi = k(A - B\cos\varphi)$$

and

$$m\omega^2 B\sin\varphi = kB\sin\varphi \ . \tag{1.28}$$

(Equivalently, we can equate separately cosine and sine terms.)

But B is nonzero: there are forces acting on the bob, so it cannot stay motionless. Moreover, except when ω is equal to the natural angular frequency of the system, $m\omega^2$ is not equal to k. From the second of equations (1.28), it follows that $\sin\varphi = 0$. That means that φ is 0 or π. By changing the sign of B if necessary, we can assume that $\varphi = 0$, and so $\cos\varphi = 1$. The first of equations (1.28) now reduces to

$$-m\omega^2 B = k(A - B) \; ;$$

thus $B = A/(1 - (m\omega^2/k))$, and the solution to the forced vibration is then

$$x = \frac{A\cos(\omega t)}{(1 - (m\omega^2/k))} \; . \tag{1.29}$$

This is not a complete solution to the forced vibration, only a solution for sinusoidal response. However the only possible *unforced* oscillations are at the natural angular frequency $\omega_0 = \sqrt{(k/m)}$; so, bearing in mind the principle of superposition, we can say that the complete solution is

$$x = \frac{A\cos(\omega t)}{(1 - (m\omega^2/k))} + C\cos(\omega_0 t + \psi) \tag{1.30}$$

for arbitrary constants C and ψ. (A is *not* arbitrary: it is the amplitude of the ceiling vibration, and is given.) The solution of the forced oscillator, then contains two terms: one due to forcing, the other a superposed vibration at the natural frequency of the oscillator.

The effect on the solution of incorporating damping is twofold. It will slightly change the amplitude of forced vibration, the extent of the change depending on the magnitude of the damping forces; and it alters the superposed natural oscillation to something like (1.22). But this natural oscillation eventually dies away, and does not represent a steady state. It is said to be *transient*. Thus, if we are interested only in steady-state vibration, the second term vanishes, leaving us with the solution in equation (1.29): a forcing (ceiling) vibration with equation $A\cos(\omega t)$ induces a forced (bob) vibration with equation $A\cos(\omega t)/(1 - (m\omega^2/k))$.

The output amplitude is proportional to the input amplitude, as you would expect from a linear system. The constant of proportionality, $1/|-(m\omega^2/k)|$, is determined by the system. Since ω can take any value, the system determines a *function of* ω:

$$X(\omega) = \frac{1}{|1 - (m\omega^2/k)|} \; . \tag{1.31}$$

It is called the *amplitude response function* or *transmissibility* of the system.

A corresponding function $X(\omega)$ can be found for forced oscillations in many oscillatory systems. In each case, its significance is the same: *a harmonic input of angular frequency ω and amplitude A forces a harmonic output of amplitude $AX(\omega)$*.

The amplitude response function of the forced extension oscillator is depicted in Fig. 1.9(a). For very slow oscillations (low ω), it is nearly 1, so that the bob moves up and down, following the ceiling almost exactly. This is what we would probably expect. For very fast oscillations (large ω), it is nearly 0, so that the bob hardly moves. And for intermediate speeds, when ω is close to ω_0, the amplitude response is very large — in fact, it tends to infinity as ω approaches ω_0. That is, if you force it at a

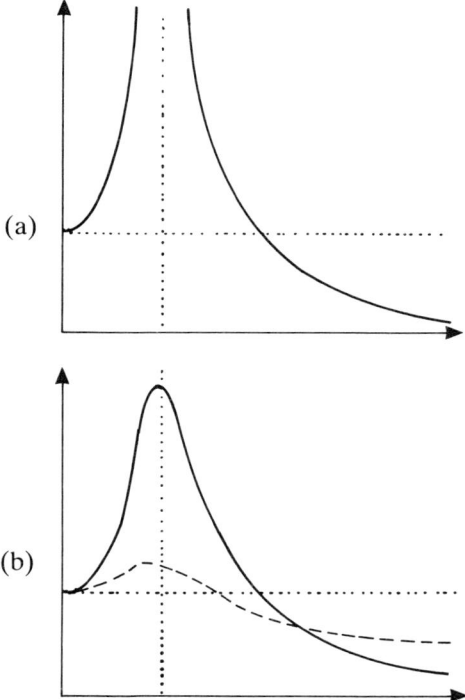

Fig. 1.9 — (a) The amplitude response function of an undamped oscillator. (b) The amplitude response function of two damped oscillators, with Q-values 3 (solid line) and 1 (dashed line).

frequency close to its natural frequency, the bob oscillates violently, even with a modest input.

For this very practical reason, finding the natural frequencies of a system can be very important. Any stray forcing vibration that happens, by chance, almost to coincide with one of them might set up violent, dangerous (and expensive) vibration of the system.

The effect of a small amount of damping on the amplitude response curve can be calculated, too. Fig. 1.9(b) represents the amplitude response to forcing vibrations of some *damped* oscillators. The major change to notice here is that the response to forcing vibration at ω_0 is now finite; in fact, it is roughly equal to Q, except when Q is very small.

1.8 The vibrating string

So far we have looked only at systems with few degrees of freedom. In this section we shall begin to analyse a more complicated system: one with infinitely many degrees of freedom. We shall see again how from the equation of motion can be extracted *harmonic* solutions, which can be superposed to give a much more general solution.

There are two main classes of infinite-degree-of-freedom systems for which analytical solutions have been obtained: those which involve Hooke's law (beams and plates), and those which do not (strings and membranes). Plates are two-dimensional analogues of beams, and membranes are two-dimensional analogues of strings. Here we shall look in detail at the easiest of these, the case of a vibrating string, of finite length, between two fixed ends; the solutions of the other cases are all more or less similar in style.

For strings, the physics is based on Newton's laws only, like the pendulum and unlike the extension oscillator. Fig. 1.10 shows the system. The string has mass M,

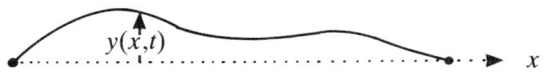

Fig. 1.10 — Representing the motion of a finite string.

which we assume to be uniformly distributed along its length l; the linear density is then $\rho = M/l$. It is also under constant tension T. It is *not* assumed to be Hookean, but it *is* assumed to be elastic; thus, it admits 'very small' (second-order) changes in length with 'very small' effect on the tension. (This condition allows the string to vibrate while maintaining the tension at a constant value.) Finally, we assume that the vibration of the string is such that the points on it move *at right angles* to the string's equilibrium line.

The state of the string is described, not by a single variable (x in the case of the extension oscillator, θ for the pendulum), but by a *function*. Thus, if the point on the string a distance x from the left-hand end is displaced vertically by $y(x)$, then the function $y(x)$ ($0 \leqslant x \leqslant l$) describes the state of the string. To describe the *motion* of the string, we need a function $y(x,t)$ of *two* variables: position along the string x, and time t. We are supposing that the vibration is small, so $y(x,t) \ll 1$; in addition, we assume that the *slope* of the string is small everywhere, so $\partial y/\partial x \ll 1$.

To find the string's equation of motion, we look at a small element of length, δx, which has mass $\rho \delta x$ (see Fig. 1.11). The elements to the left and to the right are both pulling on this element with force T, but they are pulling in slightly different directions because of the curvature of the string. The left-hand element pulls in the direction $(1, (\partial y/\partial x)(x,t))$ with force $-T$; the right-hand element, with force $+T$ in the direction $(1, (\partial y/\partial x)(x+\delta x,t))$. Since $\partial y/\partial x$ is small, these forces are $(-T, -T(\partial y/dx)(x,t))$ and $(T, T(\partial y/\partial x)(x+\delta x,t))$, to second order. The net force on the element is then

$$T\left(\frac{\partial y}{\partial x}(x+\delta x,t) - \frac{\partial y}{\partial x}(x,t)\right) ,$$

acting (almost) vertically upwards.

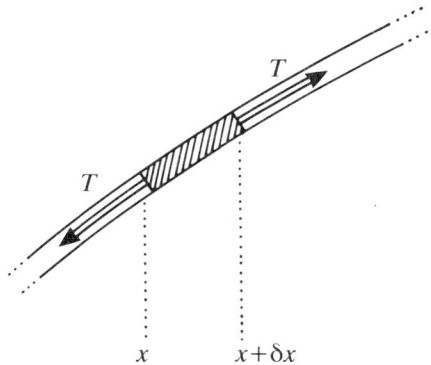

Fig. 1.11 — The force on a string element.

The vertical acceleration of the element is $\partial^2 y/\partial t^2$. Applying Newton's second law gives us

$$T\left(\frac{\partial y}{\partial x}(x+\delta x, t) - \frac{\partial y}{\partial x}(x,t)\right) = \rho \delta x \frac{\partial^2 y}{\partial t^2} .$$

Dividing both sides by δx and taking limits as $\delta x \to 0$, we end up with the equation

$$T\frac{\partial^2 y}{\partial x^2} = \rho \frac{\partial^2 y}{\partial t^2} . \qquad (1.32)$$

This, then, is the equation of motion for the string. Owing to our approximations, it is linear, so that we can freely use the principle of superposition.

Now, when we solve this partial differential equation, we shall not be able to find the exact motion of any particular string system until we have its initial conditions. That means being given the value of $y(x,0)$ and of $(\partial y/\partial t)(x,0)$ for all x. We also need to take into account the fact that the string has fixed ends; this latter criterion reduces to the statement that

$$y(0,t) = y(l,t) = 0 \quad \text{for all } t . \qquad (1.33)$$

(These are called the *boundary conditions* of the problem.)

To solve (1.32), we take a similar approach to the one we used for the double oscillator: we look for *harmonic* solutions. A harmonic solution is one in which each point on the string makes sinusoidal oscillations, at the same frequency, and in phase. The amplitude of the oscillations is allowed to vary from point to point, however (see Fig. 1.12). Thus we are restricting ourselves to solutions of the form

$$y(x,t) = A(x)\cos(\omega t + \varphi) . \qquad (1.34)$$

(Notice here that the two independent variables x and t appear separately now. The technique of *separation of variables* is widely used in solving partial differential equations.)

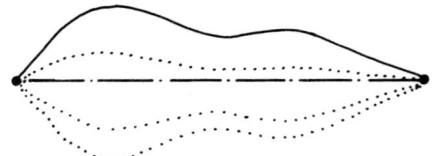

Fig. 1.12 — Harmonic solutions to the string's motion.

Now with $y(x,t)$ of this form, we have

$$\frac{\partial^2 y}{\partial t^2} = -\omega^2 A(x)\cos(\omega t + \varphi)$$

and

$$\frac{\partial^2 y}{\partial x^2} = \frac{\partial^2 A}{\partial x^2} \cos(\omega t + \varphi) \ .$$

Equation (1.32) then becomes

$$T \frac{\partial^2 A}{dx^2} \cos(\omega t + \varphi) = -\rho\omega^2 A \cos(\omega t + \varphi) \ .$$

The time-dependent part can be cancelled from both sides to leave

$$T \frac{\partial^2 A}{\partial x^2} = -\rho\omega^2 A \ .$$

But this is just the harmonic equation! If we put $\Omega = \omega\sqrt{(\rho/T)}$, then we have

$$\frac{\partial^2 A}{\partial x^2} = -\Omega^2 A \tag{1.35}$$

which we solve to give

$$A(x) = B\cos(\Omega x + \psi) \ .$$

It follows that

$$y = B\cos(\Omega x + \psi)\cos(\omega t + \varphi) \tag{1.36}$$

is the general harmonic solution to equation (1.32).

The quantity $c = \sqrt{(T/\rho)}$ is the *wave speed* on the string. Thus $\Omega = \omega/c$, and the string's equation of motion (1.32) can be rewritten as

$$\frac{\partial^2 y}{\partial x^2} = \frac{1}{c^2} \frac{\partial^2 y}{\partial t^2} \ . \tag{1.32'}$$

Not all such functions are solutions for the *string*, however. We now have to take into account the boundary conditions, equations (1.33). Putting $x = 0$ and $x = l$ along with $t = -\varphi/\omega$, we have $B\cos(\psi) = B\cos(\Omega l + \psi) = 0$. If B is nonzero, then both ψ

Ch. 1] Harmonic motion 39

and $\Omega l + \psi$ must be odd multiples of $\pi/2$. We can choose ψ to lie between 0 and 2π, and so to be $\pi/2$ of $3\pi/2$; and in fact, we can make $\psi = 3\pi/2$ by changing the sign of B if necessary. It also follows that the product Ωl is a multiple of π; say $n\pi$, for some integer n. Since l is fixed, we must have $\Omega = n\pi/l$, for some n, and thus $\omega = nc\pi/l$. Thus ω cannot be chosen arbitrarily. The string cannot vibrate at arbitrary angular frequencies — only at multiples of $c\pi/l$.

If we substitute these expressions for ω and Ω into equation (1.36), and use the fact that $\cos(\theta + 3\pi/2) = \sin\theta$, we arrive at the equation

$$y = B\sin(n\pi x/l)\cos(nc\pi t/l + \varphi) \quad \text{for some } n . \tag{1.37}$$

These are the harmonic solutions, or natural modes of vibration, of the string with fixed ends.

Oscillations corresponding to small values of n are shown diagrammatically in Fig. 1.13. There are infinitely many modes. This is to be expected, since the system

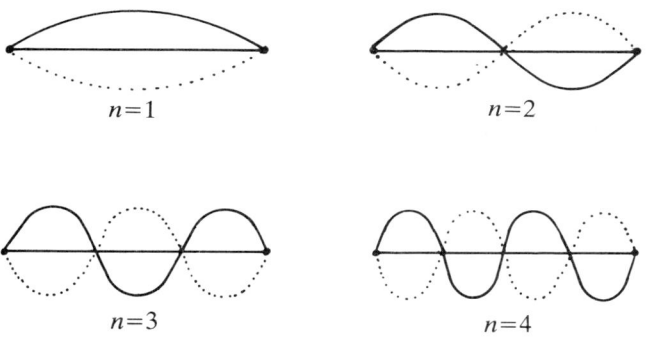

Fig. 1.13 — Four natural modes of the string.

has infinitely many degrees of freedom. In each mode, the string shape is sinusoidal at all times; the restriction on Ω and ψ (to make the endpoints fit the physical assumptions) can now be understood geometrically. The frequency of the oscillations is lowest for the case $n = 1$, at $c/2l$ (corresponding to an angular frequency of $c\pi/l$); this mode is called the *fundamental* of the string's vibration. With $n = 2, 3$, and so on, the frequency is doubled, trebled, or multiplied by larger integers; these modes are the *second, third,* and higher *harmonics*.

Example. A typical string in a grand piano is 2 m long. Suppose such a string, with a mass of 1 g, is used to sound the note A above middle C (the frequency standard, 440 Hz). What tension needs to be applied to the string?

The linear density of the string is $1/2$ gm^{-1}, or 5×10^{-4} kgm^{-1}. Suppose the tension, in newtons, is T. The fundamental (lowest) natural frequency of the string, from the previous analysis, is then $c/4 = 440$, so that $c = 1760$. Since

$c = \sqrt{(T/\rho)}$, $T = \rho c^2$; thus the tension in the string is $5 \times 10^{-4} \times 1760^2$ N, or about 1550 N.

Since equation (1.32) is linear, we can apply the principle of superposition: any sum of solutions is also a solution. For the string with fixed ends, this means sums of functions like equation (1.37). The new solutions will not be harmonic (see Fig. 1.14).

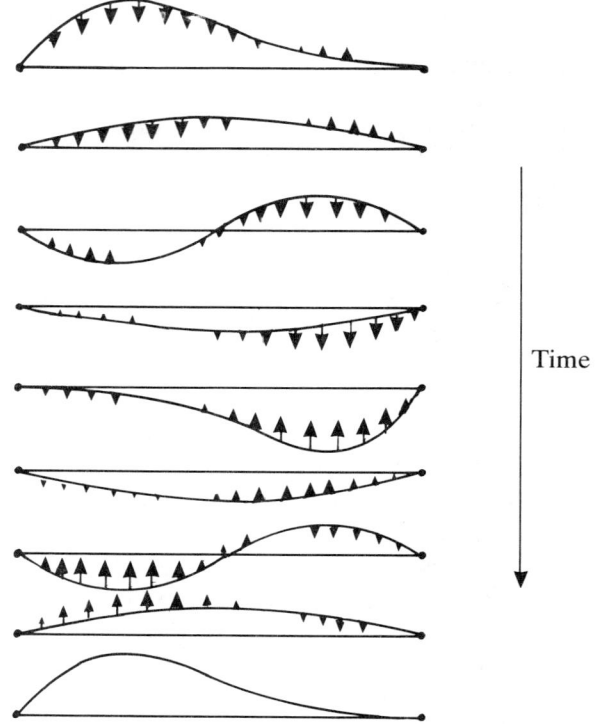

Fig. 1.14 — A nonharmonic string vibration.

But the string has infinitely many harmonic modes of vibration, at infinitely many natural frequencies. Thus we must consider sums, not just of two or three harmonic solutions, but of infinitely many. These are the *series solutions*:

$$y(x,t) = \sum_{n=1}^{\infty} B_n \sin(n\pi x/l)\cos(nc\pi t/l + \varphi_n) \ . \tag{1.38}$$

Here B_n and φ_n are infinite sequences of coefficients.

Infinite series require careful treatment. In order for (1.38) actually to make sense, for it to be a valid solution, the constants B_n and φ_n need to be such that the series *converges*, for every x and t. Not only that, but the resulting function $y(x,t)$ has to be twice partially differentiable (in order to satisfy equation (1.32)).

Equation (1.38), our candidate for the general solution of the vibrating string, is an example of a *Fourier series*: a sum of harmonic functions, all at freqencies which are multiples of the fundamental frequency. We shall study these series in the next chapter.

EXERCISES

1. Show that the function $y(x) = C\cos(\omega x + \varphi)$ is a solution of the harmonic equation $d^2y/dx^2 = -\omega^2 y$, for any values of C and φ.

2. If $y = A\cos(\omega x) + B\sin(\omega x)$ for certain constants A and B, find values of C and φ for which $y = C\cos(\omega x + \varphi)$.

3. Fig. 1.15 shows an 'inverted' extension oscillator on a Hookean spring. Assum-

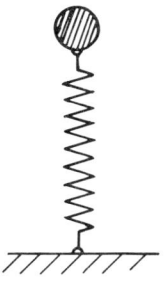

Fig. 1.15.

ing that the mass moves only vertically — that is, ignoring the question of the system's lateral stability — show that the motion of the body is harmonic. What is the body's equilibrium position? What is its angular frequency of vibration?

4. How long must a pendulum be for its period to be 1 s?

5. (Harder) Suppose a pendulum of length l oscillates with 'amplitude' C which is small enough for C^4 to be negligible, but not small enough for C^3 to be negligible. Show that one solution is

$$\theta(t) = C\cos(\omega t) - (C^3/216)\cos(3\omega t) ,$$

where $\omega^2 = (g/l)(1 - C^2/8)$, by using the approximation $\sin\theta \simeq \theta - (\theta^3/6)$ in equation (1.8). (Hence the pendulum begins to swing slightly *more slowly* as its amplitude increases.)

6. Use the results of the previous question. A pendulum is designed to oscillate with a period of 1 s with very small oscillations. How large may its oscillations become if it is desired to keep it accurate to within

(a) 1% (14 min in a day)?
(b) 0.01% (9 s in a day)?

Assume, of course, that there is no other source of variability.

7. In the double oscillator of section 1.4, find expressions for the two angular frequencies ω of oscillation, and their corresponding amplitude ratios $\gamma = C_2/C_1$. Show that $\gamma > 1$ for the lower value of ω, and that $\gamma < 1$ for the higher.

8. Verify that equation (1.18) gives a solution to the double oscillator.

9. Solve equation (1.23), for general Q and for small Q. Compare your answer with (1.22).

10. Solve explicitly the equation for the damped oscillator when Q is

(a) $\frac{1}{2} + \varepsilon^2$ \qquad (b) $\frac{1}{2} - \varepsilon^2$

for small ε.
Discuss your results.

11. Find and solve the equation of motion of a forced *damped* oscillator. Show that the amplitude response function that arises may be expressed in terms of the two parameters $\omega_0 = \sqrt{(k/m)}$ and $Q = \sqrt{(km)}/\eta$ only. (Note that for the *undamped* oscillator, the amplitude response function, (1.31), may be rewritten as $X(\omega) = 1/|1 - (\omega^2/\omega_0^2)|$.)

In particular, find expressions, in terms of Q, for the amplitude 'gain' of the forced oscillator at (i) $\omega = 0$. (ii) $\omega = \omega_0$. (iii) $\omega \to \infty$.

12. Show that the principle of superposition applies to the equation of motion of the string, (1.32). That is if $y(x,t)$ and $z(x,t)$ are solutions, show that $y + z$ is a solution.

13. Two equal masses, mass m each, are connected by a spring of spring constant k in free space. What is the natural angular frequency of the system? What happens if the masses are unequal (m_1 and m_2)?

14. Show that if f and g are any twice-differentiable functions, then

$$y(x,t) = f(x - ct) + g(x + ct)$$

is a solution to the string equation of motion (1.32′). (This is *d'Alembert's solution*.)

Project. A tower block of height h is made out of a Hookean material, with Young's modulus E and density ρ; it has a uniform cross-section A. Let us investigate *vertical* vibration: denote by $y(x,t)$ the vertical displacement from its 'natural' position of the storey at height x above ground level, at time t. By finding the force on an element δx, as is done in section 1.8 for the vibrating string. show that the equation of motion of the tower is

$$\rho \frac{\partial^2 y}{\partial t^2} = E \frac{\partial^2 y}{\partial x^2} + \rho g \ . \qquad (1.39)$$

Find the steady-state solution $Y(x)$ to this equation (that is, the solution for which $\partial Y/\partial t = 0$). Put $z = y - Y$; find a differential equation for z. Follow the method for strings to find harmonic solutions for z. (The hard part here is in finding the boundary condition at $x = h$, at the top of the tower. It is not $z = 0$, but $\partial z/\partial x = 0$ — why?) Does this situation show the 'fundamental and harmonics' features of the vibrating string?

As an approximation, we might regard all the mass of the tower as being at the top. Thus we have an extension oscillator, with a body of mass $\rho A h$ atop a spring of spring constant EA/h. How does the natural frequency of oscillation of this extension oscillator model compare with the *lowest* natural frequency in the more sophisticated calculation?

A better approximation puts half the mass at the top and half the mass at height $h/2$. This is then a double oscillator, with two masses $\rho A h/2$ connected by two springs $2EA/h$. How do the *two* natural frequencies of this double oscillator compare with the *lowest two* of the actual tower?

2
Fourier series

A. USING FOURIER SERIES
2.1 Calculating Fourier coefficients

The series, equation (1.38), that we obtained as a 'solution' to the vibrating string at the end of the previous chapter must be convergent if it is to be a genuine solution. In other words, the solutions we have found to the vibrating string problem are those functions $y(x,t)$ which *can be expressed* as a series of this form. Such an expression of a function, as the limit of a trigonometric series, is called a *Fourier expansion* of the function.

It is a natural question to ask: how general a representation is this? Which functions have Fourier expansions? This is a deep question of mathematical analysis, which we shall only begin to address in this chapter. The answer is that a very wide range of functions can be so expressed; thus, equation (1.38) gives a really rather general solution to the string problem.

First, though, let us look at the practical problem of finding a Fourier expansion of a given function. For functions of one variable defined on a finite interval, the following analysis is the most widely used.

Suppose $f(x)$ is a function, defined for $0 \leqslant x \leqslant l$. A *classical Fourier expansion* of f is an expression of the form

$$f(x) = K + \sum_{n=1}^{\infty} A_n \cos(2n\pi x/l) + \sum_{n=1}^{\infty} B_n \sin(2n\pi x/l) \tag{2.1}$$

for some constants K, A_n and B_n. Notice that the harmonic terms here all have angular frequencies which are a multiple of $2\pi/l$. This gives them periods which are submultiples of l, so that each of the terms fits a whole number of wavelengths into the interval $0 \leq x \leq l$.

Now the harmonic functions $\cos(2n\pi x/l)$ and $\sin(2n\pi x/l)$ satisfy the following relations, for all integers m and n:

$$\int_0^l \sin(2m\pi x/l)\sin(2n\pi x/l)\,dx = \begin{cases} 0 & (\text{if } m \neq n) \\ l/2 & (\text{if } m = n); \end{cases}$$

$$\int_0^l \cos(2m\pi x/l)\cos(2n\pi x/l)\,dx = \begin{cases} 0 & (\text{if } m \neq n) \\ l/2 & (\text{if } m = n); \end{cases}$$

$$\int_0^l \sin(2m\pi x/l)\cos(2n\pi x/l)\,dx = 0;$$

$$\int_0^l \sin(2m\pi x/l)\,dx = 0;$$

and

$$\int_0^l \cos(2m\pi x/l)\,dx = \begin{cases} 0 & (\text{if } m \neq 0) \\ l & (\text{if } m = 0). \end{cases} \tag{2.2}$$

They are collectively referred to as the *orthogonality relations*. These relations can be used to 'pull apart' the Fourier expansion of f, to provide explicit formulae for the coefficients. For if we multiply (2.1) throughout by $\cos(2m\pi x/l)$ and integrate between 0 and l, we obtain

$$\int_0^l \cos(2m\pi x/l) f(x)\,dx$$

$$= \int_0^l \cos(2m\pi x/l)\left\{K + \sum_{n=1}^\infty A_n \cos(2n\pi x/l) + \sum_{n=1}^\infty B_n \sin(2n\pi x/l)\right\} dx$$

$$= \int_0^l K\cos(2m\pi x/l)\,dx + \sum_{n=1}^\infty \left[\int_0^l A_n \cos(2m\pi x/l)\cos(2n\pi x/l)\,dx\right]$$

$$+ \sum_{n=1}^\infty \left[\int_0^l B_n \cos(2m\pi x/l)\sin(2n\pi x/l)\right] dx$$

$$= (l/2)A_m \quad \text{(using (2.2))}.$$

Similarly, multiplying by $\sin(2m\pi x/l)$ and integrating gives

$$\int_0^l \sin(2m\pi x/l) f(x)\,dx = (l/2)B_m\,;$$

and simply integrating equation (2.1) as it stands,

$$\int_0^l f(x) \, dx = Kl.$$

Rearranging these gives

$$A_m = (2/l) \int_0^l \cos(2m\pi x/l) f(x) \, dx$$

$$B_m = (2/l) \int_0^l \sin(2m\pi x/l) f(x) \, dx$$

$$K = (1/l) \int_0^l f(x) \, dx. \tag{2.3}$$

These formulae show that the coefficients are determined uniquely by the function f; thus, if a function has a Fourier expansion, it is unique.

(Formally, the swapping of the sum and integral signs in this calculation needs careful justification, since the sum is infinite. We take it here that, for all appropriate functions f, it may be done.)

All this applies to functions which have Fourier expansions. But the integrals in (2.3) may be calculated for a much wider range of functions. Thus, for any integrable function $f(x)$ defined on $0 \leq x \leq l$, equations (2.3) yield a sequence of values A_n, B_n, and K; and these generate a formal Fourier *series*, written

$$f(x) \to K + \sum_{n=1}^{\infty} A_n \cos(2n\pi x/l) + \sum_{n=1}^{\infty} B_n \sin(2n\pi x/l).$$

If $f(x)$ has a Fourier expansion, this is it. But there are functions $f(x)$ whose Fourier series fails to converge to $f(x)$; these functions do not have a Fourier expansion. This dichotomy, between a function *generating* a Fourier *series* and *possessing* a Fourier *expansion*, is one about which we shall say more below.

2.2 Guitar strings

This calculation applies to classical Fourier series, but the methods may be applied more widely. Thus, the solution to the vibrating string problem (1.38) involves a non-classical Fourier expansion, whose coefficients can be determined in a similar manner.

As an example, consider what happens when a guitar string is plucked in the middle. Fig. 2.1 shows the shape adopted by the string at the moment of plucking.

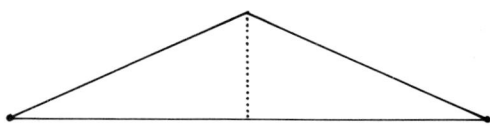

Fig. 2.1 — Guitar string at the moment of plucking.

The string has mass M and length l, the tension in it is T, and the ends are fixed. At the moment of plucking, at time $t=0$, it is drawn sideways a distance ε, and is stationary. The starting position is therefore

$$y(x,0) = \begin{cases} (2\varepsilon x/l) & \text{(if } x \leqslant l/2) \\ (2\varepsilon(1-(x/l))) & \text{(if } x \geqslant l/2). \end{cases} \tag{2.4}$$

This is the actual situation. The analysis of the previous chapter involved harmonic behaviour, giving solutions of the form

$$B_n \sin(n\pi x/l)\cos((nc\pi t/l) + \varphi_n)$$

(equation (1.37)). For this to represent a string which is stationary at $t=0$, φ_n must be 0. Summing harmonic solutions of this type gives

$$y(x,t) = \sum_{n=1}^{\infty} B_n \sin(n\pi x/l)\cos(nc\pi t/l); \tag{2.5}$$

the starting position is then

$$y(x,0) = \sum_{n=1}^{\infty} B_n \sin(n\pi x/l)$$

which is a non-classical Fourier series.

Now, in order to use Fourier methods on the guitar string, we have to match this and equation (2.4); that is, to find the coefficients B_n of a non-classical Fourier expansion of the function (2.4) in this form. As in the classical case, we use integration and orthogonality relations. For a function $f(x) = \sum_{n=1}^{\infty} B_n \sin(n\pi x/l)$, we have

$$\int_0^l \sin(m\pi x/l)f(x)\,dx = \int_0^l \sin(m\pi x/l)\left[\sum_{n=1}^{\infty} B_n \sin(n\pi x/l)\right]dx$$

$$= \sum_{n=1}^{\infty} B_n\left[\int_0^l \sin(m\pi x/l)\sin(n\pi x/l)\,dx\right]$$

$$= B_m(l/2),$$

since

$$\int_0^l \sin(m\pi x/l)\sin(n\pi x/l)\,dx = \begin{cases} 0 & (m \neq n) \\ l/2 & (m = n). \end{cases}$$

With $f(x) = y(x,0)$, as in equation (2.4), this becomes

$$B_m = (2/l)\int_0^{l/2} \sin(m\pi x/l)2\varepsilon(x/l)\,dx + (2/l)\int_{l/2}^{l} \sin(m\pi x/l)2\varepsilon(1-(x/l))\,dx$$

which can be integrated by parts to yield

$$B_m = \begin{cases} (-1)^{(m-1)/2}(8\varepsilon/\pi^2 m^2) & (m \text{ odd}) \\ 0 & (m \text{ even}). \end{cases} \qquad (2.6)$$

Thus

$$y(x,0) \to (8\varepsilon/\pi^2)\{\sin(\pi x/l) - (1/9)\sin(3\pi x/l) + (1/25)\sin(5\pi x/l) \ldots\}. \qquad (2.7)$$

This, then, is the (non-classical) Fourier series generated by the plucked string, equation (2.4).

In this case, the Fourier series converges, and is indeed a Fourier expansion of (2.4). Fig. 2.2 shows how successive partial sums of (2.7) approximate $y(x,0)$ more and more closely.

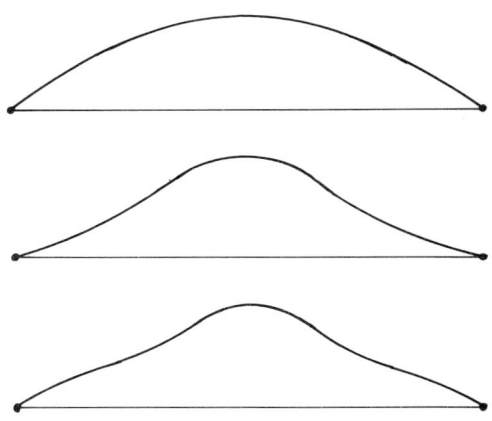

Fig. 2.2 — Three approximations to the position of a plucked string. Top: the fundamental component. Centre: the effect of adding in the $n = 3$ component. Bottom: the sum of the first three components ($n = 1$, $n = 3$ and $n = 5$).

Physically, we can understand this as follows. According to equation (2.7), the vibration of (and therefore the sound from) a plucked string can be thought of as a sum of *pure tones*, each corresponding to one of the natural frequencies of the string. (Equation (1.37) gives these natural frequencies as $nc/2l$.) These tones are present in the sound in varying strengths, and it is this which gives the sound its distinctive quality, or 'timbre'. The strength of each tone in the sound is proportional to the amplitude $|B_n|$ of the corresponding vibrational harmonic, and is therefore given by equation (2.6).

We can represent these amplitudes on a diagram (Fig. 2.3), called the *amplitude spectrum* of the function $y(x,0)$. Here, the x-axis is labelled by the mode number n; the y-axis is for amplitude. This spectrum is therefore a description of the sound

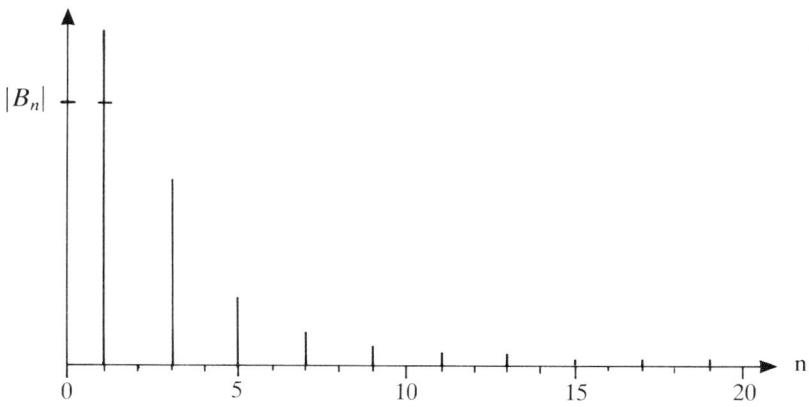

Fig. 2.3 — The amplitude spectrum of the plucked string, Fig. 2.1.

emerging from the plucked string. The largest component is the fundamental, $n = 1$. This is the note to which the string is tuned. In addition, there are, in decreasing amounts, successive (odd) higher harmonics.

Let us take a slightly more complex example. Consider the difference between plucking a guitar string with a plectrum, and plucking it with (the fleshy part of) a finger. The *sounds* made in these ways are quite distinct; and the difference between the two cases lies in the motion of the string.

When the string is plucked with something small and rigid like a plectrum, the 'kink' in the string at the plectrum is quite sharp. In this case, Fig. 2.3 is probably a fair approximation of the resulting spectrum. On the other hand, the flesh of the finger is much softer, so that the 'kink' will be less sharp and more rounded, like Fig. 2.4(a). If we assume the rounded part of the string is a parabola extending over 2% of the string's length, the new starting position is

$$y(x,0) = \begin{cases} 2\varepsilon x/l & \text{(if } x \leq 0.49l\text{)} \\ 0.99\varepsilon - 100\varepsilon((x/l) - \tfrac{1}{2})^2 & \text{(if } 0.49l \leq x \leq 0.51l\text{)} \\ 2\varepsilon(1 - (x/l)) & \text{(if } x \geq 0.51l\text{)} \end{cases} \quad (2.8)$$

for $0 \leq x \leq l$. The calculation of the coefficients is left as an exercise; the amplitude spectrum of this new function looks like Fig. 2.4(b). (For clarity, the amplitudes are plotted on a logarithmic scale.)

The difference can be seen immediately. The spectrum for the finger drops towards zero amplitude much faster than the spectrum for the plectrum. In consequence, the sound made by the finger has much less in it of the high-mode-number, high-frequency harmonics.

2.3 Periodic functions

We now return to classical Fourier series.

A function $f(x)$ is called *periodic* (of *period c*) if there is a positive constant c such that $f(x + c) = f(x)$, for all values of x. We can deduce that $f(x + nc) = f(x)$ for any

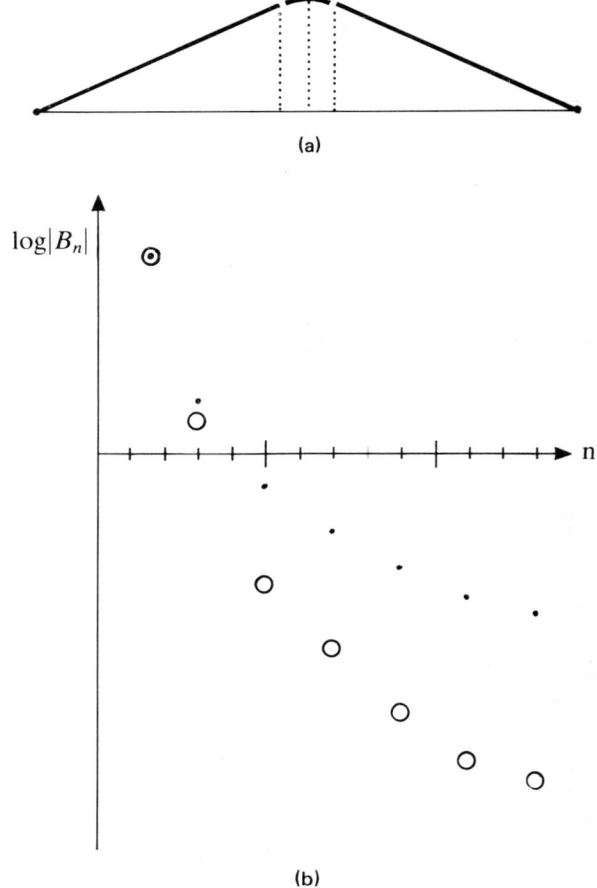

Fig. 2.4 — Plucking a guitar string with the thumb, rather than a plectrum. (a) The string's new starting position. (b) The amplitude spectrum of this curve (open circles) compared with that for Fig. 2.1 (dots). The vertical scale is logarithmic.

integer n, positive or negative. Thus the graph of f repeats itself every c units along the x-axis.

All harmonic functions are periodic. A harmonic function of angular frequency ω has period $2\pi/\omega$ (see section 1.2). But there are many periodic functions that are not harmonic; for example, the function whose graph is shown in Fig. 2.5:

$$\text{Sq}(x) = \begin{cases} 1 & (\text{if } 2n \leq x < 2n+1) \\ -1 & (\text{if } 2n+1 \leq x < 2n+2). \end{cases} \quad (2.9)$$

This function (the *square wave*) is periodic, of period 2. It is not harmonic; it is not even continuous.

A function f which is known to have period c is determined by its values over the range $0 \leq x < c$. Given these, we can find the value of $f(x)$ for any x. It follows that a

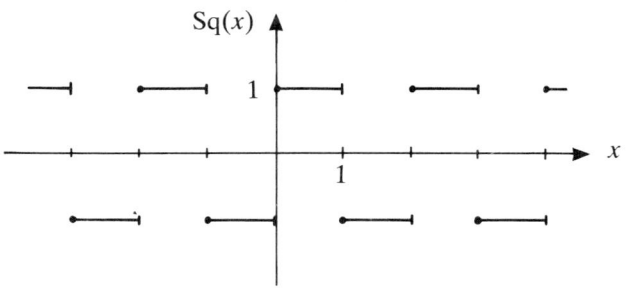

Fig. 2.5 — The square wave, Sq(x).

function of period c is the natural extension of a function on an interval of length c; conversely, given a periodic function, we can limit attention to a single period and treat it as a function on a finite interval.

Because of this close connection, we can use our analyses of functions on a finite interval to extend Fourier theory to periodic functions. Suppose $f(x)$ is periodic of period c, and the function *between 0 and c* is represented by a (classical) Fourier series

$$f(x) = K + \sum_{n=1}^{\infty} A_n \cos(2n\pi x/c) + \sum_{n=1}^{\infty} B_n \sin(2n\pi x/c). \qquad (2.10)$$

The functions involved here — $f(x)$, the constant term K, $\cos(2n\pi x/c)$, and $\sin(2n\pi x/c)$ — are all periodic of period c, so adding a multiple of c to x leaves both sides of the equation unchanged. Thus, equation (2.10) is valid for all x; it is a Fourier expansion for the periodic function $f(x)$.

A periodic function has no preferred 'starting point'. It can be restricted to *any* interval of length one period. The square wave (2.9), for instance, can be thought of as the function

$$f(x) = \begin{cases} 1 & (\text{if } 0 \leq x < 1) \\ -1 & (\text{if } 1 \leq x < 2) \end{cases}$$

repeated along the length of the x axis; or, equally well, as the function

$$f(x) = \begin{cases} -1 & (\text{if } -1 \leq x < 0) \\ 1 & (\text{if } 0 \leq x < 1) \end{cases}$$

repeated. It follows that the Fourier coefficients of a periodic function may be calculated by an integration over *any* entire period; it is not necessary to use the range $0 \leq x \leq c$. Thus, equations (2.3) may be replaced by the expressions

$$A_m = (2/c) \int_a^{a+c} \cos(2m\pi x/c) f(x) \, dx$$

$$B_m = (2/c) \int_a^{a+c} \sin(2m\pi x/c) f(x) \, dx$$

$$K = (1/c) \int_a^{a+c} f(x) \, dx \tag{2.3'}$$

for any constant a.

Except where explicitly stated otherwise, all Fourier series considered in this chapter will be classical Fourier series.

2.4 Amplitude and phase

In the last chapter we discussed the solutions of the harmonic equation, and found two different 'general forms': $A\cos(\omega x) + B\sin(\omega x)$ and $C\cos(\omega x + \varphi)$. This can be applied to the classical Fourier series. Thus there are two different ways of writing the series, equally valid, and different only in their expression: we can write

$$f(x) = K + \sum_{n=1}^{\infty} A_n \cos(2n\pi x/c) + \sum_{n=1}^{\infty} B_n \sin(2n\pi x/c),$$

as we did before, in equation (2.1) (the *cos–sin* expression); or we can write

$$f(x) = K + \sum_{n=1}^{\infty} C_n \cos((2n\pi x/c) + \varphi_n) \tag{2.11}$$

(the *amplitude–phase* expression). C_n is the *amplitude*, and φ_n is the *phase (angle)*, of the nth component of the expansion. C_n can always be taken to be positive, by replacing φ_n by $\varphi_n + \pi$ if necessary.

The double-angle expansion

$$\cos((2n\pi x/c) + \varphi_n) = \cos(\varphi_n)\cos(2n\pi x/c) - \sin(\varphi_n)\sin(2n\pi x/c),$$

may be used to connect the two forms of the series. We have

$$A_n = C_n \cos(\varphi_n)$$

and

$$B_n = -C_n \sin(\varphi_n); \tag{2.12a}$$

conversely,

$$C_n = \sqrt{(A_n^2 + B_n^2)}$$

and

$$\varphi_n = \cos^{-1}(A_n/C_n). \tag{2.12b}$$

The graphical representation of Fourier series is often desirable. There are two methods of doing this, corresponding to the two forms of the written series. One way is to plot a graph of A_n and a graph of B_n, against n; the other is to plot a graph of C_n and a graph of φ_n. These graphs-by-component are called *spectra*. Since the amplitude is usually the most important piece of information about a mode, the amplitude–phase representation is more common; indeed, the phase spectrum is often ignored altogether, and the amplitude spectrum presented alone. (Cf. the nonclassical spectra of section 2.2, Figs 2.3 and 2.4.)

B. SOME USEFUL FUNCTIONS

2.5 Odd and even functions

A function $f(x)$ (on the real numbers) is called *even* if $f(x) = f(-x)$ for all x. This means that its graph is symmetrical under a reflection through the line $x = 0$. A function is *odd* if $f(x) = -f(-x)$ for all x. Most functions are neither even nor odd. Even functions include $\cos x$, x^2 (and indeed all even powers of x), and the *unity function* $1(x)$ which takes the value 1 for all x; odd functions include $\sin x$, x, and all odd powers of x.

If an even function $f(x)$ is also periodic, then it generates a Fourier series

$$f(x) = K + \sum_{n=1}^{\infty} A_n \cos(2n\pi x/c) + \sum_{n=1}^{\infty} B_n \sin(2n\pi x/c),$$

where the coefficients are given by equations (2.3'). In particular,

$$B_m = (2/c) \int_{-c/2}^{c/2} \sin(2m\pi x/c) f(x) \, dx$$

$$= (2/c) \left\{ \int_{0}^{c/2} \sin(2m\pi x/c) f(x) \, dx + \int_{-c/2}^{0} \sin(2m\pi x/c) f(x) \, dx \right\}$$

$$= (2/c) \left\{ \int_{0}^{c/2} \sin(2m\pi x/c) f(x) \, dx - \int_{0}^{c/2} \sin(-2m\pi x/l) f(-x) \, (-dx) \right\}$$

$$= 0.$$

That is, the Fourier expansion of an even function contains no sine terms.

Similarly, if f is an odd periodic function, then there are no cosine terms *and* no constant term. That is, $A_m = 0$ for each m, and $K = 0$.

For any function $f(x)$, the function $f_e(x) = \frac{1}{2}\{f(x) + f(-x)\}$ is even and the function $f_o(x) = \frac{1}{2}\{f(x) - f(-x)\}$ is odd. Moreover, $f(x) = f_e(x) + f_o(x)$. It follows that the Fourier expansion of $f_e(x)$ consists of exactly the constant and cosine terms in the expansion of f; and the expansion of $f_o(x)$ contains exactly the sine terms of f.

2.6 The triangle wave
The function

$$\text{Tr}(x) = \begin{cases} 2n - x - \frac{1}{2} & (2n - 1 \leq x \leq 2n) \\ x - 2n - \frac{1}{2} & (2n \leq x \leq 2n + 1) \end{cases} \quad (2.13)$$

is called the *triangle wave* (see Fig. 2.6). It is related to the starting position of the plucked guitar string. Since it is a periodic function (of period 2), it generates a Fourier series.

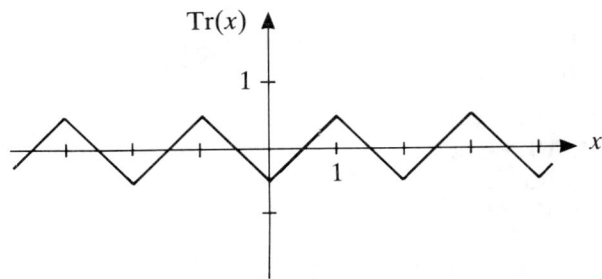

Fig. 2.6 — The triangle wave, Tr(x).

The period $-1 \leq x \leq 1$ is convenient for calculation; then, from equations (2.3'), we have

$$A_m = \int_{-1}^{1} \cos(m\pi x) \text{Tr}(x) \, dx$$

$$= \int_{-1}^{0} (-x - \tfrac{1}{2})\cos(m\pi x) \, dx + \int_{0}^{1} (x - \tfrac{1}{2})\cos(m\pi x) \, dx$$

$$= \begin{cases} -4/m^2\pi^2 & (m \text{ odd}) \\ 0 & (m \text{ even}). \end{cases}$$

Since Tr is an even function, $B_m = 0$ for all m; and

$$K = \int_{-1}^{0} (x + \tfrac{1}{2}) \, dx + \int_{0}^{1} (x - \tfrac{1}{2}) \, dx = 0.$$

Thus the Fourier series generated by the triangle wave is

$$\text{Tr}(x) \to \sum_{n=1}^{\infty} [-4/\pi^2(2n-1)^2] \cos((2n-1)\pi x). \tag{2.14}$$

This Fourier series, in fact, is convergent, and converges to Tr(x); thus (2.14) is a Fourier expansion of Tr(x). (See section 2.16 for a justification of this statement.)

2.7 The square wave

We now turn to an example of a function which demonstrates what we said before: that a function may generate a Fourier series without having a Fourier expansion. This example shows something else too; namely, that even in cases where the Fourier series does not quite work, it may still be useful, with care.

The function in question is the square wave, Fig. 2.5:

$$\text{Sq}(x) = \begin{cases} 1 & (\text{if } 2n \leq x < 2n+1 \\ -1 & (\text{if } 2n+1 \leq x < 2n+2). \end{cases}$$

This has period 2, so from equations (2.3)

$$B_m = \int_0^1 \sin(m\pi x)\, dx + \int_1^2 (-1)\sin(m\pi x)\, dx = \begin{cases} 4/m\pi & (m \text{ odd}) \\ 0 & (m \text{ even}). \end{cases}$$

Since the square wave is an odd function, we have $A_m = 0$ for all m, and $K = 0$. Thus the function Sq generates the Fourier series

$$\text{Sq}(x) \to \sum_{n=1}^{\infty} [4/(2n-1)\pi]\sin((2n-1)\pi x). \tag{2.15}$$

As in the case of the triangle wave, this series converges, although proving that it does is not trivial. But, unlike the series generated by the triangle wave, it does not converge to its generating function Sq(x). Consider the point $x = 0$, for example; here all the terms in the series vanish, so the series has sum 0, but Sq(0) = 1. (Indeed, the same problem arises at every integer value of x.)

The points at which the 'expansion' fails are peculiar, because these are precisely the points at which Sq(x) is discontinuous. It is, perhaps, not surprising that Fourier series — which are sums of the highly continuous harmonic functions — should be unable to deal wholly with discontinuities. However, these are the only 'problem' points in the expansion of Sq(x). The first few partial sums of the series are shown in Fig. 2.7, and you can see that the series does indeed appear to converge to Sq(x) at all *noninteger* values of x. The arrow in (2.15) may be replaced by an equality, except only where x is an integer.

For most applications, we may overlook this failure, and say that we have found the Fourier expansion of the square wave.

2.8 Impulsive 'functions'

Another limiting case that proves to be highly useful in Fourier analysis is the *impulse 'function'*, also called the *(Dirac) delta function*, and given the symbol $\delta(x)$. It is intended to represent an infinitesimally narrow, but infinitely sharp, pulse at $x = 0$. It takes the value 0 everywhere except at $x = 0$, where it is infinite; and its *integral* is 1. It is used in the modelling of very narrow pulses which have a positive effect.

The delta function is not really a function, for it has no finite value at $x = 0$, and we must therefore be careful in using it. (It is an example of the mathematical concept of the *generalized functions*.) Fig. 2.8 illustrates the delta function; since it cannot be plotted like a normal function, it is usually represented by a bold arrow of height 1. (Correspondingly, the function $n\delta(x)$ is drawn as a bold arrow of height n.)

(One way of realizing the delta 'function' is as the 'limit' of the functions

$$\delta_n(x) = \begin{cases} 0 & (x < -1/n) \\ \tfrac{1}{2}n & (-1/n \leq x \leq 1/n) \\ 0 & (x > 1/n) \end{cases} \tag{2.16}$$

as $n \to \infty$. In fact, $\delta(x)$ is usable as a limit of just about any sequence of functions which satisfy its defining properties; so $\delta(x)$ is also the 'limit' of, for example, the sequence $f_n(x) = n/\pi(1 + n^2 x^2)$. This 'construction' is explored in the project at the end of the following chapter.)

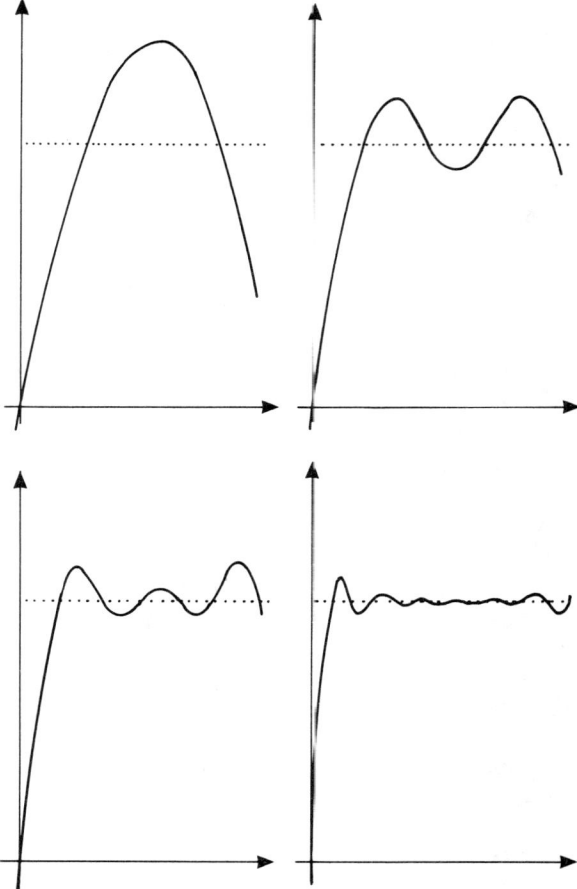

Fig. 2.7 — Partial sums of the Fourier series generated by Sq(x), including one (upper left), two (upper right), three (lower left) and seven (lower right) non-zero terms.

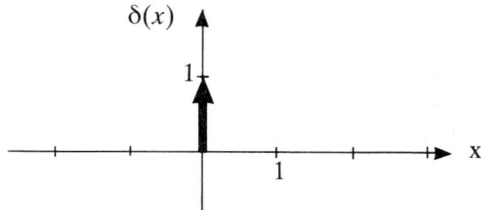

Fig. 2.8 — Representation of the delta 'function', $\delta(x)$.

The delta function has some interesting properties. Most important is the following: for any function $f(x)$ which is continuous at $x = 0$,

$$\int_{-\infty}^{\infty} f(x)\delta(x) \, dx = f(0). \tag{2.17}$$

The delta function is certainly a bizarre construct; and yet, because it is 'integrable', it is actually amenable to Fourier analysis, in a limited and qualified way. It cannot itself have a Fourier series, because it is not periodic. The simplest periodic function involving impulses is the *sampling 'function'*, often symbolized $III(x)$, using the Cyrillic letter III 'shah'; its definition is

$$III(x) = \sum_{n=-\infty}^{\infty} \delta(x - n), \tag{2.18}$$

and its 'graph' is shown in Fig. 2.9. It is zero almost everywhere, with a delta function

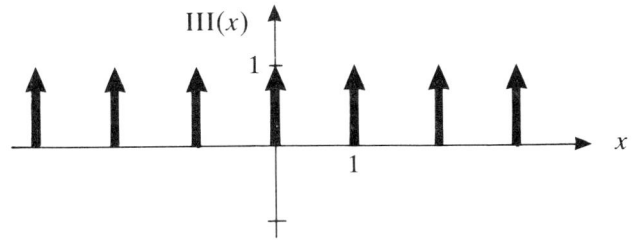

Fig. 2.9 — The sampling 'function', $III(x)$.

pulse at each integer. So the sampling function consists of infinitely many, infinitely high, infinitesimally wide spikes!

The Fourier series generated by $III(x)$ has coefficients as in equation (2.3'). In this case, the period c is 1; with $a = -\frac{1}{2}$, we have

$$B_m = 2 \int_{-1/2}^{1/2} \sin(2m\pi x) III(x) \, dx = 0$$

$$A_m = 2 \int_{-1/2}^{1/2} \cos(2m\pi x) III(x) \, dx = 2$$

and

$$K = \int_{-1/2}^{1/2} III(x) \, dx = 1.$$

The series is

$$\text{III}(x) \to 1 + \sum_{n=1}^{\infty} 2\cos(2n\pi x). \tag{2.19}$$

Fig. 2.10 shows the first few partial sums of this series. The integral under each of

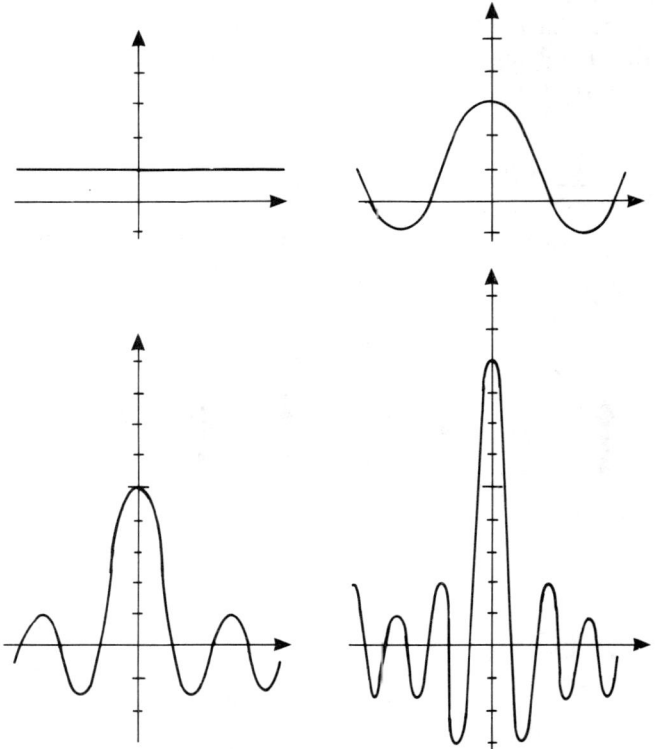

Fig. 2.10 — Partial sums of the Fourier series generated by III(x). Note that the oscillations between integer values of x do not die away.

the 'peaks' (between, say, $-\frac{1}{2}$ and $\frac{1}{2}$) is indeed 1 for each partial sum. But the series fails to converge, for any x. The value of the partial sums at *integer* values of x does tend to infinity (as it should), while at all other x the partial sums oscillate positive and negative.

This is considerably more serious than the square wave problem, where the Fourier series does at least converge. Nevertheless, the expression we have obtained is not useless, as the following example demonstrates. It concerns the forced extension oscillator of section 1.7; but this time, the ceiling vibration is no longer sinusoidal.

The new situation is this. A bob of mass m is suspended from a ceiling, as before, by a Hookean string of spring constant k. On the floor above there is a steam hammer, operating continuously, and making regular strikes once a second. As a result of this the ceiling moves: every second, on the hammer stroke, the ceiling dips briefly, and then recovers.

It does not matter exactly how the ceiling moves. Provided it is back in its rest position very quickly, we can approximate the movement of the ceiling by

$$y(t) = -\alpha \text{III}(t). \tag{2.20}$$

The value of α is determined by the time integral of the displacement of the ceiling during a stroke. (If, for example, the ceiling is shifted by 1 mm for 0.01 s, then $\alpha = 10^{-5}$ m s.) Using the Fourier series for y, we have

$$y(t) \rightarrow \alpha + \sum_{n=1}^{\infty}(2\alpha)\cos(2n\pi t).$$

Now, we solved the case of a *sinusoidally* forced oscillator in the previous chapter (equation (1.29)), so we know that the forcing component $(2\alpha)\cos(2\pi nt)$ on its own would excite the bob to move according to the equation $x(t) = (2\alpha)\cos(2\pi nt)/(1 - (m(2\pi n)^2/k))$. Also, the equation of motion of the extension oscillator is linear, so that we can apply the principle of superposition. The effect, then, of the ceiling moving according to equation (2.20) is that the bob moves according to the equation

$$x(t) = \alpha + \sum_{n=1}^{\infty} \frac{(2\alpha)\cos(2\pi nt)}{1 - (m(2\pi n)^2/k)}. \tag{2.21}$$

It is not obvious whether or not this formula can be put into any simple 'closed' form, but we can in any case use it to draw diagrams of the bob motion. Fig. 2.11 shows a range of solutions. (In fact in this case we *can* put the formula into closed form, by using a more direct analysis; I leave this as an exercise for the reader.)

Since the Fourier series for $\text{III}(x)$ fails to converge, this analysis is, strictly, logically invalid. But a more careful analysis, in which $y(t)$ is treated as a limit, would give exactly the same formula for $x(t)$. What has happened is that the (nonconvergent) series of III has been modified by the oscillator's response function, $1/(1 - m\omega^2/k)$, and the series in (2.21) *is* convergent. Thus, although the Fourier series generated by III fails to converge, it is nevertheless 'good enough' so that the result of this fallacious analysis is actually true.

C. PROPERTIES OF FOURIER SERIES

2.9 Arithmetic properties

Let the function $f(x)$ be periodic of period c, and generate the Fourier series

$$f(x) \rightarrow K + \sum_{n=1}^{\infty} A_n \cos(2n\pi x/c) + \sum_{n=1}^{\infty} B_n \sin(2n\pi x/c)$$

$$= K + \sum_{n=1}^{\infty} C_n \cos((2n\pi x/c) + \varphi_n).$$

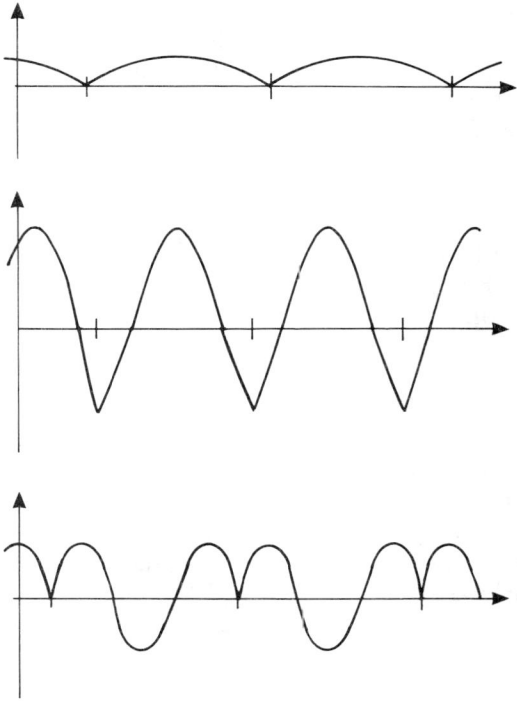

Fig. 2.11 — Movement of the tapped oscillator, for three values of k/m.

Let λ be a constant, and define the functions $r(x)$, $s(x)$, $t(x)$ as

$$r(x) = \lambda f(x); \quad s(x) = f(x + \lambda); \quad t(x) = f(\lambda x).$$

(In graphical terms, the graph of r is that of f stretched vertically by a factor of λ; the graph of s is that of f displaced to the left by λ; and the graph of t is that of f compressed horizontally by a factor of λ.) Then the functions r, s, and t are periodic too, with periods c, c, and c/λ respectively; they generate the Fourier series:

$$r(x) \to \lambda K + \sum_{n=1}^{\infty}(\lambda A_n)\cos(2n\pi x/c) + \sum_{n=1}^{\infty}(\lambda B_n)\sin(2n\pi x/c)$$

$$= \lambda K + \sum_{n=1}^{\infty}(\lambda C_n)\cos((2n\pi x/c) + \varphi_n);$$

$$s(x) \to K + \sum_{n=1}^{\infty} C_n \cos((2n\pi x/c) + [\varphi_n + (2n\pi\lambda/c)]);$$

and

$$t(x) \to K + \sum_{n=1}^{\infty} A_n \cos(2n\pi x/(c/\lambda)) + \sum_{n=1}^{\infty} B_n \sin(2n\pi x/(c/\lambda))$$

$$= K + \sum_{n=1}^{\infty} C_n \cos((2n\pi x/(c/\lambda)) + \varphi_n). \qquad (2.22)$$

(The cos–sin expression for $s(x)$ is not so simple, but it may be found from the expression given here and equations (2.12a).) Moreover, if the series for f is a Fourier *expansion*, so are the series for r, s and t.

Suppose now that $g(x)$ is a second periodic function, of the *same* period c as f, which generates the series

$$g(x) \to L + \sum_{n=1}^{\infty} D_n \cos(2n\pi x/c) + \sum_{n=1}^{\infty} E_n \sin(2n\pi x/c).$$

Then the sum, $h(x) = f(x) + g(x)$, is periodic of period c, and

$$h(x) \to (K + L) + \sum_{n=1}^{\infty} (A_n + D_n)\cos(2n\pi x/c) + \sum_{n=1}^{\infty} (B_n + E_n)\sin(2n\pi x/c).$$
$$(2.23)$$

Again, if the series from f and g are both Fourier *expansions*, then so is the series from h. (In this case, it is the amplitude–phase expression which is not simple; but again, it may be derived from (2.12b).)

There is no simple expression (of either sort) for the Fourier series generated by the product $k(x) = f(x)g(x)$. In fact, it can happen that f and g both have Fourier expansions, but fg does not.

These properties apply to functions on finite intervals as well as periodic functions, with the appropriate changes. They apply to nonclassical Fourier series as well as to classical series.

2.10 Differentiation and integration

It sometimes happens that we need to find the Fourier series for a periodic function $f(x)$ and also for its derivatives df/dx. (Note that if f is periodic and differentiable, the df/dx is also periodic, with the same period as f.) This may be achieved easily: if

$$f(x) \to K + \sum_{n=1}^{\infty} A_n \cos(2n\pi x/c) + \sum_{n=1}^{\infty} B_n \sin(2n\pi x/c)$$

is the Fourier series generated by a periodic function f, then

$$\frac{df(x)}{dx} \to \frac{d}{dx}\left\{ K + \sum_{n=1}^{\infty} A_n \cos(2n\pi x/c) + \sum_{n=1}^{\infty} B_n \sin(2n\pi x/c) \right\}$$

$$= \sum_{n=1}^{\infty} A_n \frac{d}{dx}(\cos(2n\pi x/c)) + \sum_{n=1}^{\infty} B_n \frac{d}{dx}(\sin(2n\pi x/c))$$

$$= \sum_{n=1}^{\infty} (-2A_n n\pi/c)\sin(2n\pi x/c) + \sum_{n=1}^{\infty} (2B_n n\pi/c)\cos(2n\pi x/c) \quad (2.24)$$

is the Fourier series for df/dx. In the amplitude–phase form,

$$f(x) \to K + \sum_{n=1}^{\infty} C_n \cos((2n\pi x/c) + \varphi_n)$$

yields

$$\frac{df(x)}{dx} \to (d/dx)\left\{ K + \sum_{n=1}^{\infty} C_n \cos((2n\pi x/c) + \varphi_n) \right\}$$

$$= \sum_{n=1}^{\infty} (-2C_n n\pi/c)\sin((2n\pi x/c) + \varphi_n)$$

$$= \sum_{n=1}^{\infty} (2C_n n\pi/c)\cos((2n\pi x/c) + \varphi_n + \tfrac{1}{2}\pi) \quad (2.25)$$

To prove these, we note that the coefficients of the series generated by df/dx are given by (2.3). For example, the coefficient of $\cos(2n\pi x/c)$ is

$$(2/c) \int_0^c \cos(2\pi nx/c)\,(df/dx)\,dx.$$ But, integrating by parts,

$$\int_0^c \cos(2\pi nx/c) \frac{df}{dx} dx = [\cos(2\pi nx/c)f(x)]_0^c + (2\pi n/c)\int_0^c \sin(2\pi nx/c)f(x)\,dx$$

$$= [\cos(2\pi nx/c)f(x)]_0^c + \pi n B_n.$$

The expression in the brackets is periodic of period c, and so this term vanishes. Thus the coefficient of $\cos(2n\pi x/c)$ in the series for df/dx is just $(2n\pi/c)B_n$. The other coefficients can be calculated similarly.

Warning: it does *not* follow automatically that if f has a Fourier *expansion*, then df/dx does too.

There is a converse procedure for integration, which follows the same pattern. If we put $F(x) = \int_0^x f(t)\,dt$, then

$$F(x) \to \int_0^x \left\{ K + \sum_{n=1}^{\infty} A_n \cos(2n\pi x/c) + \sum_{n=1}^{\infty} B_n \sin(2n\pi x/c) \right\} dx$$

$$= Kx + \sum_{n=1}^{\infty}(A_n c/2n\pi)\sin(2n\pi x/c) + \sum_{n=1}^{\infty}(B_n c/2n\pi)\cos(2n\pi x/c). \quad (2.26)$$

Note here that if F is to be periodic, then K must be zero, and equation (2.26) then gives the Fourier series generated by F.

2.11 Parseval's equation

We saw in section 2.1 how the orthogonality of the functions $\sin(2m\pi x/c)$ and $\cos(2m\pi x/c)$ is important in the calculation of Fourier coefficients. Thus, in order to obtain the coefficients in the expansion

$$f(x) = K + \sum_{n=1}^{\infty} A_n \cos(2n\pi x/c) + \sum_{n=1}^{\infty} B_n \sin(2n\pi x/c),$$

we multiplied both sides of the equation by each of these functions in turn, and integrated. It was their mutual orthogonality that rendered all but one of the integrals zero, and gave us an equation involving only one of the coefficients.

If we try this process again, multiplying both sides by $f(x)$ itself and integrating over a period, we obtain a curious and important relation:

$$\int_0^c f(x)^2 \, dx = \int_0^c f(x) \left\{ K + \sum_{n=1}^{\infty} A_n \cos(2n\pi x/c) + \sum_{n=1}^{\infty} B_n \sin(2n\pi x/c) \right\} dx$$

$$= K \int_0^c f(x) \, dx + \sum_{n=1}^{\infty} A_n \left(\int_0^c f(x)\cos(2n\pi x/c) \, dx \right)$$

$$+ \sum_{n=1}^{\infty} B_n \left(\int_0^c f(x)\sin(2n\pi x/c) \, dx \right)$$

$$= cK^2 + (c/2)\sum_{n=1}^{\infty} A_n^2 + (c/2)\sum_{n=1}^{\infty} B_n^2. \quad (2.27)$$

This equation is called *Parseval's equation*.

There is a physical interpretation of these equations. The *energy* (*density*) of a wave is related to the square of its amplitude. One way of expressing equation (2.27), then, is that *the energy density of a periodic function is the sum of the energy densities of its Fourier components*.

Parseval's equation can be generalized. Suppose that $f(x)$ and $g(x)$ are two separate periodic functions, of the *same period c*; let them have Fourier expansions

$$f(x) = K + \sum_{n=1}^{\infty} A_n \cos(2n\pi x/c) + \sum_{n=1}^{\infty} B_n \sin(2n\pi x/c)$$

and

$$g(x) = L + \sum_{n=1}^{\infty} C_n \cos(2n\pi x/c) + \sum_{n=1}^{\infty} D_n \sin(2n\pi x/c).$$

Now multiply the first of these equations by $g(x)$, and integrate from 0 to c, to get

$$\int_0^c f(x)g(x)\,dx = \int_0^c g(x)\left\{K + \sum_{n=1}^{\infty} A_n \cos(2n\pi x/c) + \sum_{n=1}^{\infty} B_n \sin(2n\pi x/c)\right\} dx$$

$$= K \int_0^c g(x)\,dx + \sum_{n=1}^{\infty} A_n \left(\int_0^c g(x)\cos(2n\pi x/c)\,dx\right)$$

$$+ \sum_{n=1}^{\infty} B_n \left(\int_0^c g(x)\sin(2n\pi x/c)\,dx\right)$$

$$= cKL + (c/2)\sum_{n=1}^{\infty} A_n C_n + (c/2)\sum_{n=1}^{\infty} B_n D_n. \tag{2.28}$$

Parseval's equation is just the special case of this, in the case where the two functions f and g are the same.

2.12 Decay of coefficients at large n

When we calculated the Fourier series generated by the sampling function, we found that it failed to converge. This is because its coefficients do not tend to zero at large mode numbers, with the result that the addition of additional terms of the series continues to have a significant effect on successive approximations. Conversely, for a function f to have a valid Fourier expansion

$$f(x) = K + \sum_{n=1}^{\infty} A_n \cos(2n\pi x/c) + \sum_{n=1}^{\infty} B_n \sin(2n\pi x/c),$$

we must necessarily have

$$\lim_{n\to\infty} A_n = \lim_{n\to\infty} B_n = 0. \tag{2.29}$$

That many functions have this property is the content of the *Riemann–Lebesgue lemma*. It is true, for instance, of any function f which is continuously differentiable. (We prove this in section 2.16 below.)

One way to see that this is so is from Parseval's equation, (2.27); for the integral $\int_0^c f(x)^2\,dx$ is finite, and so the sums $\sum_{n=1}^{\infty} A_n^2$ and $\sum_{n=1}^{\infty} B_n^2$ must converge. But these are sums of *positive* terms, so the terms must tend to zero with increasing n.

The Riemann–Lebesgue lemma can be combined with the results we noted concerning the derivatives of functions (2.24). If f is *twice* continuously differentiable, then (2.24) holds. It follows that df/dx is (once) continuously differentiable, and its Fourier coefficients are $2\pi n B_n/c$ and $-2\pi n A_n/c$. These tend to zero, by (2.29); and as $2\pi/c$ is a constant, we have the stronger statement that

$$\lim_{n \to \infty} nA_n = \lim_{n \to \infty} nB_n = 0.$$

That is, not only do A_n and B_n tend to zero, but they do so faster than $1/n$.

This argument may be applied any number of times. Thus, if f is $k+1$ times continuously differentiable, then $\lim_{n \to \infty} n^k A_n = \lim_{n \to \infty} n^k B_n = 0$, so that A_n and B_n tend to zero faster than $1/n^k$.

There is a corollary to this, which is that the eventual behaviour of the Fourier coefficients depends on the most 'awkward' points of the function f. The fact that the coefficients of the square wave decay no faster than $1/n$ is due to the isolated points of discontinuity; apart from at these points, the function is many times differentiable. Similarly, the plucked guitar string assumes a continuous shape, which fails to be many times differentiable only at the kink; thus the Fourier coefficients decay only as $1/n^2$.

D. THEORY OF FOURIER SERIES

2.13 Formulation using complex numbers

The formulae involved in the classical Fourier series, as we have presented them here, lack a certain elegance. In order to make the theory more mathematically appealing, we can reformulate it using complex numbers. The important relation for this is the definition

$$e^{i\theta} = \cos\theta + i\sin\theta$$

of the complex exponential function.

We define:

$$\begin{aligned} Z_m &= (1/c) \int_0^c e^{-2m\pi i x/c} f(x) \, dx \\ &= (1/c) \int_0^c \cos(2m\pi x/c) f(x) \, dx - i(1/c) \int_0^c \sin(2m\pi x/c) f(x) \, dx \\ &= \tfrac{1}{2}(A_m - iB_m). \end{aligned} \quad (2.30)$$

for $m = 1, 2, \ldots$ Moreover, for $m = 0$, $e^{-2im\pi x/c} = e^0 = 1$, so that

$$\begin{aligned} Z_0 &= (1/c) \int_0^c f(x) \, dx \\ &= K. \end{aligned} \quad (2.31)$$

Note also that the definition in equation (2.30) still makes sense when m is *negative*. Since $e^{-i\theta}$ and $e^{i\theta}$ are complex conjugates, we have

$$Z_{-m} = \overline{Z_m} = \tfrac{1}{2}(A_m + iB_m). \quad (2.32)$$

We therefore have that
$$A_m = Z_m + Z_{-m}$$
and (2.33)
$$iB_m = Z_m - Z_{-m}.$$

The classical Fourier series generated by a function f is
$$f(x) \to K + \sum_{n=1}^{\infty} A_n \cos(2n\pi x/c) + \sum_{n=1}^{\infty} B_n \sin(2n\pi x/c)$$
$$= K + \sum_{n=1}^{\infty} [A_n \cos(2n\pi x/c) + B_n \sin(2n\pi x/c)].$$

But notice that $A_n \cos(2n\pi x/c) + B_n \sin(2n\pi x/c)$ is the real part of the complex product
$$(A_n - iB_n)(\cos(2n\pi x/c) + i\sin(2n\pi x/c)) = 2Z_n e^{2\pi nix/c},$$
so that
$$f(x) \to K + \sum_{n=1}^{\infty} \mathrm{Re}[2Z_n e^{2\pi nix/c}]. \tag{2.34}$$

Since, for any complex number z, $\mathrm{Re}\, z = \tfrac{1}{2}z + \tfrac{1}{2}\overline{z}$, we have
$$\mathrm{Re}[2Z_n e^{2\pi nix/c}] = [Z_n e^{2\pi nix/c}] + [\overline{Z}_n e^{-2\pi nix/c}]$$
$$= Z_n e^{2\pi nix/c} + Z_{-n} e^{2\pi(-n)ix/c},$$
and therefore
$$A_n \cos(2n\pi x/c) + B_n \sin(2n\pi x/c) = Z_n e^{2\pi nix/c} + Z_{-n} e^{2\pi(-n)ix/c}. \tag{2.35}$$

Thus the standard series is the same as
$$f(x) \to Z_0 + \sum_{n=1}^{\infty} (Z_n e^{2\pi nix/c} + Z_{-n} e^{2\pi(-n)ix/c})$$
$$= \sum_{n=-\infty}^{\infty} Z_n e^{2\pi nix/c}. \tag{2.36}$$

Notice the minus sign in the definition of Z_n.

There is a close connection between the complex formulation of the Fourier series and the amplitude–phase formulation,
$$f(x) = K + \sum_{n=1}^{\infty} C_n \cos((2n\pi x/c) + \varphi_n).$$

Just as $Z_n = \tfrac{1}{2}(A_n - iB_n)$, so $Z_n = \tfrac{1}{2}C_n \exp(i\varphi_n)$; it follows that

$$C_n = 2|Z_n|$$

and

$$\varphi_n = \arg(Z_n). \tag{2.37}$$

Parseval's equation, (2.27), becomes neater if we use the complex formulation of (2.36). If the Fourier series is a Fourier expansion, then

$$\int_0^c f(x)^2 \, dx = c \sum_{n=-\infty}^{\infty} |Z_n|^2. \tag{2.38}$$

2.14 Approximation of functions

The Fourier series gives us a way of expressing many periodic functions *exactly* as an infinite sum of sinusoidal components. By truncating the series, we can express such functions *approximately* as a *finite* sum of harmonics. Thus, if we have a function $f(x)$ which has a valid Fourier expansion

$$f(x) = K + \sum_{n=1}^{\infty} A_n \cos(2n\pi x/c) + \sum_{n=1}^{\infty} B_n \sin(2n\pi x/c),$$

and we define

$$f_N(x) = K + \sum_{n=1}^{N} A_n \cos(2n\pi x/c) + \sum_{n=1}^{N} B_n \sin(2n\pi x/c), \tag{2.39}$$

then $f(x) \simeq f_N(x)$ for large enough N.

Although truncated series are easier to calculate with, they have an important limitation: it is not easy to tell, in advance, how many terms are needed for accurate approximation. Indeed, it is not quite clear what the phrase 'accurate approximation' means. Sometimes we require that $f(x) - f_N(x)$ must be between the same limits $\pm \varepsilon$ for all x; then $f_N(x)$ is said to approximate $f(x)$ *uniformly*. At other times we ask that the 'average error', the integral $\int_0^c [f(x) - f_N(x)]^2 \, dx$, be small; then $f_N(x)$ is said to approximate $f(x)$ *in the mean*. Which definition we choose depends on our purposes. But there is a simple rule of thumb that will work adequately for many practical cases. If the smallest detail of $f(x)$ we want to capture in the approximation extends over an x-range of δ, then we need to take the series at least as far as sinusoids with wavelength δ. Thus we need to take $N \geq c/\delta$. If we want this detail represented *accurately*, we will probably need a significantly larger N. As a guide to how this works, look again at the successive approximations in Figs 2.2, 2.7, and 2.10.

Curiously, to find the truncations $f_N(x)$ it is not necessary to calculate the coefficients separately. There is a 'short cut'.

By definition,

$$f_N(x) = K + \sum_{n=1}^{N} [A_n \cos(2n\pi x/c) + B_n \sin(2n\pi x/c)]$$

$$= \sum_{n=-N}^{N} Z_n e^{2n\pi ix/c}, \qquad (2.40)$$

from equation (2.35). Inserting the integral definition of Z_n, we have

$$f_N(x) = \sum_{n=-N}^{N} \left\{ (1/c) \int_0^c e^{-2n\pi iy/c} f(y) \, dy \right\} e^{2n\pi ix/c}$$

$$= (1/c) \int_0^c f(y) \left[\sum_{n=-N}^{N} e^{2n\pi i(x-y)/c} \right] dy$$

$$= (1/c) \int_0^c f(y) D_N[2\pi(x-y)/c] \, dy, \qquad (2.41)$$

where

$$D_N(x) = \sum_{n=-N}^{N} e^{nix}. \qquad (2.42)$$

The function $D_N(x)$ is called the *Dirichlet kernel of order N*. Once we have calculated the Dirichlet kernel, we only need to perform a *single* integration to get $f_N(x)$.

The formula of (2.42) can be simplified further, for the right-hand side is a finite geometric series of $2N+1$ terms, with starting point e^{Nix} and constant ratio e^{-ix}. It follows that

$$D_N(x) = e^{Nix}[1 - e^{-(2N+1)ix}]/[1 - e^{-ix}]$$
$$= [e^{(2N+1)ix/2} - e^{-(2N+1)ix/2}]/[e^{ix/2} - e^{-ix/2}]$$
$$= \sin[(N+\tfrac{1}{2})x]/\sin[\tfrac{1}{2}x]. \qquad (2.43)$$

Thus equation (2.41) becomes

$$f_N(x) = (1/c) \int_0^c f(y) \frac{\sin[(2N+1)\pi(x-y)/c]}{\sin[\pi(x-y)/c]} \, dy. \qquad (2.44)$$

Fig. 2.12 shows the graph of $D_N(x)$ for $-\pi \leq x \leq \pi$.

Notice that the integrand is periodic of period c. Thus the integration may be made over any interval $[a, a+c]$ to yield the same result.

2.15 Discontinuities and the Gibbs phenomenon

We saw in section 2.7 that the square wave $Sq(x)$ does not have a Fourier expansion, owing to the presence of discontinuities at integer values of x. Yet the Fourier series it generates is convergent, and the limit of the series is very close to Sq. The series converges even at integer x. This is because the discontinuities are *simple*: there exist right- and left-hand limits, and finite right- and left-hand derivatives, at each point of discontinuity.

If $f(x)$ is a periodic function with a simple discontinuity at $x = x_0$, which is

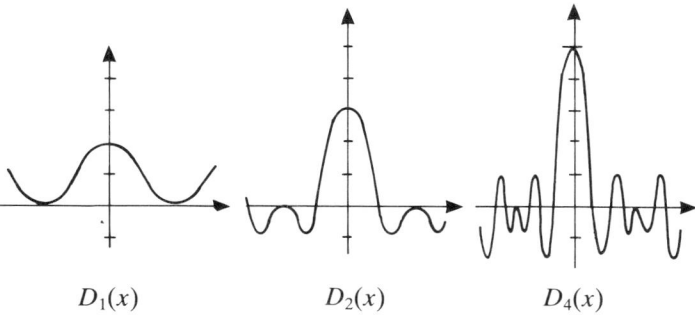

Fig. 2.12 — The Dirichlet kernels of orders 1, 2 and 4.

continuously differentiable near x_0, then it can be proved that the Fourier series generated by f converges, even at $x = x_0$; and that the limit of the series lies halfway between the left-hand limit and the right-hand limit:

$$f_N(x_0) \to \tfrac{1}{2}(f(x_0+) + f(x_0-)) \quad \text{as } N \to \infty. \tag{2.45}$$

In the case of Sq(x), the one-sided limits at $x = 0$ are Sq($0+$) = 1, Sq($0-$) = 1, so that the Fourier series tends to $\tfrac{1}{2}(1 + (-1)) = 0$; as indeed we found.

A more interesting property of a Fourier approximation $f_N(x)$ near a discontinuity concerns the 'overshoot'. Fig. 2.7 shows how Sq$_N(x)$ rises from near -1 (left of $x = 0$) to near $+1$ (right of $x = 0$): rather than approaching steadily and monotonically, it overshoots, then oscillates, before settling down. As N becomes large, the oscillation and settling happen faster and faster, so that Sq$_N(x) \to$ Sq(x) for non-integral x, as we have seen. However, the *amount* of overshoot does *not* tend to zero with increasing N.

The function Sq$_{2N}(x)$ is given, by equation (2.39) and section 2.7, by

$$\text{Sq}_{2N}(x) = \text{Sq}_{2N-1}(x) = \sum_{n=1}^{N} (4/\pi(2n-1))\sin((2n-1)\pi x).$$

Differentiating this gives

$$\frac{d}{dx}(\text{Sq}_{2N}(x)) = \sum_{n=1}^{N} 4\cos((2n-1)\pi x)$$

$$= \frac{2\sin(2N\pi x)}{\sin(\pi x)}. \tag{2.46}$$

Thus, Sq$_{2N}(x)$ has its turning points where $\sin(2N\pi x) = 0$: at integer multiples of $1/2N$ (other than $x = 0$). Its first turning point is the point of maximum overshoot, and occurs at $x = 1/2N$.

$Sq_{2N}(x)$ is also given by equation (2.41). For the square wave, $c = 2$; so changing variables to $z = x - y$ gives

$$Sq_{2N}(x) = \tfrac{1}{2} \int_{-1}^{1} Sq(x-z) D_{2N}[\pi z] \, dz. \tag{2.47}$$

But D_{2N} is even, while Sq is odd. Thus much of the integral cancels out, and we are left with

$$Sq_{2N}(1/2N) = \left\{ \int_{0}^{1/2N} - \int_{1}^{1+(1/2N)} \right\} D_{2N}[\pi z] \, dz. \tag{2.48}$$

Since D_N is bounded by $\sqrt{2}$ between $\pi/2$ and $3\pi/2$, the second integral tends to zero as $N \to \infty$. Thus

$$\lim_{N \to \infty} Sq_{2N}(1/2N) = \lim_{N \to \infty} \int_{0}^{1/2N} \frac{\sin[(2N+\tfrac{1}{2})\pi z]}{\sin[\tfrac{1}{2}\pi z]} \, dz. \tag{2.49}$$

Putting $u = (2N + \tfrac{1}{2})\pi z$, and using the fact that $\sin t/t \to 1$ as $t \to 0$,

$$\lim_{N \to \infty} Sq_{2N}(1/2N) = \lim_{N \to \infty} (1/(2N+\tfrac{1}{2})\pi) \int_{0}^{\pi(1+1/4N)} \frac{\sin u}{\sin[u/(4N+1)]} \, du$$

$$= (2/\pi) \int_{0}^{\pi} (\sin u/u) \, du$$
$$\simeq 1.179. \tag{2.50}$$

This compares with $Sq(0+) = 1$. Thus, for very large N, the Fourier approximation to Sq near the discontinuity at $x = 0$ still overshoots by about 9% of the total jump (of 2, from -1 to $+1$). This, like equation (2.45), is a very general property of simple discontinuities; it is called the *Gibbs phenomenon*.

2.16 Convergence of Fourier series

Finally in this chapter, we sketch a proof that a large range of functions f have Fourier expansions: Fourier series which converge to $f(x)$.

The class of functions for which this proof is valid is the class of continuously differentiable ('smooth') functions, though with a little extra effort it can be expanded to include a greater range (for instance, *piecewise* smooth functions, which are smooth except at a finite number of simple discontinuities). The conclusion is *not* true for general continuous functions: there are continuous functions whose Fourier series fail to converge at infinitely many points. (These examples are not easy to construct, however, and are certainly beyond the scope of this book.)

The proof proceeds in three steps.

Step 1.
We begin by proving the Riemann–Lebesgue lemma, (2.29), for a smooth function $f(x)$. For constants a, b, and α, integration by parts yields

$$\int_a^b f(x)\cos(\alpha x)\, dx = [(1/\alpha)\sin(\alpha x)f(x)]_a^b - \int_a^b (1/\alpha)\sin(\alpha x)\frac{df(x)}{dx}\, dx$$

$$= (1/\alpha)\Bigg\{f(b)\sin(\alpha b) - f(a)\sin(\alpha a)$$

$$- \int_a^b \sin(\alpha x)\frac{df(x)}{dx}\, dx\Bigg\}.$$

Now df/dx is continuous on $[a,b]$, so $|df/dx|$ is integrable. It follows that

$$\left|\int_a^b f(x)\cos(\alpha x)\, dx\right| \leq (1/\alpha)\left\{|f(b)| + |f(a)| + \int_a^b \left|\frac{df(x)}{dx}\right|\, dx\right\}.$$

In particular,

$$\lim_{\alpha \to \infty} \int_a^b f(x)\cos(\alpha x)\, dx = 0. \tag{2.51}$$

Similarly, $\lim_{\alpha \to \infty} \int_a^b f(x)\sin(\alpha x)\, dx = 0$. Equation (2.29) is obtained from this by putting $a = 0$, $b = c$, and $\alpha = 2\pi n/c$.

(In fact this argument proves a stronger result: namely, that for smooth functions, the set $\{nA_n\}$ is bounded.)

Step 2.

The next step is the evaluation of a second integral limit, $\lim_{\alpha \to \infty} \int_{-a}^b f(x)(\sin(\alpha x)/x)\, dx$. We assume a and b are positive, and that f is smooth. Then we can divide up the integral as $\int_{-a}^{-\varepsilon} + \int_{-\varepsilon}^{\varepsilon} + \int_{\varepsilon}^b$, for any $\varepsilon > 0$. Since $f(x)/x$ is smooth over the interval $[\varepsilon, b]$, the Riemann–Lebesgue lemma tells us that $\lim_{\alpha \to \infty} \int_\varepsilon^b f(x)(\sin(\alpha x)/x)\, dx = 0$; similarly with the interval $[-a, -\varepsilon]$. Thus

$$\lim_{\alpha \to \infty} \int_{-a}^b f(x)(\sin(\alpha x)/x)\, dx = \lim_{\alpha \to \infty} \int_{-\varepsilon}^{\varepsilon} f(x)(\sin(\alpha x)/x)\, dx$$

for any $\varepsilon > 0$.

Now f is differentiable at 0, with derivative $df(0)/dx = \beta$, say; so, by definition, for any $\delta > 0$ there is an $\varepsilon > 0$ such that $|(f(x)/x) - (f(0)/x) - \beta| \leq \delta$ whenever $-\varepsilon \leq x \leq \varepsilon$.

It follows that $f(x)(\sin(\alpha x)/x)$ always lies between $f(0)(\sin(\alpha x)/x) - \beta\sin(\alpha x) \pm \delta$; and therefore on integration, we have that $\lim_{\alpha \to \infty} \int_{-a}^{b} f(x)(\sin(\alpha x)/x) \, dx$ lies between

$$\lim_{\alpha \to \infty} \left\{ f(0) \int_{-\varepsilon}^{\varepsilon} (\sin(\alpha x)/x) \, dx - \beta \int_{-\varepsilon}^{\varepsilon} \sin(\alpha x) \, dx \right\} \pm 2\delta\varepsilon.$$

But as $\delta \to 0$, the last term tends to zero; and the β term simply vanishes. We are left with the equation

$$\lim_{\alpha \to \infty} \int_{-a}^{b} f(x)(\sin(\alpha x)/x) \, dx = f(0) \lim_{\alpha \to \infty} \int_{-\varepsilon}^{\varepsilon} (\sin(\alpha x)/x) \, dx.$$

Putting $y = \alpha x$, the last integral becomes $\lim_{\alpha \to \infty} \int_{-\alpha\varepsilon}^{\alpha\varepsilon} (\sin y/y) \, dy$, or just

$\int_{-\infty}^{\infty} (\sin y/y) \, dy$. This is a standard integral, with value π. Thus

$$\lim_{\alpha \to \infty} \int_{-a}^{b} f(x)(\sin(\alpha x)/x) \, dx = \pi f(0). \tag{2.52}$$

Step 3.
We now have enough to prove our convergence theorem. For suppose $f(x)$ is a smooth, periodic function, and suppose $f_N(x)$ is its Nth approximation (equation (2.39)). Fix x between 0 and c, and define

$$g(y) = (1/c)f(y)\frac{[\pi(x-y)/c]}{[\sin(\pi(x-y)/c)]}. \tag{2.53}$$

Since $t/\sin t$ is smooth for $-\pi < t < \pi$, $g(x)$ is smooth for $x - c < y < x + c$, and so certainly for $0 \leq y \leq c$. Then equation (2.44) gives the expression

$$f_N(x) = (1/c) \int_0^c f(y) \frac{\sin[(2N+1)\pi(x-y)/c]}{\sin[\pi(x-y)/c]} \, dy$$

$$= \int_0^c g(y) \frac{\sin[(2N+1)\pi(x-y)/c]}{[\pi(x-y)/c]} \, dy.$$

Make the substitution $z = 2\pi(x-y)/c$, so $y = zc/2\pi + x$, and this becomes

$$f_N(x) = \int_{2\pi(x/c-1)}^{2\pi x/c} g(x + zc/2\pi) \, (\sin[(N+\tfrac{1}{2})z]/\tfrac{1}{2}z) \, (c/2\pi) \, dz;$$

then, from (2.52), we have

$$\lim_{N \to \infty} f_N(x) = \pi(c/\pi) g(x+0) = f(x).$$

So at each point x in the range $0 < x < c$, $f_N(x)$ converges to $f(x)$.

Notice in this proof that the endpoints, $x=0$ and $x=c$, are excluded. This restriction can be removed, since the function f is smooth at these points, by changing the period to $[-\frac{1}{2}c,\frac{1}{2}c]$. Functions defined on a finite interval, however, may be smooth *within* the interval, but yield periodic functions which are not smooth *at* the endpoints. For such functions, the restriction is important.

Notice also that, provided f is integrable merely, $f_N(x)$ exists. The proof given may then be restructured to give the stronger theorem: if f is continuously differentiable *at* $x=a$, $0<a<c$, then $f_N(a) \to f(a)$ as $N \to \infty$.

EXERCISES

1. We saw how plucking a guitar string in the middle gives a response whose spectrum contains a substantial component at the string's fundamental frequency. What happens if the string is 'plucked' at two points simultaneously, so that the starting position is as in Fig. 2.13?

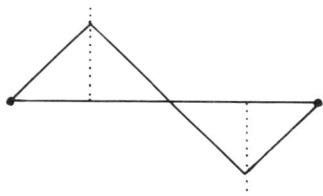

Fig. 2.13.

2. Find the (non-classical) Fourier series generated by the function $f(x) = (x-1)^2$, $0 \leq x \leq 2$, in the form $f(x) = \sum_{n=1}^{\infty} B_n \sin(\frac{1}{2}n\pi x)$. (Note that the function $f(x)$ is not zero at the endpoints of its interval; this is not a solution for the vibrating string.)

3. Find explicitly the coefficients of the Fourier series of the 'finger-plucked' guitar string, equation (2.8). Compare them with those for the 'plectrum-plucked' string, equation (2.6), in the light of section 2.12.

4. The *sawtooth wave* is defined as

$$\text{Saw}(x) = x - 2n \quad (2n-1 \leq x < 2n+1) \tag{2.54}$$

(Fig. 2.14). What are the coefficients of the Fourier series it generates?

Does this series converge at $x=1$? If so, to what? How does this fit in with equation (2.45)?

5. Use section 2.5 to find the Fourier series generated by the function

$$f(x) = \begin{cases} 0 & (2n-1 \leq x < 2n) \\ x - 2n & (2n \leq x < 2n+1). \end{cases}$$

Sketch the graph of this function.

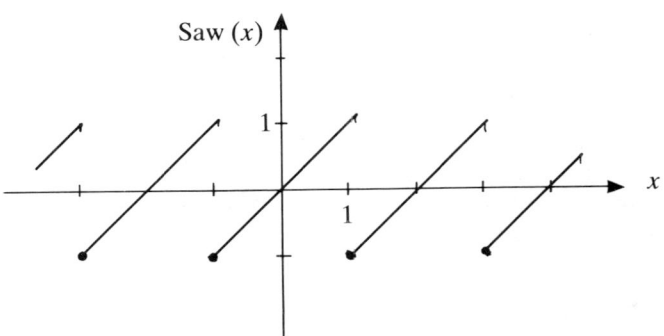

Fig. 2.14 — The sawtooth wave, Saw(x).

6. Use section 2.10 to connect the Fourier series generated by Tr(x) and Sq(x).
7. Use section 2.10 to connect the Fourier series generated by Saw(x) and III(x).
8. Apply the derivative theorem to the function of Question 2.
9. Analyse directly the problem given in section 2.8 of an extension oscillator forced by a ceiling vibration of III(t).
10. Find the Fourier expansion of the *fully rectified wave* function $f(x) = |\sin(\pi x)|$ (Fig. 2.15a).

Show how this can be deduced from the result of the preceding question, by making a suitable choice of m, k, and α.

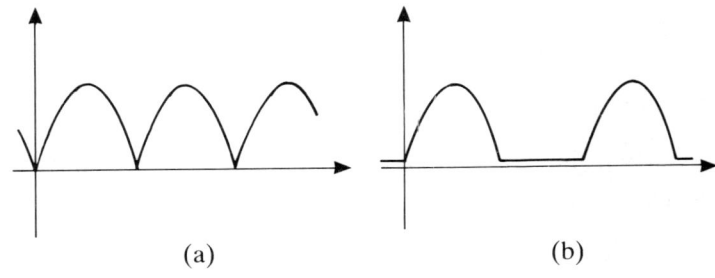

Fig. 2.15.

11. The *half-rectified wave* function is the function

$$g(x) = \begin{cases} \sin(\pi x) & (\sin(\pi x) \geq 0) \\ 0 & (\sin(\pi x) < 0) \end{cases}$$

(shown in Fig. 2.15b). Find the even and odd parts of this function, $g_e(x)$ and $g_o(x)$. Using the results of the previous question, find the Fourier expansion of g.

12. Find an expression for the generalized Parseval's equation, (2.28), in terms of the complex formulation of Fourier series.

13. Use the techniques of section 2.14 to find expressions for the Nth order truncations of the Fourier series for

    (a) the sawtooth wave, Saw(x).
    (b) the sampling function, III(x).

14. (Harder) Show that the sawtooth wave also shows Gibbs' phenomenon: that for large N, Saw$_N(x)$ overshoots Saw(x) near the discontinuities by about 9% of the total jump.

15. What is the Fourier series generated by the function

$$f_N(x) = \begin{cases} 2N & (n - 1/N \leq x \leq n + 1/N) \\ 0 & \text{(otherwise)}? \end{cases}$$

What happens as $N \to \infty$?

16. Using the appropriate orthogonality relations, find an analogue of Parseval's equation, (2.27), for nonclassical Fourier expansions of the form

$$y(x) = \sum_{n=1}^{\infty} B_n \sin(n\pi x/l) \text{ (see section 2.2).}$$

17. A string of linear density ρ and length l vibrates as $y(x,t)$. Show that, to second order, its kinetic energy at time t is

$$T(t) = \tfrac{1}{2}\rho \int_0^l \left[\frac{\partial y}{\partial t}\right]^2 dx.$$

If y has a Fourier expansion $y(x,t) = \sum_{n=1}^{\infty} B_n \sin(n\pi x/l)\cos(nc\pi t/l + \varphi_n)$, find an expression for $T(t)$ in terms of the Fourier coefficients. Compare your answer with the Parseval's equation for nonclassical Fourier series found in the previous question.

Note that the kinetic energy of this string is not constant. What has happened to the conservation of energy?

18. Find the complex formulation of the Fourier series generated by

    (a) Tr(x).   (b) Sq(x).   (c) III(x).

Project: There are, as has been mentioned, cases of continuous periodic functions generating Fourier series which do not converge at some points. That is, there exist continuous periodic functions f for which $f_N(x)$ does not converge to $f(x)$ everywhere.

For any integrable periodic f, define $\tilde{f}_N(x) = (1/N)\sum_{n=0}^{N-1} f_N(x)$. Using equations (2.41)–(2.44), find an integral expression for $\tilde{f}_N$ in the form

$$\tilde{f}_N(x) = \int_0^c f(y) F_N[2\pi(x-y)/c] \, dy$$

(cf. (2.42)). Show that $F_N(x)$ can be expressed compactly as

$$F_N(x) = \sin^2[\tfrac{1}{2}Nx]/N\sin^2[\tfrac{1}{2}x].$$

$F_N(x)$ is the *Fejér kernel* of order N. Note that $F_N(x)$ is a nonnegative function — unlike $D_N(x)$, which oscillates positive and negative.

Prove that $F_N(x) \simeq (4/N)(\sin^2[\tfrac{1}{2}Nx]/x^2)$ near $x = 0$. Using an argument analogous to Step 2 in section 2.16, show that

$$\lim_{\alpha \to \infty} \int_{-a}^{b} f(x)\,(\sin^2(\alpha x)/\alpha x^2)\,dx = f(0)\int_{-\infty}^{\infty} \sin^2 y/y^2\, dy$$

$$= \pi f(0)$$

for any bounded f continuous at 0.

Prove also that $\left(\int_{-a}^{-\varepsilon} + \int_{\varepsilon}^{b}\right) F_N(x)\,dx \to 0$ as $N \to \infty$. Hence show that for any continuous periodic (and therefore bounded) function f, $\tilde{f}_N(x) \to f(x)$ everywhere. Hence this simple process of averaging the first N approximations f_N produces a sequence of new approximations $\tilde{f}_N$ guaranteed to converge for any merely continuous f, even though the conditions needed for f_N to converge are much stronger.

3
Fourier transforms

A. DEFINITION OF FOURIER TRANSFORMS

3.1 The Fourier transform operator

The major drawback with Fourier series theory is that it deals only with periodic functions (or, equivalently, functions on a finite interval). However, we may, for any function $f(x)$ which is integrable on the *whole* of the real numbers, define the following integrals:

$$Cf(\omega) = \int_{-\infty}^{\infty} \cos(\omega x) f(x) \, dx$$

$$Sf(\omega) = \int_{-\infty}^{\infty} \sin(\omega x) f(x) \, dx \tag{3.1}$$

These are analogous to the integrals that define the *coefficients* A_n and B_n of Fourier series, equations (2.3). In (3.1), though, the range of integration is from $-\infty$ to $+\infty$, rather than over an interval of length c; and instead of the arguments $2\pi nx/c$, we allow any angular frequency ω.

Because any ω is allowed, equations (3.1) define *functions* of ω. The two functions Cf and Sf are called, respectively, the *(two-sided Fourier) cosine transform* and the *(two-sided Fourier) sine transform* of f. (Note that C and S act as *operators*, which are applied to the function f to yield new functions.)

As with Fourier series, there is a complex formulation of Fourier transformation which allows the cosine and sine formulae to be collected into a single expression. With Fourier series we defined

$$Z_n = (1/c) \int_0^c e^{-2\pi i n x/c} f(x) \, dx$$

(equation (2.30)); for Fourier transforms we write

$$Ff(\omega) = \int_{-\infty}^{\infty} e^{-i\omega x} f(x) \, dx. \tag{3.2}$$

The function $Ff(\omega)$ is the (*complex*) *Fourier transform* of f, and F is the Fourier transform operator. By expanding the term $e^{-i\omega x}$, we can relate this to the cosine and sine transforms:

$$Ff(\omega) = Cf(\omega) - iSf(\omega). \tag{3.3}$$

(Note: the minus sign appears here, and in the exponent of equation (3.2), in the definition used most frequently by mathematicians. Sometimes it is dropped — physicists often favour this version — but then care has to be taken that all the formulae that are used are appropriately switched. Engineers frequently use yet another formulation, which we shall discuss at the end of the chapter.)

The argument f of the Fourier transform operator need not be a real-valued function. In many important cases, we shall need to take Fourier transforms of *complex*-valued functions. There is no new theory required for this.

The theory and practice of Fourier transforms is very similar to that of Fourier series, and this and the following chapter are largely concerned with results analogous to those in Chapter 2.

3.2. Inversion

The Fourier transform is defined by analogy with the *coefficients* of the classical Fourier series. The most important thing about Fourier series is that, for many periodic functions, the Fourier *series* is actually a Fourier *expansion*, and we can write, as in equation (2.36),

$$f(x) = \sum_{n=-\infty}^{\infty} Z_n e^{2n\pi xi/c}.$$

For nonperiodic functions, there is a corresponding expression. However, in this case there is, instead of a *sequence* of coefficients Z_n, a *function* $Ff(\omega)$, with a value for each possible angular frequency ω. Thus where the Fourier series is a sum over n, for Fourier transforms we have an *integral* over ω.

$$f(x) = (1/2\pi) \int_{-\infty}^{\infty} Ff(\omega) e^{i\omega x} d\omega. \tag{3.4}$$

(Note the factor of $1/2\pi$.) Equation (3.4) is called the *inversion theorem*.

Just as not all periodic functions have Fourier series which are Fourier *expansions*, so not all Fourier transforms are *invertible*. That is, although any integrable function f has a Fourier transform Ff, the function f cannot always be recovered from Ff, using equation (3.4) (see section 3.5 below for an example of this). Indeed, the question of which functions satisfy the inversion theorem is yet unsolved, and is an important question of mathematical analysis. However, equation (3.4) can be proved for a wide enough variety of functions to be useful. For instance, if $f(x)$ is a *smooth* (continuously differentiable) function, and the function $|f(x)|$ is integrable over $(-\infty, \infty)$, then its transform is invertible. (A sketch proof of the inversion theorem for such functions in given in section 3.16 below.)

There is a temptation, in dealing with Fourier transforms, to treat the forward transform operator F as acting on 'basic' functions to produce 'derived' functions, and to regard the inversion theorem as dealing with the 'recovery' of 'basic' functions from their transforms. In many physical situations, indeed, this is a natural viewpoint: a function of time seems more fundamental than a function of frequency. But there is no mathematical justification for it. Let us see why this is.

The Fourier transform operator is defined by the formula

$$Ff(\omega) = \int_{-\infty}^{\infty} e^{-i\omega x} f(x) \, dx$$

(equation (3.2)). Let us define a second operator G by

$$Gf(\omega) = (1/2\pi) \int_{-\infty}^{\infty} e^{i\omega x} f(x) \, dx.$$

Clearly, this is closely related to F; indeed,

$$Gf(\omega) = (1/2\pi) Ff(-\omega). \tag{3.5}$$

Thus, if Ff exists, then Gf does too, and vice versa.

But the action of G is precisely that involved in the Fourier inversion theorem. We can reformulate equation (3.4) as

$$G[Ff](x) = f(x),$$

or more simply,

$$GFf(x) = f(x), \tag{3.6}$$

for any suitable function f.

If now the G in equation (3.6) is expanded using (3.5), we get the surprising relation

$$f(x) = (1/2\pi) FFf(-x). \tag{3.7}$$

Also, we can calculate that

$$FGf(x) = (1/2\pi) FFf(-x)$$
$$= f(x).$$

The operator G is said to be the *inverse* of F, and we write Gf as $F^{-1}f$. Equation (3.5) now becomes

$$F^{-1}f(\omega) = (1/2\pi) Ff(-\omega). \tag{3.8}$$

Thus, if $g(\omega)$ is the transform of $f(x)$, then $(1/2\pi) f(-\omega)$ is the transform of $g(x)$.

Because 'forward' transformation and inverse transformation are so closely related, one sometimes refers, not to g being the transform of f, but to f and g being a *transform pair* of functions.

3.3 Spectra

We saw in the previous chapter how Fourier series may be represented graphically, by means of spectra. The same technique is used for Fourier transforms. As with

Fourier series, there are two methods of representing a transform: the cos–sin method and the amplitude–phase method.

In the cos–sin method, the cosine transform and the sine transform are plotted. This is possible only for the transforms of real-valued functions. Since $Ff(\omega) = Cf(\omega) - iSf(\omega)$, this method gives the real and imaginary parts of the complex Fourier transform, although the imaginary part is 'upside down' by virtue of the minus sign in this formula.

In the amplitude–phase method, the two functions plotted are the modulus and the argument of Ff:

$$Af(\omega) = |Ff(\omega)|$$

and

$$\Phi f(\omega) = \arg(Ff(\omega)). \qquad (3.9)$$

These functions (and their graphs) are the *amplitude spectrum* and the *phase spectrum*, respectively. Alternative to the amplitude spectrum is the *power spectrum*,

$$Pf(\omega) = (Af(\omega))^2 = |Ff(\omega)|^2; \qquad (3.10)$$

this is used for technical reasons, some of which we shall examine in due course.

The amplitude–phase representation of Ff is more common.

3.4 Power spectral density functions

In many practical applications of Fourier theory, the function $f(x)$ to be transformed is not integrable from $-\infty$ to ∞; instead, the function behaves in much the same way across the whole range of x (or at least, over a range so large as to be effectively infinite). Functions of this kind include meteorological records: the temperature pattern over the past week is much the same as the week before, or last year, or a millenium ago.

We say that $f(x)$ is a *function of constant character* if its Fourier structure is stable over large intervals. More precisely, f has constant character if the limit

$$pf(\omega) = \lim_{R \to \infty} (1/2R) \left| \int_{-R}^{R} f(x) e^{-i\omega x} dx \right|^2 \qquad (3.11)$$

exists; if it does, pf is called the *power spectral density function* (PSDF) of f.

The power spectral density operator p behaves very similarly to the power spectrum operator P. Among other points to note, it is not possible to invert a PSDF: there is no 'phase information' in it.

B. SOME USEFUL FUNCTIONS

3.5 The bell curve

The *bell curve* is the function

$$N(x) = \exp(-\tfrac{1}{2}x^2) \qquad (3.12)$$

(see Fig. 3.1). It arises frequently in statistics, where it is used as the basis for the so-called *normal distribution*.

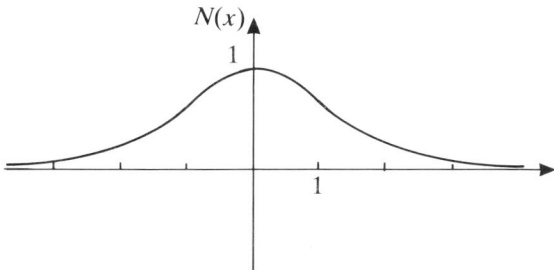

Fig. 3.1 — The bell curve, $N(x)$.

The bell curve is smooth, indeed infinitely many times differentiable. It is positive and integrable, with integral $\int_{-\infty}^{\infty} N(x)\,dx = \sqrt{(2\pi)}$. Hence it has a Fourier transform, which is invertible.

The transform may be calculated using 'contour integration', as

$$FN(\omega) = \sqrt{(2\pi)}\exp(-\tfrac{1}{2}\omega^2) = \sqrt{(2\pi)}\,N(\omega). \tag{3.13}$$

That is, apart from a constant of $\sqrt{(2\pi)}$, the bell curve is its own Fourier transform. This has an important consequence for statistics, which we shall see in Chapter 10.

3.6 Step functions

A function $f(x)$ is a *step function* if there exist finitely many values $x_0, \ldots, x_n$ of x such that (i) $f(x)$ is constant over each interval $x_{i-1} < x < x_i$, and (ii) $f(x) = 0$ for $x < x_0$ and for $x > x_n$. The points x_i are the only points at which $f(x)$ changes value, and at these points $f(x)$ makes a discontinuous jump. Thus the graph of a step function consists of a finite number of finite horizontal 'steps'. Step functions arise naturally as voltage outputs of digital electronic circuits, for instance.

Basic among the step functions are the *characteristic functions of finite intervals*. For an interval I (which may include one, or both, or neither, of its endpoints), the function χ_I is defined as

$$\chi_I(x) = \begin{cases} 1 & (x \in I) \\ 0 & (x \notin I) \end{cases} \tag{3.14}$$

Thus, for example, the function $\chi_{[0,1)}(x)$ takes the value 0 for $x < 0$, 1 for $0 \le x < 1$, and 0 again for $x \ge 1$. Any step function can be expressed as a sum of multiples of characteristic functions of intervals.

Characteristic functions of intervals have an important property when combined with integration. Suppose we have an arbitrary function $f(x)$, and $\chi_{(a,b)}(x)$ is the characteristic function of the interval (a,b). Then the product $f\chi_{(a,b)}$ has the form

$$f(x)\chi_{(a,b)}(x) = \begin{cases} 0 & (x \le a) \\ f(x) & (a < x < b) \\ 0 & (x \ge b) \end{cases}$$

Thus, multiplication by $\chi_{(a,b)}$ *truncates* the function $f(x)$ at $x = a$ and $x = b$. Upon integration, this yields

$$\int_{-\infty}^{\infty} f(x)\chi_{(a,b)}(x)\,dx = \int_{-\infty}^{a} 0\,dx + \int_{a}^{b} f(x)\,dx + \int_{b}^{\infty} 0\,dx$$

$$= \int_{a}^{b} f(x)\,dx. \tag{3.15}$$

Multiplying a function by $\chi_{(a,b)}$ and integrating from $-\infty$ to ∞ has the same effect as integrating the function from a and b. Similar formulae hold for the functions $\chi_{[a,b)}$, $\chi_{(a,b]}$, and $\chi_{[a,b]}$, in which one or both endpoints are included in the interval.

All step functions are integrable; therefore they all have Fourier transforms. Indeed

$$F\chi_{(a,b)}(\omega) = \int_{-\infty}^{\infty} \chi_{(a,b)}(x)e^{-i\omega x}\,dx$$

$$= \int_{a}^{b} e^{-i\omega x}\,dx$$

$$= i(e^{-i\omega b} - e^{-i\omega a})/\omega. \tag{3.16}$$

A useful particular case arises when $a = -b$. In this case, the interval is symmetrical about $x = 0$, and the function is even; thus its transform is real. For we have

$$F\chi_{(-b,b)}(\omega) = i(e^{-i\omega b} - e^{i\omega b})/\omega$$

$$= 2\sin(\omega b)/\omega. \tag{3.17}$$

3.7 Limiting arguments and the function $\sin x/x$

The transforms expressed in (3.16) and (3.17) are not themselves absolutely integrable functions. Thus the inverse transform cannot be applied to them; the functions χ_I do not satisfy the inversion theorem. This is not too surprising, for χ_I is not smooth.

This behaviour is reminiscent of the failure of the square wave $Sq(x)$ to have a Fourier series expansion, which we saw in section 2.7. As in that case, the failure is slight: it may be shown that

$$\lim_{R\to\infty} (1/2\pi) \int_{-R}^{R} F\chi_{(a,b)}(\omega)e^{i\omega x}\,d\omega = \begin{cases} 0 & (x < a) \\ \tfrac{1}{2} & (x = a) \\ 1 & (a < x < b) \\ \tfrac{1}{2} & (x = b) \\ 0 & (x > b). \end{cases} \tag{3.18}$$

The left-hand limit exists, and is equal to $\chi_{(a,b)}(x)$ everywhere except at its discontinuities, a and b.

Conversely, consider the function $f(x) = \sin x/x$. It does not have a Fourier transform, since $|f(x)|$ does not have a finite integral. However, we can regard $f(x)$ as the limit of functions $f_R(x) = f(x)\chi_{(-R,R)}(x)$, as $R \to \infty$; and then we can say that

$$\lim_{R\to\infty} Ff_R(\omega) = \lim_{R\to\infty} \int_{-R}^{R} (\sin x/x)e^{-i\omega x}\,dx = \begin{cases} 0 & (\omega < -1) \\ \pi & (\omega = -1) \\ 2\pi & (-1 < \omega < 1) \\ \pi & (\omega = 1) \\ 0 & (\omega > 1). \end{cases} \quad (3.19)$$

Thus, the limit of the transforms $Ff_R(\omega)$ as $R \to \infty$ is (apart from at the discontinuities) equal to $\chi_{(-1,1)}(\omega)$, so that the functions $\sin x/x$ and $\chi_{(-1,1)}(x)$ are (almost) a transform pair. (See Fig. 3.2.)

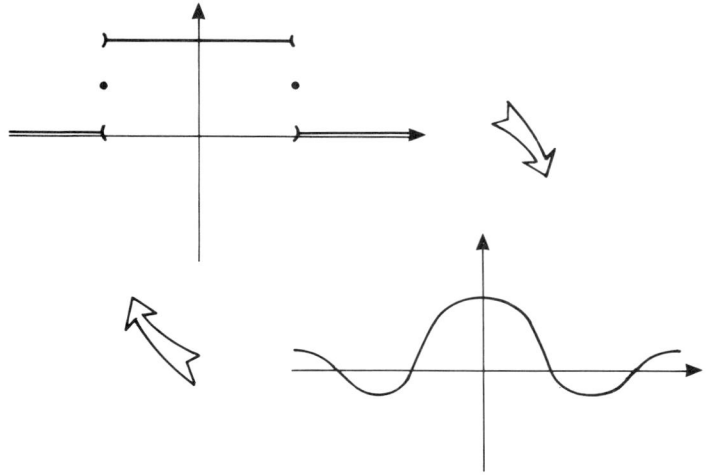

Fig. 3.2.

This kind of limiting argument is widely applicable. Many useful functions are, in fact, not integrable, and so do not have 'genuine' Fourier transforms. In some of these cases, we may for practical purposes define 'improper' Fourier transforms by adopting a limiting argument in the definition. Certainly, for an *integrable* function $f(x)$,

$$Ff(\omega) = \int_{-\infty}^{\infty} e^{-i\omega x} f(x)\, dx$$

$$= \lim_{R\to\infty} \int_{-R}^{R} e^{-i\omega x} f(x)\, dx;$$

$$= \lim_{R\to\infty} F[f\chi_{(-R,R)}](\omega).$$

In circumstances where f is not integrable, $f\chi_{(-R,R)}$ may still be. The limit may or may not exist. When it does exist, it may fairly be called the transform of f; this is the case with $f(x) = \sin x/x$, and we treat $\chi_{(-1,1)}(x)$ and $\sin x/x$ as a transform pair. When the limit does not exist, the transforms of the truncated functions $f\chi_{(-R,R)}$ can still be useful.

Related to step functions is the *Heaviside function*, denoted by $H(x)$ and defined by the formula:

$$H(x) = \begin{cases} 0 & (x<0) \\ \tfrac{1}{2} & (x=0) \\ 1 & (x>0). \end{cases} \qquad (3.20)$$

Its graph is shown in Fig. 3.3. The Heaviside function, like the function $\sin x/x$, does

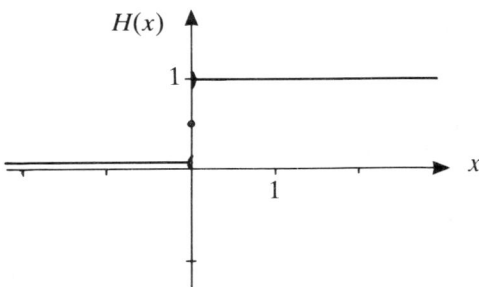

Fig. 3.3 — The Heaviside function, $H(x)$.

not have a Fourier transform, for it is not integrable. $H(x)$ is the limit of the functions $H(x)\chi_{(-R,R)}(x) = \chi_{(0,R)}(x)$, as $R \to \infty$. The transform of this truncation is $F\chi_{(0,R)}(\omega) = (ie^{-i\omega R}/\omega) - (i/\omega)$. But these functions do not tend to a limit as $R \to \infty$, and no 'transform' of $H(x)$ can be found in this way. (However, see Question 12 at the end of this chapter.)

The function $\sin(\pi x)/\pi x$ is sometimes written $\operatorname{sinc}(x)$.

3.8 More on impulsive 'functions'

In the previous chapter we found that the Fourier series generated by the sampling 'function', $III(x)$, was $1 + \sum_{n=1}^{\infty} 2\exp(2\pi inx)$: all the Fourier coefficients are 2. Similarly, the Fourier transform of the impulse function, $\delta(x)$, is given by

$$F\delta(\omega) = \int_{-\infty}^{\infty} e^{-i\omega x} \delta(x) \, dx$$
$$= e^{-i\omega 0}$$
$$= 1 \qquad (3.21)$$

(see equation (2.17)).

Again we can be more rigorous by using a limiting argument. The delta function is the limit of (for example) the step functions $\delta_n(x) = \tfrac{1}{2}n\chi_{(-1/n, 1/n)}(x)$, as $n \to \infty$. From equation (3.17), we can write

$$F\delta_n(\omega) = \frac{\sin(\omega/n)}{(\omega/n)}. \qquad (3.22)$$

As $n \to \infty$, $\omega/n \to 0$, and $\sin(\omega/n)/(\omega/n) \to 1$. Thus, the limiting case of δ_n is δ, and the limit of $F\delta_n$ is 1 (see Fig. 3.4).

We can try to argue similarly in reverse, for the unity function $1(x)$ is the limit of the step functions $1_n(x) = \chi_{(-n,n)}(x)$. Here Fourier transformation yields

$$F1_n(\omega) = 2\sin(n\omega)/\omega$$
$$= 2n\sin(n\omega)/(n\omega). \qquad (3.23)$$

These functions, however, have no proper limit as $n \to \infty$. There is a spike around $\omega = 0$, of height $2n$. The integral of the function is always 2π. But the functions do not tend to $2\pi\delta(\omega)$, as we would expect from (3.21) if the inversion theorem applied to $\delta(x)$. For, just as with the partial sums of the Fourier series for $III(x)$, the function $2n\sin(n\omega)/(n\omega)$ *oscillates* with increasing n, between $\pm 2/\omega$. (Fig. 3.5 shows this function for some values of n.)

3.9 Periodic functions

A nonzero periodic function can never be integrable over $(-\infty, \infty)$. Thus the Fourier transform of a periodic function cannot strictly exist. However, the concept of impulsive 'functions' like $\delta(x)$ allows us to extend the theory of Fourier transforms to incorporate periodic functions. Under this extension, the Fourier transform of a periodic function is an impulsive 'function'.

A smooth periodic function $f(\omega)$ of period c has a Fourier series of the form

$$f(\omega) = \sum_{n=-\infty}^{\infty} Z_n \exp(2n\pi\omega i/c).$$

If $g(x)$ is to be the Fourier *inverse* transform of $f(\omega)$, then we must have the

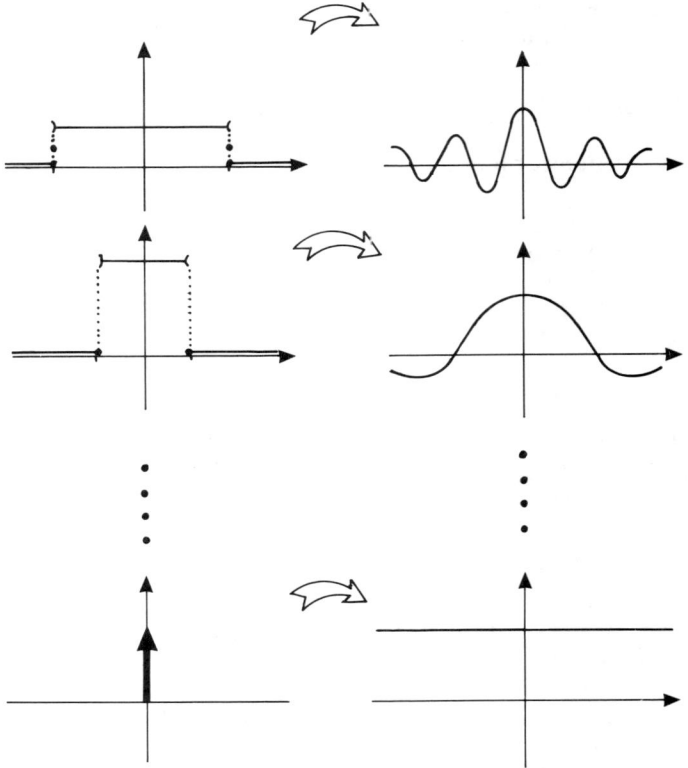

Fig. 3.4 — Transform of the delta function, as a limit.

expression $f(\omega) = Fg(\omega) = \int_{-\infty}^{\infty} g(x)e^{-i\omega x} dx$. In order to pick out discrete harmonic terms from this integral, we need to use impulses; in fact, it may be readily checked that the 'function'

$$g(x) = \sum_{n=-\infty}^{\infty} Z_n \delta(x + 2n\pi/c) \tag{3.24}$$

will satisfy this equation.

This gives the *inverse* transform. But there is a close relation between forward and inverse Fourier transforms (section 3.2). Thus, although $f(x)$ is not integrable, it is frequently useful to regard its *forward* transform as being

$$Ff(\omega) = 2\pi g(-\omega) = \sum_{n=-\infty}^{\infty} (2\pi Z_n)\delta(\omega - 2n\pi/c). \tag{3.25}$$

In particular, we have

Ch. 3] Fourier transforms

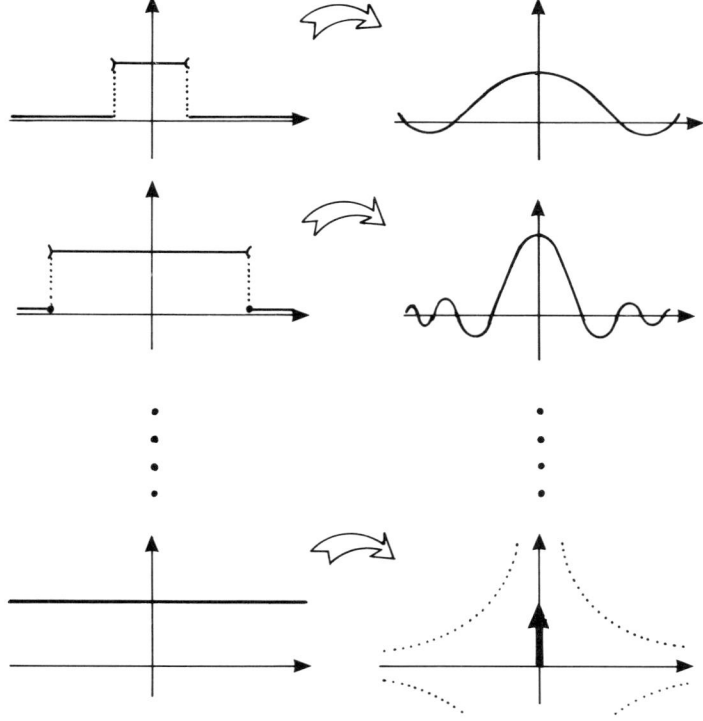

Fig. 3.5 — Transform of the unity function, as a limit.

$F[e^{-i\alpha x}](\omega) = 2\pi\delta(\omega + \alpha)$, while $F^{-1}[e^{-i\alpha x}](x) = \delta(x - \alpha)$;

$F[\cos(\alpha x)](\omega) = \pi(\delta(\omega + \alpha) + \delta(\omega - \alpha))$, while $F^{-1}[\cos(\alpha\omega)](x) = \frac{1}{2}(\delta(x + \alpha) + \delta(x - \alpha))$;

$F[\sin(\alpha x)](\omega) = \pi i(\delta(\omega - \alpha) - \delta(\omega + \alpha))$, while $F^{-1}[\sin(\alpha\omega)](x) = \frac{1}{2}i(\delta(x + \alpha) - \delta(x - \alpha))$.

(3.26)

Again, using (3.25) and sections 2.6–2.8, we can calculate the Fourier 'transforms' of the periodic functions Tr, Sq, and III as

$$F\text{Tr}(\omega) = \sum_{n=-\infty}^{\infty} (-4/\pi(2n-1)^2)\delta(\omega - (2n-1)\pi)$$

$$FSq(\omega) = \sum_{n=-\infty}^{\infty} (-4i/(2n-1))\delta(\omega - (2n-1)\pi)$$

and

$$FIII(\omega) = \sum_{n=-\infty}^{\infty} (2\pi)\delta(\omega - (2n\pi))$$

$$= III(\omega/2\pi). \qquad (3.27)$$

The amplitude spectra of these three functions are shown in Figs 3.6–3.8. Note in particular that III is (almost) its own transform.

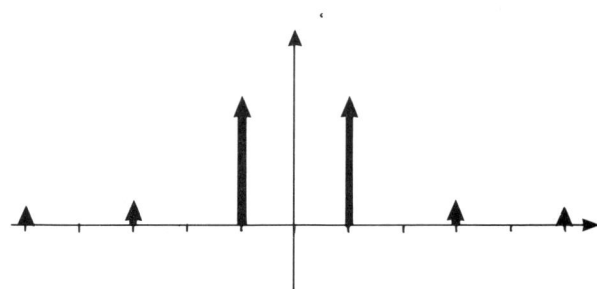

Fig. 3.6 — Transform of the triangle wave.

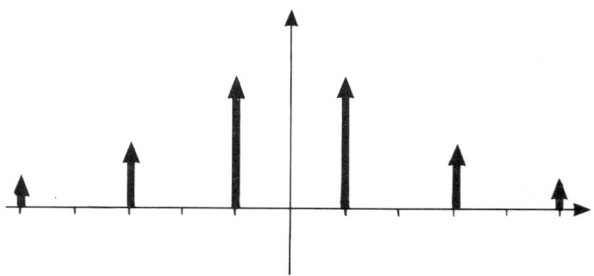

Fig. 3.7 — Transform of the square wave.

Periodic functions are not integrable, but have 'improper' Fourier transforms. Likewise, periodic functions are not of constant character, although functions of constant character are a kind of generalization of periodic functions. Thus, periodic

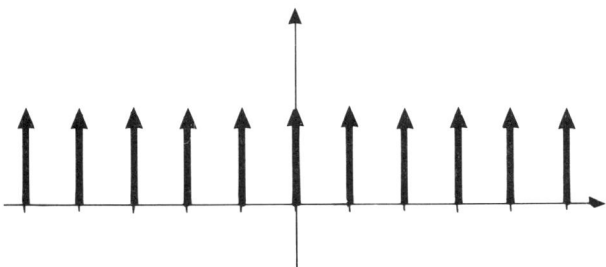

Fig. 3.8 — Transform of the sampling function.

functions have improper, impulsive PSDFs too. More precisely, suppose $f(x)$ is a periodic function of period c, generating the Fourier series $\sum_{n=-\infty}^{\infty} Z_n e^{2\pi i n x/c}$. Then it may be shown that

$$pf(\omega) = \sum_{n=-\infty}^{\infty} (2\pi Z_n^2)\delta(\omega - 2\pi n/c). \qquad (3.28)$$

In particular,

$$p[e^{-i\alpha x}](\omega) = 2\pi\delta(\omega + \alpha)$$
$$p[\cos(\alpha x)](\omega) = \tfrac{1}{2}\pi(\delta(\omega + \alpha) + \delta(\omega - \alpha))$$
$$p[\sin(\alpha x)](\omega) = \tfrac{1}{2}\pi(\delta(\omega + \alpha) + \delta(\omega - \alpha))$$

$$p\mathrm{Tr}(\omega) = \sum_{n=-\infty}^{\infty} (8/\pi^3(2n-1)^4)\,\delta(\omega - (2n-1)\pi)$$

$$p\mathrm{Sq}(\omega) = \sum_{n=-\infty}^{\infty} (8/\pi(2n-1)^2)\,\delta(\omega - (2n-1)\pi)$$

and

$$p\mathrm{III}(\omega) = \sum_{n=-\infty}^{\infty} 2\pi\delta(\omega - 2n\pi) = \mathrm{III}(\omega/2\pi). \qquad (3.29)$$

Note that $\cos(\alpha x)$ and $\sin(\alpha x)$ have the same PSDF. This is because the PSDF contains no phase information; and $\cos(\alpha x)$ and $\sin(\alpha x)$ differ only in phase.

Table 3.1 shows some functions and their Fourier transforms.

C. PROPERTIES OF FOURIER TRANSFORMS
3.10 Arithmetic properties
Fourier transforms have many of the same properties as Fourier series. First, the

Table 3.1 — Some functions and their Fourier transforms. Shown are functions f and their transforms Ff. For the transforms of functions in the second list, use the inversion theorem: if $g(x) = Ff(x)$, then $Fg(\omega) = 2\pi f(-x)$

f	Ff
$\Pi(x)$	$\sin\omega/\omega$
$\chi_{(a,b)}(x)$	$i(e^{-i\omega b} - e^{-i\omega a})/\omega$
$\Lambda(x)$	$4\sin^2(\tfrac{1}{2}\omega)/\omega^2$
$e^{-|x|}$	$2/(1+\omega^2)$
$e^{-x}H(x)$	$1/(1+i\omega)$
$N(x)$	$\sqrt{(2\pi)}N(\omega)$
$xN(x)$	$i\sqrt{(2\pi)}\omega N(\omega)$
$\cos(\alpha x)N(x)$	$\sqrt{(\pi/2)}(N(\omega+\alpha) + N(\omega-\alpha))$
$J_0(x)$	$2\Pi(\omega)/\sqrt{(1-\omega^2)}$
$\delta(x)$	$1(\omega)$
$\cos x$	$\pi(\delta(\omega+1) + \delta(\omega-1))$
$\sin x$	$\pi i(\delta(\omega+1) - \delta(\omega-1))$
$\mathrm{III}(x)$	$\mathrm{III}(\omega/2\pi)$
$H(x)$	$\pi\delta(\omega) - i/\omega$
$\lambda g(x)$	$\lambda Fg(\omega)$
$g(x+\lambda)$	$e^{i\omega\lambda}Fg(\omega)$
$g(\lambda x)$	$(1/|\lambda|)Fg(\omega/\lambda)$
$\overline{g}(x)$	$\overline{Ff(-\omega)}$
$g'(x)$	$i\omega Fg(\omega)$
$(g*h)(x)$	$Fg(\omega)Fh(\omega)$ (see Chapter 5)

Fourier transform operator is a *linear* operator: for functions f and g and constant α, we have

$$F(f+g)(\omega) = Ff(\omega) + Fg(\omega)$$

and

$$F(\alpha f)(\omega) = \alpha Ff(\omega). \tag{3.30}$$

Similar formulae hold separately for the Fourier cosine and Fourier sine transforms. However, the amplitude, phase, power, and power density transforms A, Φ, P, and p are *not* linear.

Again, if $f(x)$ is a function whose Fourier transform exists and λ is a constant, and the functions r and s are defined as

$$r(x) = f(\lambda x) \quad \text{and} \quad s(x) = f(x+\lambda),$$

then r and s have transforms which are given by

$$Fr(\omega) = (1/|\lambda|)Ff(\omega/\lambda)$$

and

$$Fs(\omega) = e^{i\omega\lambda}Ff(\omega). \qquad (3.31)$$

(Compare equations (2.22).)

A further property of Fourier transformation has to do with the case of negative ω. It arises as a consequence of the result $e^{-it} = \overline{e^{it}}$. If $f(x)$ is a *real*-valued function, then

$$Ff(-\omega) = \int_{-\infty}^{\infty} f(x)e^{i\omega x}\,dx$$

$$= \int_{-\infty}^{\infty} f(x)\overline{\exp(-i\omega x)}\,dx$$

$$= \overline{Ff(\omega)}. \qquad (3.32)$$

The case of *complex*-valued functions $f(x)$ is only slightly more complicated; in this case,

$$F\bar{f}(-\omega) = \overline{Ff(\omega)}, \qquad (3.33)$$

where $\bar{f}$ is the complex conjugate of f, the function defined by $\bar{f}(x) = \overline{f(x)}$.

An important special case of (3.30) is the case where $r(x) = f(-x)$, so that the graph of $r(x)$ is just the graph of $f(x)$ reflected about the vertical line $x = 0$. Here we have $\lambda = -1$, so that

$$Fr(\omega) = Ff(-\omega). \qquad (3.34)$$

If $f(x)$ is an *even*, real-valued function, $f(x) = f(-x) = r(x)$. It follows that $Ff(\omega) = Ff(-\omega)$, so $Ff(\omega)$ is even. Similarly, if f is odd and real-valued, Ff is odd. But we can also apply (3.32): so, if f is even and real-valued then $Ff(\omega) = Ff(-\omega) = \overline{Ff(\omega)}$. Thus, Ff is also real-valued. If f is odd and real-valued, then $Ff(\omega) = -\overline{Ff(\omega)}$; so Ff takes pure imaginary values.

Again, the situation for complex-valued functions is slightly more complicated. For an even function f, we have $Ff(\omega) = Ff(-\omega) = \overline{F\bar{f}(\omega)}$; for odd f, $Ff(\omega) = -\overline{F\bar{f}(\omega)}$.

All the properties discussed in this section apply to the Fourier transforms of all integrable functions, whether or not those transforms are invertible.

3.11 Differentiation and integration

As with Fourier series, the Fourier transforms of derivatives and integrals have simple forms. Suppose that $f(x)$ is smooth integrable function, such that $f(x) \to 0$ as $x \to \pm\infty$; and suppose df/dx is also integrable over $(-\infty, \infty)$. Then

$$F\frac{df}{dx}(\omega) = \int_{-\infty}^{\infty} e^{-i\omega x}\frac{df(x)}{dx}\,dx$$

$$= [e^{-i\omega x}f(x)]_{-\infty}^{\infty} - \int_{-\infty}^{\infty}\left(\frac{d}{dx}e^{-i\omega x}\right)f(x)\,dx,$$

upon integration by parts. But $e^{-i\omega x}$ has absolute value 1 for all x, whereas (by assumption) $f(x)$ tends to zero as $x \to \pm \infty$. The term in square brackets is therefore zero, and we are left with:

$$F\frac{df}{dx}(\omega) = -\int_{-\infty}^{\infty} \left(\frac{d}{dx} e^{-i\omega x}\right) f(x) \, dx$$

$$= -\int_{-\infty}^{\infty} (-i\omega e^{-i\omega x}) f(x) \, dx$$

$$= i\omega F f(\omega). \qquad (3.35)$$

Thus the transform of df/dx is just $i\omega$ times the transform for f. This is sometimes called the *derivative theorem*.

There is a converse to this result: if f is a smooth function, with $xf(x) \to 0$ as $x \to \pm \infty$, and if $xf(x)$ has a finite integral over $(-\infty, \infty)$, then

$$F(xf)(\omega) = i\frac{d}{d\omega}(Ff(\omega)). \qquad (3.36)$$

For we have

$$\frac{d}{d\omega} Ff(\omega) = \frac{d}{d\omega} \left\{ \int_{-\infty}^{\infty} e^{-i\omega x} f(x) \, dx \right\}$$

$$= \int_{-\infty}^{\infty} (\partial/\partial\omega)(e^{-i\omega x} f(x)) \, dx$$

$$= \int_{-\infty}^{\infty} (-ix) e^{-i\omega x} f(x) \, dx$$

$$= -i \int_{-\infty}^{\infty} e^{-i\omega x} xf(x) \, dx$$

$$= -i F(xf)(\omega).$$

Equations (3.35) and (3.36) are closely connected. Indeed, each can be deduced from the other, using the inversion theorem, equation (3.8) (see Question 12 at the end of this chapter). Remember, though, that the inversion theorem does not apply to all functions; this is why the criteria for (3.36) are different from those for (3.35).

It follows by repeated application of equations (3.35) and (3.36) that

$$F\frac{d^n f}{dx^n}(\omega) = (i\omega)^n Ff(\omega)$$

and

$$F(x^n f)(\omega) = i^n \frac{d^n}{d\omega^n} Ff(\omega). \tag{3.37}$$

3.12 Other properties

In the previous chapter we discussed two further properties of Fourier series, namely Parseval's equation and the Riemann–Lebesgue lemma. Both of these may be adapted for the Fourier transform.

The transform analogue of Parseval's equation is an extremely powerful result called the *power theorem*. We shall be discussing this in Chapter 5, for it is related to some important new theory.

The Riemann–Lebesgue lemma, on the other hand, comes over to transform theory very simply. In its transform version, it says that for any integrable function $f(x)$,

$$Ff(\omega) \to 0 \quad \text{as} \quad \omega \to \pm\infty; \tag{3.38}$$

it can be proved, in the case where f is smooth, by adapting the proof of equation (2.51) as given in section 2.16. (This proof is left as an exercise.)

We can combine this with the derivative theorem, equation (3.35). Thus, if $f(x)$ is k times differentiable and each derivative is integrable, then $\omega^k Ff(\omega) \to 0$ as $\omega \to \infty$; that is, $Ff(\omega)$ tends to zero faster than $1/\omega^k$.

There is one final property of Fourier transforms that we should mention, which has no counterpart in the series theory, namely continuity. If $f(x)$ is an integrable function for which $\int_{-\infty}^{\infty} |f(x)| \, dx$ is finite, then the Fourier transform $Ff(\omega)$ of f is a continuous function of ω. Here is a sketch of the proof.

For any positive ε, choose R such that

$$\int_{-\infty}^{-R} |f(x)| \, dx + \int_{R}^{\infty} |f(x)| \, dx \leq \tfrac{1}{2}\varepsilon;$$

that is, such that the interval $(-R, R)$ contains 'most' of the function $|f|$. Now, for given ω, let δ be such that

$$|\delta| \leq \varepsilon / \left\{ 4R \int_{-\infty}^{\infty} |f(y)| \, dy \right\};$$

then, whenever $-R \leq x \leq R$,

$$|e^{i\omega x} - e^{i(\omega+\delta)x}| = |e^{i\omega x}||1 - e^{i\delta x}|$$
$$\leq |\delta| R$$

$$\leq \varepsilon / \left\{ 4 \int_{-\infty}^{\infty} |f(y)| \, dy \right\},$$

and it follows that

$$\left| \int_{-R}^{R} f(x) \, (e^{i\omega x} - e^{i(\omega + \delta)x}) \, dx \right| \leq \int_{-R}^{R} |f(x)| \, |e^{i\omega x} - e^{i(\omega + \delta)x}| \, dx$$

$$\leq \tfrac{1}{2}\varepsilon.$$

Therefore

$$|Ff(\omega + \delta) - Ff(\omega)| = \left| \int_{-\infty}^{\infty} f(x) e^{i\omega x} \, dx - \int_{-\infty}^{\infty} f(x) e^{i(\omega + \delta)x} \, dx \right|$$

$$\leq \varepsilon.$$

for δ in this range.

D. USING FOURIER TRANSFORMS

3.13 Heat conduction

In the second half of this book we shall be looking at a range of applications of Fourier theory that are specific to particular disciplines. There are, however, some more general points that should be made here.

Thus, Fourier transforms can be used to solve problems involving differential equations, just as Fourier series were used in the previous chapter in the analysis of the vibrating string. Among these is the problem of heat conduction; which was, indeed, the original application of Fourier transform theory, studied by Fourier himself.

For an infinitely long, thin, uniform wire, Fourier's 'heat equation' is

$$\frac{\partial^2 \theta}{\partial x^2} = \kappa \frac{\partial \theta}{\partial t}, \tag{3.39}$$

where $\theta(x,t)$ is the temperature of the point x at time t, and κ is a constant related to the material of which the wire is made. Now suppose that at time $t = 0$ the temperature profile of the wire is $\theta(x,0) = I(x)$; how does the temperature of the wire evolve?

This problem can be solved, as in the case of the vibrating string, by first looking for solutions of the form $\theta(x,t) = X(x)T(t)$, in which the variables are separated. In this case, $\partial^2\theta/\partial x^2 = (d^2X/dx^2)T$, and $\partial\theta/\partial t = X(dT/dt)$. Substituting these into (3.39) and rearranging gives

$$\frac{1}{X}\frac{d^2X}{dx^2} = \frac{\kappa}{T}\frac{d/T}{dT/dt}. \tag{3.40}$$

Since the right-hand side is not x-dependent, neither is the left-hand side; thus $(1/X)\mathrm{d}^2 X/\mathrm{d}x^2 = k$ for some constant k, or $(\mathrm{d}^2 X/\mathrm{d}x^2) - kX = 0$. This is the harmonic equation, when $k = -\omega^2$, with solutions $X(x) = Z_1 e^{-i\omega x}$. (The reason for the minus sign will become apparent.) But then $(\kappa/T)\mathrm{d}T/\mathrm{d}t = -\omega^2$ too, so $(\mathrm{d}T/\mathrm{d}t) + (\omega^2/\kappa)T = 0$; and the solutions of this are $T(t) = Z_2 \exp(-\omega^2 t/\kappa)$. Putting $Z_1 Z_2 = Z$, we have

$$\theta(x,t) = Z e^{-i\omega x} \exp(-\omega^2 t/\kappa) \tag{3.41}$$

as solutions. (Note that Z_1, Z_2 and Z can be complex-valued, even though such solutions cannot arise physically.)

These solutions are like the harmonic solution of the vibrating string. The significant difference here is that there is no restriction on ω, because the wire is infinite and there are therefore no boundary conditions.

Now equation (3.39) is linear, and we may therefore apply the principle of superposition. Any sum of solutions like (3.41) is also a solution; so, by extension, are integrals of the form

$$\theta(x,t) = \int_{-\infty}^{\infty} Z(\omega) e^{-i\omega x} \exp(-\omega^2 t/\kappa)\, \mathrm{d}\omega, \tag{3.42}$$

where the constant Z of (3.41) becomes a function of ω.

This expression is clearly connected with Fourier transformation. Indeed, at time $t = 0$, we have $\theta(x,0) = \int_{-\infty}^{\infty} Z(\omega) e^{-i\omega x}\, \mathrm{d}\omega$: $\theta(x,0)$ is, in fact, exactly the Fourier transform of $Z(\omega)$. Thus, if we are given the initial state of the wire as $\theta(x,0) = I(x)$, we can say that $Z(\omega) = F^{-1} I(\omega)$, and therefore

$$\theta(x,t) = \int_{-\infty}^{\infty} F^{-1} I(\omega) e^{-i\omega x} \exp(-\omega^2 t/\kappa)\, \mathrm{d}\omega$$
$$= F[F^{-1} I(\omega) \exp(-\omega^2 t/\kappa)](x), \tag{3.43}$$

Example. A thin uniform wire follows the thermodynamics of equation (3.39). It is uniformly at temperature $\theta(x,t) = 0$ for $t < 0$. At time $t = 0$, it is touched instantaneously by an infinitely hot probe at the point $x = 0$; so that $\theta(x,0) = A\delta(x)$. What is the maximum temperature subsequently attained at the point $x = 1$?

The initial state of the wire is $I(x) = A\delta(x)$, where A is a constant related to the amount of heat energy transferred from the probe to the wire. From equation (3.43),

$$\theta(x,t) = F[F^{-1}(A\delta)(\omega)\exp(-\omega^2 t/\kappa)](x);$$

so, using the results of sections 3.5, 3.8 and 3.10,

$$\theta(x,t) = AF[(1/2\pi) \times 1 \times \exp(-\omega^2 t/\kappa)](x)$$
$$= (A/2\pi) F[\exp(-\omega^2 t/\kappa)](x)$$

$$= (A/2)\sqrt{(\kappa/\pi t)}\exp(-\kappa x^2/4t).$$

It follows that

$$\theta(1,t) = (A/2)\sqrt{(\kappa/\pi t)}\exp(-\kappa/4t).$$

To find the maximum temperature attained at this point, we differentiate with respect to t:

$$\frac{\partial \theta(1,t)}{\partial t} = \theta(1,t)(-1/2t + \kappa/4t^2),$$

which is zero when $t = \frac{1}{2}\kappa$. And $\theta(1,\frac{1}{2}\kappa) = A/\sqrt{(2\pi e)}$; so the temperature at $x = 1$ peaks at $\underline{A/\sqrt{(2\pi e)}}$.

3.14 Interpreting spectral peaks

This use of Fourier transformations in the solution of differential equations is as a technical tool. There are also more practical uses of transforms: cases where data is collected and then subjected to Fourier transformation in order to extract information about it.

Thus, for instance, a meteorologist might plot a graph of the air temperature, measured over a period of time, at a particular point in Greenwich. In doing so, he is plotting the graph of a function of time, $\theta(t)$ say. The function $\theta(t)$ is said to be in the *time domain*, or the *t-domain*. Interpreting such a graph is a familiar exercise. (See Fig. 3.9(a). The curves do not represent real data.)

If the meteorologist performs a Fourier transformation on his data, he obtains the frequency-dependent function $F\theta(\omega)$. Its graph is the *spectrum* of θ, and is a curve in the *frequency domain*, or the *ω-domain* (Fig. 3.9(b)). Interpreting the spectrum of θ is less familiar.

The spectrum of θ breaks $\theta(t)$ down into its harmonic components. If the temperature readings are made in such a way as to give the entire function $\theta(t)$, from $-\infty$ to ∞, with perfect accuracy, so that we have the entire spectrum $F\theta(\omega)$ available, then surely the first thing we would notice in the spectrum would be a large peak at an angular frequency of about $\omega = 7.27 \times 10^{-5}$ s^{-1}, corresponding to a period of $2\pi/\omega = 86400$ s. We would interpret this as indicating a significant harmonic component in $\theta(t)$ that oscillates with this period. Since 86400 s is the length of the solar day, we might deduce that there is a significant connection between the time of day and the air temperature. The cycling of day and night is a fairly constant, periodic phenomenon, and its effect on the air temperature would be expected to be periodic too.

Other peaks in the spectrum of θ would suggest other connections. There would be a peak at the much lower angular frequency $\omega \approx 1.99 \times 10^{-7}$ s^{-1}, for instance, and perhaps one at $\omega \approx 1.8 \times 10^{-8}$ s^{-1}. (Identifying these is left as a task for the reader.)

But there would be many peaks with no natural cycle to connect with. There would be one at $\omega \approx 2.26 \times 10^{-4}$ s^{-1}, for instance, which gives a cycle length of exactly 8 h. This is not the period of any important cyclic phenomenon. The clue to identifying this is that 8 h is exactly one-third of a day. A component of period 8 h could, therefore, be a harmonic of the component of period one day. This peak in the spectrum may be, in fact, part of the effect of the solar cycle.

To see how this might arise, we can use our understanding of Fourier *series*.

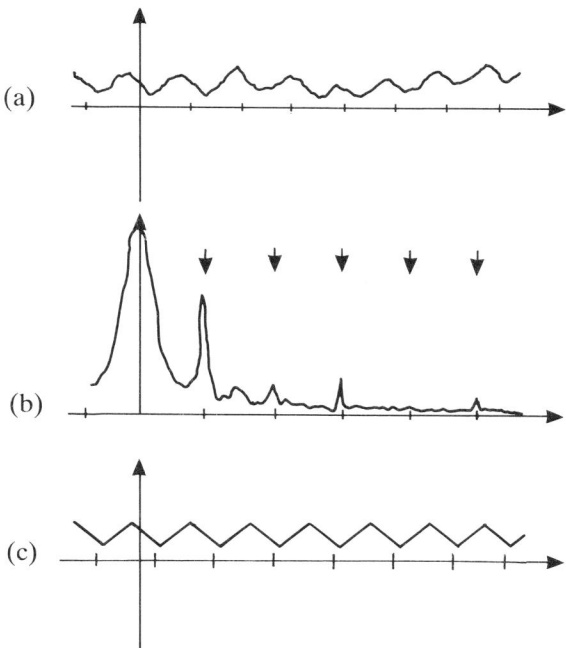

Fig. 3.9 — (a) Typical temperature record, clearly showing daily swings. (b) 'Spikes' in the transform, at frequencies of multiples of 1 day−1. (c) An underlying periodic function that might give rise to such spikes.

Suppose, for simplicity, that during the day the sun exerts a constant warming effect on the air, and that during the night there is a corresponding cooling effect; and that no heat is lost by the air through re-radiation. Ignore factors such as the variation in day length, the greater warming effect at midday than at dawn or at dusk. If there were no other influences on the air temperature, then the temperature curve would look like Fig. 3.9(c): it would be a triangle wave. In fact, there *are* other influences, which also affect the temperature; these, we assume, *superpose* their effects onto the solar effect, so that the actual curve looks like Fig. 3.9.(a).

Let the triangle wave due to the solar cycle be $\theta_s(t)$, and the remainder be $\theta_r(t)$; so $\theta(t) = \theta_s(t) + \theta_r(t)$. Fourier transformation is linear, so $F\theta(t) = F\theta_s(t) + F\theta_r(t)$. But the transform of a triangle wave contains an impulse at every odd harmonic of the fundamental frequency (equation (3.23)). Thus, in addition to a large spike at the once-a-day angular frequency 7.27×10^{-5} s^{-1}, there will also be smaller spikes at all odd multiples of that frequency, each of which is (in theory) impulsive.

In reality, the warming effect of the sun is not uniform, and the underlying solar cycle of temperature is not going to be a triangle wave, but some other non-harmonic wave. This will alter the sizes of the spikes, but not their position. In reality too, the solar heating will not be exactly periodic. This too will alter the spectrum slightly.

E. FOURIER TRANSFORM THEORY

3.15 Angular frequency versus frequency

In this chapter, we have looked at the properties and uses of the Fourier transform, $Ff(\omega) = \int_{-\infty}^{\infty} e^{-i\omega x} f(x) \, dx$. This, as we said above, is the way in which the transform is usually defined among mathematicians; physicists often use a version without the minus sign in the exponent.

There is a third formulation, sometimes encountered among engineers and other practical users. In this version, the transform is made a function of *frequency* $s = \omega/2\pi$ rather than of *angular frequency* ω. We define

$$\hat{F}f(s) = \int_{-\infty}^{\infty} e^{-2\pi i s x} f(x) \, dx. \qquad (3.44)$$

We might call this function the *Fourier frequency transform* of f. Clearly, it is related to F by the equation

$$\hat{F}f(s) = Ff(2\pi s). \qquad (3.45)$$

Although we shall, in this book, always use the function $Ff(\omega)$ in preference to $\hat{F}f(s)$, the operator $\hat{F}$ does have certain advantages. It is therefore appropriate briefly to look at the properties of $\hat{F}$.

A function has a Fourier frequency transform if and only if it has a Fourier transform. The operator $\hat{F}$ is, like F, a linear operator. Its arithmetic properties are similar: so, if we replace F by $\hat{F}$ in the equations of section 3.10, they remain true.

$\hat{F}$, like F, is invertible, for a large range of functions. Indeed, we can calculate its inverse: using equations (3.45) and (3.31),

$$F\hat{F}f(x) = (1/2\pi)FFf(x/2\pi)$$

which, by equation (3.4), gives

$$F\hat{F}f(x) = f(-x/2\pi).$$

A final application of equation (3.45) gives

$$\hat{F}\hat{F}f(x) = f(-x),$$

and therefore

$$\hat{F}^{-1}f(s) = \hat{F}f(-s). \qquad (3.46)$$

Thus, the inverse of the Fourier frequency transform does not include the constant 2π of equation (3.4). (This elegance is obtained, however, at the price of including the 2π in the definition of $\hat{F}$.)

The derivative theorem holds, in modified form, becoming

$$\hat{F}\frac{df}{dx}(s) = 2\pi i s \hat{F}f(s).$$

This is a case in which F is a more elegant operator than $\hat{F}$.

3.16 Applicability of the inversion theorem

Here is a sketch-proof that the inversion theorem, equation (3.4), holds when f is a *smooth* integrable function. It is analogous to the proof in section 2.16 of the convergence of Fourier series for smooth periodic functions.

The Fourier transform of $f(x)$ is $Ff(\omega) = \int_{-\infty}^{\infty} e^{-i\omega x} f(x) \, dx$; by the integrability of f, this exists for all ω. The inversion theorem concerns the integral $\int_{-\infty}^{\infty} e^{i\omega x} Ff(\omega) \, d\omega$, which we can expand as

$$\lim_{R \to \infty} \int_{-R}^{R} e^{i\omega x} \left\{ \int_{-\infty}^{\infty} e^{-i\omega y} f(y) \, dy \right\} d\omega. \tag{3.47}$$

Swapping the order of integration (this can be justified formally) gives

$$\int_{-\infty}^{\infty} e^{i\omega x} Ff(\omega) \, d\omega = \lim_{R \to \infty} \int_{-\infty}^{\infty} f(y) \left\{ \int_{-R}^{R} e^{i\omega(x-y)} \, d\omega \right\} dy$$

$$= \lim_{R \to \infty} \int_{-\infty}^{\infty} f(y) \left\{ (e^{i\omega(x-y)}/i(x-y))_{-R}^{R} \right\} dy$$

$$= \lim_{R \to \infty} \int_{-\infty}^{\infty} f(y) \left(\frac{2\sin(R(x-y))}{(x-y)} \right) dy$$

(putting $z = x - y$)

$$= 2 \lim_{R \to \infty} \int_{-\infty}^{\infty} f(x-z) (\sin(Rz)/z) \, dz. \tag{3.48}$$

Now we may apply (an analogue of) equation (2.52) to deduce that the limit on the right-hand side of this equation is $\pi f(x-0) = \pi f(x)$. Dividing by 2π gives the inversion theorem.

3.17 Fourier transforms and Fourier series

Over the course of the last two chapters, we have developed the theories for Fourier transforms and Fourier series independently: series for periodic functions, transforms for integrable functions. However, the theories are clearly very similar; we have, indeed, on most occasions adapted the proofs for transforms from the corresponding proofs for series. In fact, although there is no actual overlap between the theories, each can be seen to arise as a limiting case of the other.

First, let us look a little more closely at the calculation of the Fourier transform, (3.25), of a periodic function of period c. Without assuming that f has a Fourier series, we can treat its transform as a limit:

$$Ff(\omega) = \int_{-\infty}^{\infty} f(x)e^{-i\omega x} dx$$

$$= \lim_{R\to\infty} \int_{-R}^{R} f(x)e^{-i\omega x} dx.$$

In particular, when $R = (n + \tfrac{1}{2})c$ this integral becomes

$$\int_{-(n+1/2)c}^{(n+1/2)c} f(x)e^{-i\omega x} dx = \int_{-c/2}^{c/2} f(x)\left[\sum_{r=-n}^{n} e^{-i\omega(x+rc)}\right] dx.$$

Now the sum inside the right-hand integral is a geometric series. If $\tfrac{1}{2}\omega c$ is not a multiple of π, it can be written as

$$\sum_{r=-n}^{n} e^{-i\omega(x+rc)} = e^{-i\omega x} e^{i\omega nc}(1 - e^{-(2n+1)i\omega c})/(1 - e^{-i\omega c})$$

$$= e^{-i\omega x}[\sin((n+\tfrac{1}{2})\omega c)/\sin(\tfrac{1}{2}\omega c)],$$

and the integral becomes

$$\int_{-(n+1/2)c}^{(n+1/2)c} f(x)e^{-i\omega x} dx = [\sin((n+\tfrac{1}{2})\omega c)/\sin(\tfrac{1}{2}\omega c)] \int_{-c/2}^{c/2} f(x)e^{-i\omega x} dx.$$

The denominator $\sin(\tfrac{1}{2}\omega c)$ in the first term is nonzero. Since $\sin((n+\tfrac{1}{2})\omega c)$ oscillates between ± 1 as $n \to \infty$, and the integral $\int_{-c/2}^{c/2} f(x)e^{-i\omega x} dx$ is not dependent on n, the left-hand side is bounded with increasing n.

On the other hand, if $\tfrac{1}{2}\omega c = n\pi$, then $\sum_{r=-n}^{n} e^{-i\omega(x+rc)} = (2n+1)e^{-i\omega x}$, and we have

$$\int_{-(n+1/2)c}^{(n+1/2)c} f(x)e^{-i\omega x} dx = (2n+1) \int_{-c/2}^{c/2} f(x)e^{-i\omega x} dx.\text{ As } n\to\infty, \text{ this integral tends}$$

to infinity too. The Fourier 'transform' of $f(x)$ will, insofar as it exists at all, be impulsive (infinite) at angular frequencies ω which are multiples of $2\pi/c$, and be finite elsewhere. This, indeed, is the content of equation (3.25).

Thus, the Fourier transform of f has the form $Ff(\omega) = \sum_{n=-\infty}^{\infty} Y_n \delta(\omega - 2\pi n/c)$. From the inversion theorem,

$$f(x) = \sum_{n=-\infty}^{\infty} Y_n F^{-1}[\delta(\omega - 2\pi n/c)](x)$$

$$= \sum_{n=-\infty}^{\infty} Y_n e^{2\pi nx/c}.$$

(Of course this argument is not mathematically rigorous, and it needs careful justification in terms of limiting processes.)

Conversely, we may 'deduce' the Fourier inversion theorem for transforms from the Fourier convergence theorem for series. Consider a smooth integrable function $f(x)$, which decays reasonably quickly for large (positive and negative) x: say, there is some constant A for which

$$|f(x)| \leq A/x^2 \qquad (x \neq 0). \tag{3.49}$$

Define, for positive integers n, the function f_n by

$$f_n(x) = \sum_{m=-\infty}^{\infty} f(x - m2^n); \tag{3.50}$$

(3.49) ensures that this sum converges for all x, and defines a smooth function f_n. Moreover, $f_n(x) \to f(x)$ as $n \to \infty$, for all x (Fig. 3.10 shows how this works for a

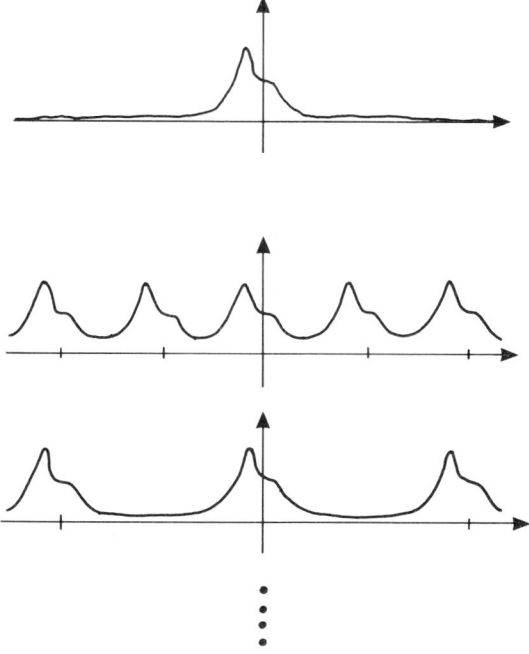

Fig. 3.10 — A function (top) yields a sequence of periodic functions (below).

typical f). In addition, f_n is periodic of period $c = 2^n$, so it has a Fourier series expansion

$$f_n(x) = \sum_{r=-\infty}^{\infty} Z_{n,r} \exp(2r\pi xi/c). \tag{3.51}$$

But $f_{n-1}(x) = f_n(x) + f_n(x + \tfrac{1}{2}c)$. It follows from equations (2.22) and (3.51) that

$$\sum_{r=-\infty}^{\infty} Z_{n-1,r} \exp(2r\pi xi/\tfrac{1}{2}c) = \sum_{r=-\infty}^{\infty} Z_{n,r} \exp(2r\pi xi/c) + \sum_{r=-\infty}^{\infty} Z_{n,r} \exp(2r\pi xi/c + r\pi).$$

Treating all these as Fourier series with $c = 2^n$, and remembering that such expressions are unique, gives us the relations $Z_{n,2r} = \tfrac{1}{2} Z_{n-1,r}$ (the terms in $Z_{n,2r-1}$ cancel out). Thus, in passing from f_{n-1} to f_n, we double the number of components (from harmonics with frequencies at multiples of $1/\tfrac{1}{2}c$ to those at multiples of $1/c$), but halve the amplitude of the components that are retained.

Now let us define the impulsive function

$$g_n(x) = \sum_{r=-\infty}^{\infty} Z_{n,r} \delta(x - 2r\pi/c). \tag{3.52}$$

Using this, we can write (3.51) as

$$f_n(x) = \int_{-\infty}^{\infty} g_n(\omega) e^{i\omega x} d\omega \tag{3.53}$$

(see equation (2.17)). Fig. 3.11 shows 'functions' $|g_n|$ that arise from the function f depicted in Fig. 3.10. The density of arrows doubles at every step, but their height halves. Remember that the height of an impulsive arrow denotes the integral, or area, under the 'spike'. The net effect of passing from g_{n-1} to g_n is thus to make the function smoother. In the limit, the density of spikes increases to infinity, and the spike height tends to zero; but the area under the curves between a and $a + h$ — the integral $\int_a^{a+h} g_n(\omega) d\omega$ — stabilizes.

What this means is that the limit of the functions g_n is a real-valued function. There remain no impulsive components. Indeed, we can calculate this limit. In virtue of (3.49), $f_n(x) \simeq f(x)$ for x in the range $-\tfrac{1}{2}c \leq x < \tfrac{1}{2}c$ and large n. It follows that the height of the rth impulsive arrow in g_n, at angular frequency $\omega = 2r\pi/c$, is

$$Z_{n,r} = (1/c) \int_{-c/2}^{c/2} f_n(x) e^{-i\omega x} dx$$

$$\simeq (1/c) \int_{-c/2}^{c/2} f(x) e^{-i\omega x} dx$$

$$\simeq (1/c) Ff(\omega).$$

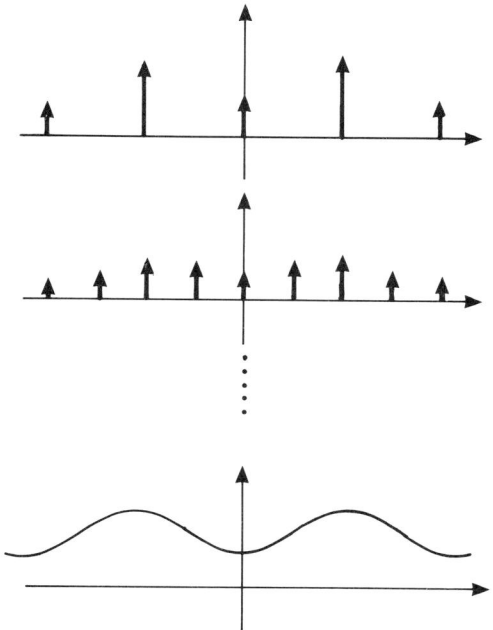

Fig. 3.11 — The 'convergence' of the impulsive transforms of the functions f_n to the transform of f.

The interval between successive impulses in g_n is $2\pi/c$. Thus, the limit of $g_n(\omega)$ as $n \to \infty$ is $(1/c)Ff(\omega)/(2\pi/c)$, or $(1/2\pi)Ff(\omega)$.

Since $f_n \to f$, and $g_n \to (1/2\pi)Ff$, as $n \to \infty$, equation (3.53) becomes, in the limit,

$$f(x) = (1/2\pi)\int_{-\infty}^{\infty} Ff(\omega)e^{i\omega x}d\omega.$$

This is the inversion theorem.

EXERCISES

1. Draw a graph of the function $f(x) = e^{-\lambda x}H(x)$. What is its (complex) Fourier transform? Sketch its amplitude and phase spectra.

2. What are the (two-sided) cosine and sine transforms of the function in the previous question?

3. Let $f(x)$ be as in Question 1. Of what function is $f(\omega)$ the transform?

4. Show how equation (3.16) may be deduced from equation (3.17) and the 'shift theorem', (3.31b).

5. Given a function $f(x)$, show that the transform of $f\chi_{(-a,a)}$ is

$$F[f\chi_{(-a,a)}](\omega) = \int_{-\infty}^{\infty} Ff(\xi) \frac{2\sin[a(\omega-\xi)]}{(\omega-\xi)} d\xi.$$

Compare this with equation (2.44).

6. Consider the forced oscillator analysed in section 1.7. The bob is at rest when, at time $t=0$, the ceiling is hit by a dropped load, If the ceiling's displacement is treated as a delta function of time, so that $y(t) = -A\delta(t)$, what is the motion of the bob?

7. (Harder) In reality, the ceiling's displacement will not be a delta function. Instead, it is likely to be a damped oscillation, with a formula like

$$y(t) = ae^{-\gamma t}\sin(\alpha t)\chi_{(0,\infty)}(t).$$

How will *this* forcing affect the bob?

8. What happens to your answer to the previous question as (i) $\gamma \to \infty$? (ii) $\gamma \to 0$?

9. Prove equations (3.26) and (3.27).

10. Prove equations (3.30) and (3.31)

⌐ 11. In the text, the delta 'function' $\delta(x)$ was treated as the limiting case of the step functions $\delta_n(x) = \frac{1}{2}n\chi_{(-1/n,1/n)}(x)$. It can also be treated as the limiting case of the bell curves $\tilde{\delta}_n(x) = (n/\sqrt{(2\pi)})\exp(-\frac{1}{2}n^2x^2)$. Prove that (i) $\lim_{n\to\infty} \tilde{\delta}_n(0) = \infty$; (ii) $\lim_{n\to\infty} \tilde{\delta}_n(x) = 0$

for $x \neq 0$; and (iii) $\int_{-\infty}^{\infty} \tilde{\delta}_n(x) dx = 1$ for all n.

Calculate $F\tilde{\delta}_n(\omega)$. How does it behave as $n \to \infty$?

12. In sections 3.7 and 3.8, the nonintegrable functions $H(x)$ and $I(x)$ were treated as limits of step functions, and we found that the Fourier transforms of these functions failed to converge. If instead we write $H_n(x) = e^{-x/n}H(x)$ and $I_n(x) = \exp(-x^2/2n^2)$, what happens?

13. The simplest *odd* step function is the function

$$f(x) = \chi_{(0,1)}(x) - \chi_{(-1,0)}(x).$$

Sketch this function. Calculate its transform. and sketch the amplitude and phase spectra.

14. Calculate and sketch the amplitude spectrum of $\chi_{(a,b)}(x)$.

15. Deduce equation (3.36) from equation (3.35) and the inversion theorem.

⌐ 16. Sketch the graph of the function $f(x) = x\exp(-\frac{1}{2}x^2)$. Is it even, odd, or neither? Find its Fourier transform, using section 3.11.

17. Prove the Riemann–Lebesgue lemma for integrals, equation (3.38), for smooth functions f.

18. (Harder) A thin uniform wire follows the thermodynamics of equation (3.39). For time $t<0$ it is at a uniform zero temperature; then, at $t=0$, a section of length 2 units is heated instantaneously so that its temperature is raised by a uniform $\frac{1}{2}A$ degrees.

Show that, after a long while, the temperature profile of the wire is approximately that of the example in section 3.9, i.e.

$$\theta(x,t) \simeq (A/2)\sqrt{(\kappa/\pi t)}\exp(-\kappa x^2/4t) \quad \text{for large } t.$$

Project. In the text, the delta 'function' $\delta(x)$ is treated as the 'limit' of the sequence of step functions $\delta_n(x) = \frac{1}{2}n\chi_{(-1/n,\,1/n)}(x)$. Above, in Question 12, we saw that $\delta(x)$ is also the 'limit' of the sequence $\tilde{\delta}_n(x) = (n/\sqrt{(2\pi)})\exp(-\frac{1}{2}n^2x^2)$. But the most important feature of the delta function is that expressed in equation (2.17)

Prove, for any integrable function $f(x)$ which is continuous at $x=0$, that

$$\lim_{n\to\infty} \int_{-\infty}^{\infty} f(x)\delta_n(x)\,dx \text{ exists and equals } f(0). \text{ Is the same true for } \tilde{\delta}_n(x), \text{ that }$$

$$\lim_{n\to\infty} \int_{-\infty}^{\infty} f(x)\tilde{\delta}_n(x)\,dx \text{ exists and equals } f(0)? \text{ If so, prove it. If not, provide a counterexample.}$$

The *lambda* (or *triangle*) *function* — not to be confused with the triangle wave — is defined as

$$\Lambda(x) = \begin{cases} 0 & (x<-1) \\ x+1 & (-1\leq x<0) \\ 1-x & (0\leq x<1) \\ 0 & (x\geq 1). \end{cases} \quad (3.54)$$

Calculate its Fourier transform. Use it to construct a third sequence of functions of which $\delta(x)$ is the 'limit', and prove that $\delta(x)$ is indeed the 'limit' of these functions in the senses explored above.

How might δ_n be related to the *pi* (or *rectangle*) *function*,

$$\Pi(x) = \begin{cases} 0 & (x<-1) \\ 1 & (-1\leq x<1) \\ 0 & (x\geq 1)? \end{cases} \quad (3.55)$$

(This is a step function, of course — indeed $\Pi(x) = \chi_{(-1,1)}(x)$.) How about $\tilde{\delta}_n$ and the bell curve $N(x)$, equation (3.12)? Can this construction be generalized? If so, how?

4
Two and more dimensions

A. FORMULATION AND PROPERTIES

4.1 Fourier transforms

We saw in Chapter 2 and in sections 3.13–3.14 that an important application of Fourier theory is in the solution of problems involving *time-dependent* differential equations, like the oscillation of a pendulum and the variation of Greenwich temperature. Similar problems involving *spatial* variables can be treated in the same way: hence our analysis of the vibrating guitar string (section 2.2), in which, although it involves time as well, the actual Fourier series analysis is on the *shape* of the string.

But the string is a simple example for, of course, space is three-dimensional. Thus, many spatial variables must be represented by functions of the form $f(x,y,z)$. Density, electric potential, temperature, and pressure are all variables of this kind: in bulk bodies, they have values at each point of space, and will therefore depend on the three coordinate values of that point. Indeed, a non-equilibrium variable like atmospheric temperature must be represented by a function of the kind $\theta(x,y,z,t)$. In order to be able to use Fourier methods in such cases, we must extend the theory to functions of several variables.

We achieve this by treating each of the independent variables separately. For simplicity, suppose $f(x,y)$ is a function of two variables; then the *(two-dimensional) Fourier transform* of f is defined by

$$F_2 f(\alpha, \beta) = \int_{-\infty}^{\infty} \left\{ \int_{-\infty}^{\infty} f(x,y) e^{-i\alpha x}\, dx \right\} e^{-i\beta y}\, dy \ . \tag{4.1}$$

In other words: we first treat y as constant, and perform a Fourier transform on $f(x,y)$ as if it were a function of x only. The result, the inner integral, is a function of α and y. Then, we treat α as constant, and perform a Fourier transform on this as a function of y. The result is a function of α and β, the Fourier transform of $f(x,y)$.

It does not matter in which order the transforms are taken. The same function $F_2 f(\alpha, \beta)$ is obtained by taking the y-transform first:

$$F_2 f(\alpha, \beta) = \int_{-\infty}^{\infty} \left\{ \int_{-\infty}^{\infty} f(x,y) e^{-i\beta y} \, dy \right\} e^{-i\alpha x} \, dx \ .$$

To see this, note that in each case the exponentials can all be brought inside the inner integration, so that

$$F_2 f(\alpha, \beta) = \int_{-\infty}^{\infty} \left\{ \int_{-\infty}^{\infty} f(x,y) e^{-i\alpha x} e^{-i\beta y} \, dx \right\} dy \ . \tag{4.2}$$

Example What is the two-dimensional Fourier transform of the function $f(x,y) = e^{-x} H(x-y) H(y)$?

The function $f(x,y)$ is sketched in Fig. 4.1. Transforming with respect to x, we

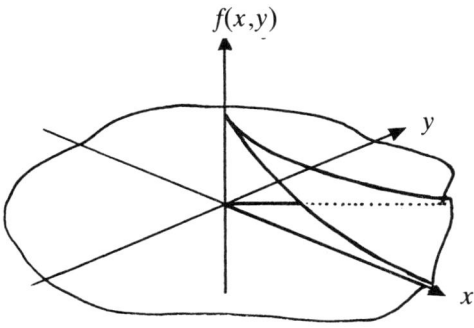

Fig. 4.1.

have

$$\int_{-\infty}^{\infty} f(x,y) e^{-i\alpha x} \, dx = H(y) \int_{-\infty}^{\infty} e^{-x} H(x-y) e^{-i\alpha x} \, dx$$

(putting $z = x - y$)

$$= H(y) \int_{-\infty}^{\infty} H(z) e^{-(i\alpha + 1)(z + y)} \, dz$$

$$= e^{-(i\alpha + 1) y} H(y) \int_{0}^{\infty} e^{-(i\alpha + 1) z} \, dz$$

$$= e^{-(i\alpha+1)y}H(y)/(i\alpha+1) .$$

Then

$$F_2 f(\alpha, \beta) = \int_{-\infty}^{\infty} \left\{ \int_{-\infty}^{\infty} f(x,y) e^{-i\alpha x} \, dx \right\} e^{-i\beta y} \, dy$$

$$= \frac{1}{i\alpha+1} \int_{-\infty}^{\infty} e^{-(i\alpha+1)y} H(y) e^{-i\beta y} \, dy$$

$$= \frac{1}{i\alpha+1} \int_{0}^{\infty} e^{-(1+i\alpha+i\beta)y} \, dy$$

$$= \frac{1}{(1+i\alpha)(1+i\alpha+i\beta)} .$$

Thus the two-dimensional transform of $f(x,y)$ is $1/(1+i\alpha)(1+i\alpha+i\beta)$.

For functions of more than two variables, the same process of repeated transformation is used. In general, a function $f(x_1,\ldots,x_n)$ of n variables has an n-dimensional Fourier transform, $F_n f(\omega_1,\ldots,\omega_n)$; this is also a function of n variables, and is defined using n consecutive transforms:

$$F_n f(\omega_1,\ldots,\omega_n)$$

$$= \int_{-\infty}^{\infty} \left\{ \ldots \left\{ \int_{-\infty}^{\infty} f(x_1,\ldots,x_n) \exp(-i\omega_1 x_1) \, dx_1 \right\} \ldots \right\} \exp(-i\omega_n x_n) \, dx_n$$

$$= \int_{-\infty}^{\infty} \left\{ \ldots \left\{ \int_{-\infty}^{\infty} f(x_1,\ldots,x_n) \exp(-i\omega_1 x_1) \ldots \exp(-i\omega_n x_n) \, dx_1 \right\} \ldots \right\} dx_n .$$
(4.3)

In this case, the x-domain and the ω-domain are both n-dimensional.

The range of functions which have n-dimensional Fourier transforms is large. If $f(x_1,\ldots,x_n)$ is any integrable function of n variables for which $|f|$ is also integrable over the whole of n-space, then f has an n-dimensional transform.

Clearly, the one-dimensional transform F_1 is identical to the Fourier transform operator F of the previous chapter.

Inversion of multidimensional transforms is possible, in view of their definition by repeated applications of the invertible operator F. Thus we have a multidimensional inversion theorem

$$f(x_1,\ldots,x_n) = F_n^{-1}[F_n f](x_1,\ldots,x_n)$$

$$= (1/2\pi)^n \int_{-\infty}^{\infty} \left\{ \ldots \left\{ \int_{-\infty}^{\infty} F_n f(\omega_1,\ldots,\omega_n)\exp(i\omega_1 x_1) \, d\omega_1 \right\} \ldots \right\} \exp(i\omega_n x_n) \, d\omega_n \, .$$
(4.4)

As with the ordinary (one-dimensional) transform, however, the inversion only applies to a restricted class of 'well-behaved' functions. Thus, for example, if the function $f(x_1,\ldots,x_n)$ has an everywhere-continuous *gradient* (vector derivative) $\nabla f = (\partial f/\partial x_1,\ldots,\partial f/\partial x_n)$, and $|f|$ has finite integral over the whole of n-space, then the multidimensional inversion theorem applies.

Since $F^{-1}f(\omega) = (1/2\pi)Ff(-\omega)$ for a one-dimensional function (equation (3.8)), for an n-dimensional function we have

$$F_n^{-1}f(\omega_1,\ldots,\omega_n) = (1/2\pi)^n F_n f(-\omega_1,\ldots,-\omega_n) \, .$$

4.2 Fourier series

Extending Fourier *series* to functions of several variables is slightly more involved, for we must first define an analogue of periodic functions (or, equivalently, of finite intervals). This is something we shall return to in section 4.7.

For the moment, let us say that a function $f(x,y)$ of two variables is (*doubly*) *periodic* if that there are *two* constants c_1 and c_2 such that

$$f(x + c_1, y) = f(x, y)$$

and

$$f(x, y + c_2) = f(x, y) \tag{4.5}$$

for all x and y. Thus, for any constant y_0, $f((x,y_0)$ is a periodic function of x, and each of these functions has period c_1; similarly, for each x_0, $f(x_0,y)$ is a periodic function of y with period c_2. The two constants c_1 and c_2 are the *x-period* and the *y-period* of f, respectively.

Each y-cross-section $f(x,y_0)$ is a periodic function of period c_1, so it generates a Fourier series

$$f(x,y_0) \to K + \sum_{n=1}^{\infty} A_n \cos(2n\pi x/c_1) + \sum_{n=1}^{\infty} B_n \sin(2n\pi x/c_1) \, .$$

The values of the Fourier coefficients K, A_n and B_n in this series will, naturally, depend on the choice of y_0. Thus, we can write

$$f(x,y) \to K + \sum_{n=1}^{\infty} A_n(y)\cos(2n\pi x/c_1) + \sum_{n=1}^{\infty} B_n(y)\sin(2n\pi x/c_1) \, , \tag{4.6}$$

where the coefficients K, A_n and B_n are now *functions of y*. From equations (2.3), these coefficient functions are given by the formulae

$$A_n(y) = \int_0^{c_1} f(x,y)\cos(2n\pi x/c_1) \, dx$$

110 **Fourier theory** [Pt. I

$$B_n(y) = \int_0^{c_1} f(x,y)\sin(2n\pi x/c_1)\, dx$$

and

$$K(y) = \int_0^{c_1} f(x,y)\, dx \ . \tag{4.7}$$

Now $f(x,y)$ is periodic with respect to y; indeed $f(x,y) = f(x,y+c_2)$. It follows from (4.7) that $A_n(y)$, $B_n(y)$, and $K(y)$ are all periodic of period c_2. Thus $A_n(y)$, $B_n(y)$, and $K(y)$ each generates a Fourier series:

$$A_n(y) \to L_n + \sum_{m=1}^{\infty} C_{m,n}\cos(2m\pi y/c_2) + \sum_{m=1}^{\infty} F_{m,n}\sin(2m\pi y/c_2)$$

$$B_n(y) \to M_n + \sum_{m=1}^{\infty} D_{m,n}\cos(2m\pi y/c_2) + \sum_{m=1}^{\infty} G_{m,n}\sin(2m\pi y/c_2)$$

$$K(y) \to N + \sum_{m=1}^{\infty} E_m\cos(2m\pi y/c_2) + \sum_{m=1}^{\infty} H_m\sin(2m\pi y/c_2) \ , \tag{4.8}$$

where

$$C_{m,n} = \int_0^{c_2} A_n(y)\cos(2m\pi y/c_1)\, dy$$

$$= \int_0^{c_2}\left(\int_0^{c_1} f(x,y)\cos(2n\pi x/c_2)\cos(2m\pi y/c_2)\, dx\right) dy \ , \tag{4.9}$$

and so on.

Putting expressions (4.8) into equation (4.6) gives

$$f(x,y) \to N + \sum_{n=1}^{\infty} L_n\cos(2n\pi x/c_1) + \sum_{n=1}^{\infty} M_n\sin(2n\pi x/c_1)$$

$$+ \sum_{m=1}^{\infty} E_m\cos(2m\pi y/c_2) + \sum_{m=1}^{\infty}\sum_{n=1}^{\infty} C_{m,n}\cos(2m\pi y/c_2)\cos(2n\pi x/c_1)$$

$$+ \sum_{m=1}^{\infty}\sum_{n=1}^{\infty} D_{m,n}\cos(2m\pi y/c_2)\sin(2n\pi x/c_1)$$

$$+ \sum_{m=1}^{\infty} H_m\sin(2m\pi y/c_2) + \sum_{m=1}^{\infty}\sum_{n=1}^{\infty} F_{m,n}\sin(2m\pi y/c_2)\cos(2n\pi x/c_1)$$

$$+ \sum_{m=1}^{\infty} \sum_{n=1}^{\infty} G_{m,n} \sin(2m\pi y/c_2)\sin(2n\pi x/c_1) \ . \qquad (4.10)$$

Thus, a general two-dimensional periodic function generates a two-dimensional Fourier series, which is a linear combination of the functions

$$1; \qquad \cos(2n\pi x/c_1); \qquad \sin(2n\pi x/c_1);$$
$$\cos(2m\pi y/c_2); \quad \cos(2m\pi y/c_2)\cos(2n\pi x/c_1); \quad \cos(2m\pi y/c_2)\sin(2n\pi x/c_1);$$
$$\sin(2m\pi y/c_2); \quad \sin(2m\pi y/c_2)\cos(2n\pi x/c_1); \quad \sin(2m\pi y/c_2)\sin(2n\pi x/c_1) \ . \qquad (4.11)$$

For a periodic function of r variables $f(x_1, \ldots, x_r)$, this process of repeated substitution must be repeated r times, yielding an r-dimensional Fourier series. The series obtained in this way are long and cumbersome, the number of terms rising exponentially with the dimension of the function.

The complexity can be obviated by the use of the complex formulation of the Fourier series, equation (2.36):

$$f(x) \rightarrow \sum_{n=-\infty}^{\infty} Z_n e^{2n\pi ix/c} \ .$$

In this case, there is only a single series of coefficients. Hence, it may be used in the construction of multidimensional series without an 'explosion of terms'. For instance, let $f(x,y)$ be a periodic function of just two variables, with x-period c_1 and y-period c_2. We begin by fixing $y = y_0$, and writing the Fourier series generated by the y-cross-section $f(x, y_0)$ in its complex form

$$f(x, y_0) \rightarrow \sum_{n=-\infty}^{\infty} Z_n \exp(2n\pi ix/c_1) \ .$$

This can be done for each such cross-section. But the value of the coefficient Z_n depends on the choice of y_0; thus we have

$$f(x,y) \rightarrow \sum_{n=-\infty}^{\infty} Z_n(y) \exp(2n\pi ix/c_1) \ , \qquad (4.12)$$

where $Z_n(y)$ is a function of y. Now by definition (equation (2.30)),

$$Z_n(y) = (1/c_1) \int_0^{c_1} f(x,y) \exp(-2n\pi ix/c_1) \, dx \ . \qquad (4.13)$$

Since f is periodic in y of period c_2, so is Z_n. As a periodic function, $Z_n(y)$ generates a Fourier series

$$Z_n(y) \rightarrow \sum_{m=-\infty}^{\infty} Y_{n,m} \exp(2m\pi iy/c_2) \ . \qquad (4.14)$$

The coefficients $Y_{m,n}$ in these series are given by equation (2.30) as

$$Y_{n,m} = (1/c_2) \int_0^{c_2} Z_n(y) \exp(-2n\pi i y/c_2) \, dy$$

$$= (1/c_1 c_2) \int_0^{c_2} \int_0^{c_1} f(x,y) \exp(-2\pi i n x/c_1) \exp(-2\pi i m y/c_2) \, dx \, dy \, . \quad (4.15)$$

Putting (4.14) into (4.12) gives:

$$f(x,y) \to \sum_{n=-\infty}^{\infty} \sum_{m=-\infty}^{\infty} Y_{n,m} \exp(2n\pi i x/c_1) \exp(2m\pi i y/c_2) \, . \quad (4.16)$$

This is the simplest expression of the two-dimensional Fourier series for $f(x,y)$.

n-fold periodic functions of more than two variables are dealt with by repeated substitutions of this kind.

Example. What is the two-dimensional Fourier series generated by the function $f(x,y) = Sq(x) \, Tr(2y)$?

The function $f(x,y)$ is sketched in Fig. 4.2. It has an x-period of 2 and a y-

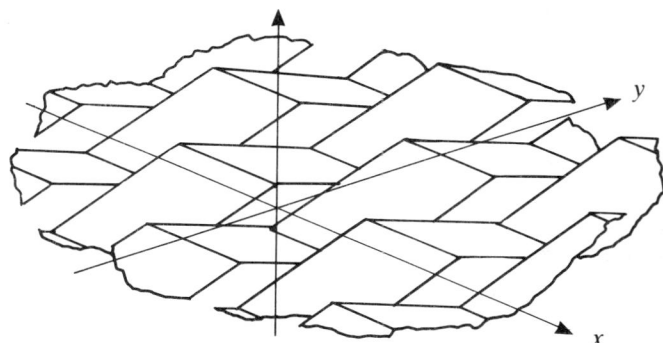

Fig. 4.2 — The function $Sq(x)Tr(2y)$.

period of 1. Fixing y, we have, from (4.13),

$$Z_n(y) = \tfrac{1}{2} \int_0^2 f(x,y) \exp(-n\pi i x) \, dx$$

$$= \tfrac{1}{2} Tr(y) \int_0^2 Sq(x) \exp(-n\pi i x) \, dx$$

$$= \begin{cases} -(2i/n\pi)\mathrm{Tr}(y) & (n \text{ odd}) \\ 0 & (n \text{ even}) \end{cases}$$

(cf. section 2.7). Then, from (4.12)

$$Y_{n,m} = \int_0^1 Z_n(y)\exp(-2m\pi i y) \, dy$$

$$= \begin{cases} -(2i/n\pi \int_0^1 \mathrm{Tr}(y)\exp(-2m\pi i y) \, dy & (n \text{ odd}) \\ 0 & (n \text{ even}) \end{cases}$$

$$= \begin{cases} -8i/m^2 n\pi^3 & (n \text{ odd}, m \text{ odd}) \\ 0 & (\text{otherwise}) \end{cases}$$

(cf. section 2.6). Hence the two-dimensional Fourier series generated by $f(x,y)$ is

$$\sum_{n \text{ odd}} \sum_{m \text{ odd}} (-8i/m^2 n\pi^3)\exp(n\pi i x)\exp(2m\pi i y) \ .$$

As with one-dimensional series, for certain 'well-behaved' functions the n-dimensional Fourier series is actually a Fourier *expansion*: the series converges to the function that generates it. This is true, for instance, of any differentiable (periodic) function $f(x_1,\ldots,x_n)$ for which the vector derivative $\nabla f = (\partial f/\partial x_1,\ldots,\partial f/\partial x_n)$ is continuous.

4.3 Vector formulation

Both multidimensional Fourier transforms and multidimensional Fourier series are rendered more appealing by the use of vector variables. This is natural, for the use of vectors to describe (in particular) functions in real two- and three-space is fundamental. Thus the temperature distribution of a room might be given by a function $\theta(\mathbf{x})$, where the vector variable $\mathbf{x} = (x,y,z)$ denotes the point in the room at which θ is measured.

If the n-dimensional x-domain is described in terms of vectors, the n-dimensional ω-domain can be too. In equation (4.3), for, example, if f is treated as a function of the vector variable $\mathbf{x} = (x_1,\ldots,x_n)$, then $\mathbf{F}_n f$ can be thought of as a function of the *angular frequency vector* $\boldsymbol{\omega} = (\omega_1,\ldots,\omega_n)$. In this way, the n-dimensional Fourier transform becomes the Fourier *vector transform*, $\mathbf{F}_n f(\boldsymbol{\omega})$.

In calculation, the vector transform is identical to the multi-dimensional transform as we have already seen it. But using vector ideas gives us a greater insight into its interpretation. Thus, we can use the *scalar product* (or *dot product*) of vectors, defined as

$$(a_1,\ldots,a_n) \cdot (b_1,\ldots,b_n) = \sum_{r=1}^n a_r b_r \ , \qquad (4.17)$$

and the associated concept of the *norm* (or *length*) of a vector, $\|\mathbf{a}\| = \sqrt{(\mathbf{a} \cdot \mathbf{a})}$. Among other useful properties of these operations is the fact that, for a pair of (spatial) vectors $\mathbf{a}$ and $\mathbf{b}$, the quotient $(\mathbf{a} \cdot \mathbf{b})/\|\mathbf{a}\|\|\mathbf{b}\|$ is the cosine of the angle between them.

Using these ideas, we can rewrite equation (4.3) as

$$\mathbf{F}_n f(\mathbf{\omega}) = \int_{-\infty}^{\infty} \left\{ \ldots \left\{ \int_{-\infty}^{\infty} f(\mathbf{x}) \exp(-i\omega_1 x_1) \ldots \exp(-i\omega_n x_n) \, dx_1 \right\} \ldots \right\} dx_n$$

$$= \int_{-\infty}^{\infty} \left\{ \ldots \left\{ \int_{-\infty}^{\infty} f(\mathbf{x}) \exp(-i(\omega_1 x_1 + \ldots + \omega_n x_n)) \, dx_1 \right\} \ldots \right\} dx_n$$

$$= \int_{-\infty}^{\infty} \left\{ \ldots \left\{ \int_{-\infty}^{\infty} f(\mathbf{x}) \exp(-i\mathbf{\omega} \cdot \mathbf{x}) \, dx_1 \right\} \ldots \right\} dx_n$$

$$= \int f(\mathbf{x}) \exp(-i\mathbf{\omega} \cdot \mathbf{x}) \, d\mathbf{x} \, , \qquad (4.18)$$

where the last integration is taken over the whole of n-space.

This is a significant advance over (4.3). In that equation, the idea was implicit that we were measuring the x_r according to some fixed coordinate system. (Consequently, the ω_r are measured, in the frequency domain, according to a corresponding coordinate system.) But in (4.18) there is no preferred coordinate system. It follows that the multidimensional Fourier transform $\mathbf{F}_n f$ of a multidimensional function $f(\mathbf{x})$ is an *inherent* 'partner' of f, and not one of many associated functions, each depending on a particular arbitrary framework.

The multidimensional inversion theorem, equation (4.4), can be cast into the same, vectorial, form. There we have

$$f(\mathbf{x}) = \mathbf{F}_n^{-1} [\mathbf{F}_n f](\mathbf{x})$$

$$= (1/2\pi)^n \int_{-\infty}^{\infty} \left\{ \ldots \left\{ \int_{-\infty}^{\infty} \mathbf{F}_n f(\mathbf{\omega}) \exp(i\mathbf{\omega} \cdot \mathbf{x}) \, d\omega_1 \right\} \ldots \right\} d\omega_n$$

$$= (1/2\pi)^n \int \mathbf{F}_n f(\mathbf{\omega}) \exp(i\mathbf{\omega} \cdot \mathbf{x}) \, d\mathbf{\omega} \, . \qquad (4.19)$$

In the last line, the integration is carried out over the whole of n-dimensional ω-space.

Multidimensional Fourier *series* can be made to benefit from such simplifications, with, however, less ease than transforms. Suppose first that $f(\mathbf{x}) = f(x_1, \ldots, x_n)$ is a periodic function of n variables, with x_r-period 1 for each r. For any vector $\mathbf{m} = (m_1, \ldots, m_n)$ with integer entries, we define

$$Z_\mathbf{m} = \int_0^1 \left\{ \ldots \left\{ \int_0^1 f(\mathbf{x}) \exp(-2\pi i m_1 x_1) \ldots \exp(-2\pi i m_n x_n) \, dx_1 \right\} \ldots \right\} dx_n$$

$$= \int_V f(\mathbf{x})\exp(-2\pi i \mathbf{m} \cdot \mathbf{x})\, d\mathbf{x} ,$$

where V is the unit volume $0 \leq x_1, \ldots, x_n \leq 1$. (Compare equation (4.15).) Then the n-dimensional Fourier series generated by the function f is

$$f(\mathbf{x}) \to \sum_{\mathbf{m}} Z_{\mathbf{m}} \exp(2\pi i m_1 x_1) \ldots \exp(2\pi i m_n x_n)$$

$$= \sum_{\mathbf{m}} Z_{\mathbf{m}} \exp(2\pi i \mathbf{m} \cdot \mathbf{x}) , \qquad (4.21)$$

where the sum is taken over all vectors $\mathbf{m}$ with integer entries.

For periodic functions $g(\mathbf{x})$ where the periods are not all 1, the variables must first be normalized. That is, if $g(\mathbf{x})$ has x_r-period c_r, we define $f(t_1, \ldots, t_n) = g(t_1/c_1, \ldots, t_n/c_n)$. Then f has t_r-period 1 for each r, and yields equation (4.21) with coefficients calculated as in (4.20). Putting $\mathbf{x} = (x_1, \ldots, x_n)$ and $\mathbf{y} = (x_1/c_1, \ldots, x_n/c_n)$, we have $f(\mathbf{x}) = g(\mathbf{y})$; so that

$$g(\mathbf{x}) \to \sum_{\mathbf{m}} Z_{\mathbf{m}} \exp(2\pi i \mathbf{m} \cdot \mathbf{y}) , \qquad (4.22)$$

where

$$Z_{\mathbf{m}} = (1/c_1, \ldots c_n) \int_V g(\mathbf{x}) \exp(-2\pi i \mathbf{m} \cdot \mathbf{y})\, d\mathbf{x} ; \qquad (4.23)$$

here, V is the volume $0 \leq x_1 \leq c_1, \ldots, 0 \leq x_n \leq c_n$.

4.4 Properties of vector transforms

All the properties of one-dimensional Fourier series and transforms have analogues for their n-dimensional counterparts. Thus, for instance, the Fourier vector transform is *linear*: for all functions f and g,

$$F_n(f+g)(\omega) = F_n f(\omega) + F_n g(\omega)$$

and

$$F_n(\alpha f)(\omega) = \alpha F_n f(\omega) . \qquad (4.24)$$

These formulae are analogous to equations (3.30). Similarly, there are n-dimensional analogues of equations (3.31); they are as follows.

If $g(\mathbf{x}) = f(\lambda \mathbf{x})$ for some scalar λ, then

$$F_n g(\omega) = 1/(|\lambda|^n) F_n f((1/\lambda)\omega) . \qquad (4.25)$$

If $g(\mathbf{x}) = f(\mathbf{x} + \mathbf{a})$ for some constant vector $\mathbf{a}$, then

$$F_n g(\omega) = e^{i\omega \cdot \mathbf{a}} F_n f(\omega) . \qquad (4.26)$$

The proofs follow directly from the one-dimensional cases.

We have already remarked upon the relationship between n-dimensional Fourier transformation and inverse transformation,

$$F_n^{-1}f(\omega) = (1/2\pi)^n F_n f(-\omega) \ . \tag{4.27}$$

Vector transforms have additional properties. Consider what happens, for instance, if we act on $\mathbf{x}$ with some invertible matrix A. This might be associated with a rotation of the coordinate axes along which we are measuring. The proof of the result is a little technical, but it is possible to show that if $g(\mathbf{x}) = f(A\mathbf{x})$, then

$$F_n g(\omega) = (1/\det A) F_n f(A^*\omega) \ , \tag{4.28}$$

where $A^* = (\bar{A}')^{-1}$ is the *inverse* of the *transpose* of the *complex conjugate* of A. ($\mathbf{x}$ and ω must of course be treated as *column* vectors for the actions $A\mathbf{x}$ and $A^*\omega$ to make sense.)

Example. What is the two-dimensional Fourier transform of the function $g(x,y) = e^{-2x}H(x-y)H(x+y)$?

The function $g(x,y)$ is sketched in Fig. 4.3.

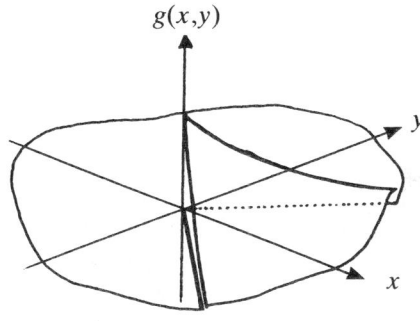

Fig. 4.3.

The transform of this function can be calculated directly, as the transform of the function $f(x,y) = e^{-x}H(x-y)H(y)$ was calculated in the Example of section 4.1, as $Ff(\alpha,\beta) = 1/(1+i\alpha)(1+i\alpha+i\beta)$. But we can also use equation (4.28). For $g(x,y) = f(2x, x+y)$; in other words, $g(\mathbf{x}) = f(A\mathbf{x})$, where

$$A = \begin{bmatrix} 2 & 0 \\ 1 & 1 \end{bmatrix} \ .$$

Then $\det A = 2$ and

$$A^* = (\bar{A}')^{-1} = \begin{bmatrix} \frac{1}{2} & -\frac{1}{2} \\ 0 & 1 \end{bmatrix} \ .$$

By (4.28), $Fg(\boldsymbol{\omega}) = (1/\det A)Ff(A^*\boldsymbol{\omega})$; so that

$$Fg(\alpha,\beta) = \tfrac{1}{2}Ff(\tfrac{1}{2}\alpha - \tfrac{1}{2}\beta, \beta)$$
$$= \tfrac{1}{2}/(1 + i(\tfrac{1}{2}\alpha - \tfrac{1}{2}\beta))\,(1 + i(\tfrac{1}{2}\alpha - \tfrac{1}{2}\beta) + i\beta)\ .$$

Thus the transform of g is $1/(2 + i\alpha - i\beta)(2 + i\alpha + i\beta)$.

Finally, it follows from the definition that the n-dimensional Fourier transform is *separable* in its variables. Thus, if the two-dimensional function $f(x,y)$ is of the form $f(x,y) = g(x)h(y)$, then $\boldsymbol{F}_2 f(\alpha,\beta) = Fg(\alpha)Fh(\beta)$. More generally, if

$$f(x_1,\ldots,x_n) = g(x_1,\ldots,x_r)\,h(x_{r+1},\ldots,x_n)\ ,$$

then

$$\boldsymbol{F}_n f(\omega_1,\ldots,\omega_n) = \boldsymbol{F}_r g(\omega_1,\ldots,\omega_r)\,\boldsymbol{F}_{n-r} h(\omega_{r+1},\ldots,\omega_n)\ . \quad (4.29)$$

B. SOME USEFUL FUNCTIONS

4.5 The *n*-dimensional bell curve

The standard n-dimensional bell-curve is the function

$$N_n(x_1,\ldots,x_n) = \exp(-\tfrac{1}{2}x_1^2)\ldots\exp(-\tfrac{1}{2}x_n^2)$$
$$= N(x_1)\ldots N(x_n) \quad (4.30)$$

(see section 3.5). In vector notation,

$$N_n(\mathbf{x}) = \exp(-\tfrac{1}{2}\mathbf{x}\cdot\mathbf{x}) = \exp(-\tfrac{1}{2}\|\mathbf{x}\|^2)\ .$$

Fig. 4.4 shows N_2.

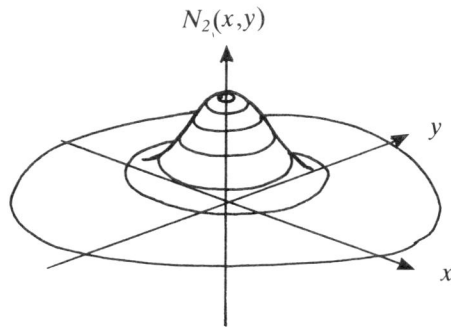

Fig. 4.4 — The two-dimensional bell curve, $N_2(x)$.

The transform of N_n follows from equations (3.13) and (4.29):

$$F_n N_n(\omega) = (2\pi)^{n/2}\exp(-\tfrac{1}{2}\|\omega\|^2) = (2\pi)^{n/2} N_n(\omega) \ . \tag{4.31}$$

Like its one-dimensional counterpart, the n-dimensional bell curve is, apart from a constant factor of $(2\pi)^{n/2}$, its own (n-dimensional) Fourier transform.

4.6 Impulsive 'functions' in n dimensions

Functions derived from the delta function $\delta(x)$ hold an important position in the canon of n-dimensional Fourier transforms. There are several categories of n-dimensional impulsive functions, corresponding roughly to the number of dimensions in which the function is impulsive.

For example, there is the two-dimensional impulse function $\delta_2(x,y)$, which is zero everywhere apart from at the origin (0,0), where it is infinite; and its total integral

$$\int_{-\infty}^{\infty}\int_{-\infty}^{\infty} \delta_2(x,y)\,dx\,dy = 1.$$

Like $\delta(x)$, such a function cannot exist strictly; so δ_2 is a generalized function, realizable only as a limiting case. For example, $\delta_2(x,y)$ is the 'limit' of the sequence of functions $(t^2/2\pi) N_2(tx,ty)$, as $t \to \infty$ (see Fig. 4.5).

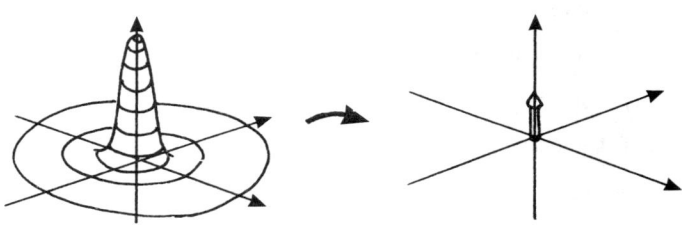

Fig. 4.5 — The two-dimensional delta 'function' $\delta_2(x,y)$ may be thought of as the 'limit' of compacted bell curves.

The function $\delta_2(x,y)$ has useful integral properties. For instance,

$$\int_{-\infty}^{\infty}\int_{-\infty}^{\infty} f(x,y)\delta_2(x,y)\,dx\,dy = f(0,0) \tag{4.32}$$

for any function $f(x,y)$. (Compare equation (2.17).) This formula means that we can sensibly write

$$\delta_2(x,y) = \delta(x)\delta(y) \ . \tag{4.33}$$

Comparable functions can be defined in three and more dimensions. The n-dimensional impulse function $\delta_n(\mathbf{x}) = \delta_n(x_1,\ldots,x_n)$ is zero except at the origin $\mathbf{x} = \mathbf{0}$ and infinite at that point, and has total integral 1.

The (two-dimensional) Fourier transform of δ_2 is easy to find: $F_2\delta_2(\omega) = 1$

everywhere. The transform of δ_2 is thus the *two-dimensional unity function*, which we write as $I_2(\omega)$. (The proof of this is left as an exercise.) More generally, the transform of the n-dimensional delta function $\delta_n(x)$ is $I_n(\omega)$, the n-dimensional unity function. (Conversely, using the relationship between F_n and its inverse (equation (4.27)), the transform of the unity function is $F_n I_n(\omega) = (2\pi)^n \delta_n(\omega)$.)

Similarly, the function $\delta_n(\mathbf{x} - \mathbf{c})$ consists of an impulsive spike (of unit magnitude) at the point $\mathbf{x} = \mathbf{c}$. The transform of this follows from equation (4.26):

$$F_n[\delta_n(\mathbf{x} - \mathbf{c})](\omega) = e^{-i\omega \cdot \mathbf{c}} F_n[\delta_n(\mathbf{x})](\omega)$$
$$= e^{-i\omega \cdot \mathbf{c}} . \quad (4.34)$$

Conversely, $F_n[e^{i\mathbf{x} \cdot \mathbf{c}}](\omega) = (2\pi)^n \delta_n(\omega - \mathbf{c})$.

Composed of such impulses is the two-dimensional sampling function, or the 'bed-of-nails' function, $III_2(x,y)$. This is defined as

$$III_2(x,y) = \sum_{m,n=-\infty}^{\infty} \delta_2(x-m, y-n) ; \quad (4.35)$$

its graph is shown in Fig. 4.6. In view of (4.33), we can write

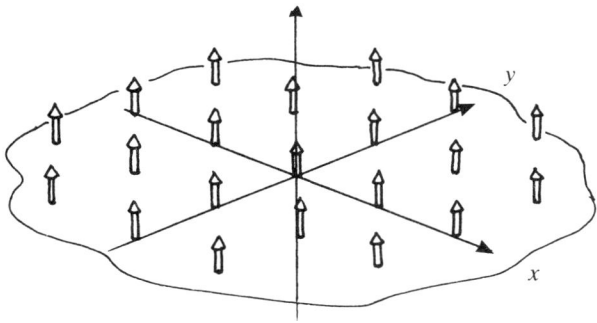

Fig. 4.6 — The two-dimensional sampling 'function' $III_2(x,y)$.

$$III_2(x,y) = III(x)III(y) . \quad (4.36)$$

It follows from the separability of Fourier transforms (equation (4.29)) and equations (3.27) that

$$F_2 III_2(\alpha, \beta) = III(\alpha/2\pi)III(\beta/2\pi)$$
$$= III^2(\alpha/2\pi, \beta/2\pi) .$$

In vectorial notation,

$$F_2 III_2(\omega) = III^2((1/2\pi)\omega) .$$

More generally, the transform of the n-dimensional sampling function, defined as
$$III_n(\mathbf{x}) = \sum_{\mathbf{m}} \delta_n(\boldsymbol{\omega} - \mathbf{m})$$ (the sum being over all vectors with integer entries), is

$$F_n III_n(\boldsymbol{\omega}) = III_n((1/2\pi)\boldsymbol{\omega}) \ . \tag{4.37}$$

Thus, like the n-dimensional bell curve, the sampling function is (almost) its own transform.

The 'function' δ_2 is an example of a function of two variables which is, as it were, impulsive in both dimensions. Contrast this with the function $f(x,y) = \delta(x)N(y)$, which is impulsive in only one dimension (Fig. 4.7). The transform of this function is, from (4.29), $F_2 f(\alpha, \beta) = \sqrt{2\pi}N(\beta)$.

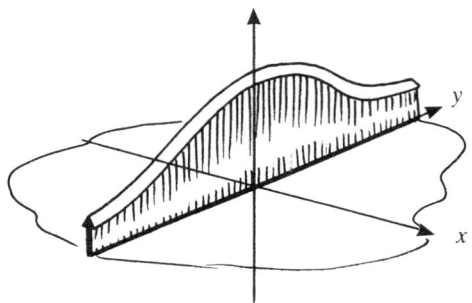

Fig. 4.7 — The function $\delta(x)N(y)$. This is an example of a function which is impulsive in only one dimension.

Of this second kind are the functions

$$\delta_x(x,y) = \delta(x) \quad \text{and} \quad \delta_y(x,y) = \delta(y) \ ;$$
$$III_x(x,y) = III(x) \quad \text{and} \quad III_y(x,y) = III(y) \ ;$$
$$\text{and} \quad \delta_r(x,y) = (1/2\pi)\delta(\sqrt{(x^2+y^2)}-1), \tag{4.38}$$

of which the graphs are shown in Fig. 4.8. Since $\delta_x(x,y) = \delta(x)I(y)$, we can use separability (equation (4.29)) to obtain

$$F_2 \delta_x(\alpha, \beta) = F\delta(\alpha) F_I(\beta)$$
$$= I(\alpha) 2\pi\delta(\beta)$$
$$= 2\pi\delta_y(\alpha, \beta) \ ; \tag{4.39}$$

conversely, $F_2 \delta_y(\alpha, \beta) = 2\pi\delta_x(\alpha, \beta)$. Thus, δ_x and δ_y constitute a transform pair. Similarly, $III_x(x,y) = III(x)I(y)$, so

$$F_2 III_x(\alpha, \beta) = FIII(\alpha) FI(\beta)$$

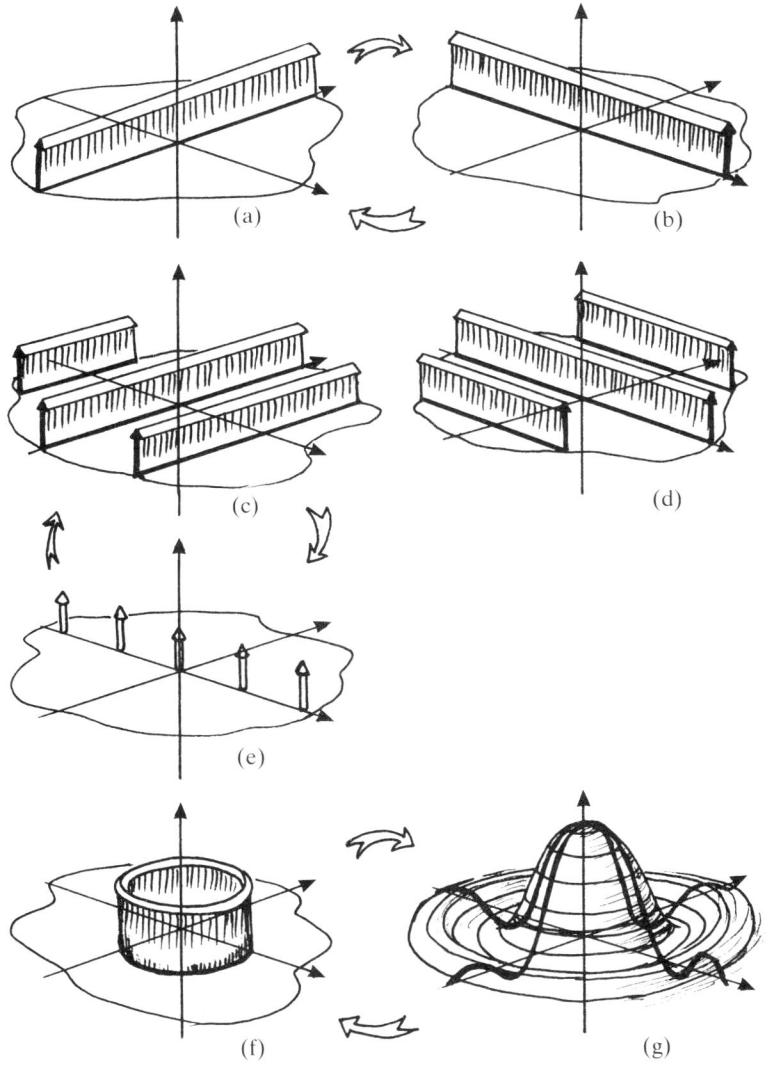

Fig. 4.8 — Some impulsive functions in two variables, and their transforms.

$$= III(\alpha/2\pi)2\pi\delta(\beta)$$
$$= III_x(\alpha/2\pi, \beta/2\pi)\delta_y(\alpha/2\pi, \beta/2\pi) \qquad (4.40)$$

(Fig. 4.8(e)); while $F_2 III_y(\alpha, \beta) = \delta_x(\alpha/2\pi, \beta/2\pi) III_y(\alpha/2\pi, \beta/2\pi)$. (The functions III_x and III_y are sometimes called 'grating' functions, while $III_x\delta_y$ and $III_y\delta_x$ are 'row-of-spikes' functions.)

The transform of the function δ_r is less easy to calculate. It turns out, in fact, to be a *Bessel function*. Bessel functions are discussed further in section 8.2; for now, we simply note that

$$F_2\delta_r(\omega) = J_0(\|\omega\|) \ . \tag{4.41}$$

Fig. 4.8(g) shows this function.

4.7 Periodic functions

As we saw in section 4.2, the functions $f(x_1,\ldots x_n)$ that generate n-dimensional Fourier series must satisfy stringent conditions of periodicity; namely, they must be independently periodic in *each* of the variables x_r. So, for instance, a function having x_r-period 1 for each r yields a Fourier series $f(\mathbf{x}) \to \sum_m Z_m \exp(2\pi i \mathbf{m} \cdot \mathbf{x})$, where

$$Z_\mathbf{m} = \int_V f(\mathbf{x})\exp(-2\pi i \mathbf{m} \cdot \mathbf{x})d\mathbf{x} \quad (V \text{ being the unit volume, } 0 \leq x_1,\ldots,x_n \leq 1; \text{ cf.}$$

equations (4.20) and (4.21)). If f is smooth, this Fourier series is a Fourier expansion, and the arrow may be replaced by an equality.

Such functions will have (improper) impulsive Fourier transforms. Thus, if $g(\omega)$ is to be the Fourier *inverse* transform of $f(\omega)$, then we require that $f(\mathbf{x}) = Fg(\mathbf{x}) = \int g(\omega)\exp(i\omega \cdot \mathbf{x}) \, d\omega$; since $f(\mathbf{x}) = \sum_m Z_\mathbf{m}\exp(2\pi i \mathbf{m} \cdot \mathbf{x})$, we might take $g(\omega) = \sum_m Z_\mathbf{m}\delta_n(\omega + 2\pi \mathbf{m})$. Using the relationship between forward and inverse transforms, we must then have

$$F_n f(\omega) = (2\pi)^n g(-\omega) = \sum_m ((2\pi)^n Z_\mathbf{m})\delta_n(\omega - 2\pi \mathbf{m}) \ . \tag{4.42}$$

Compare this with equations (3.24) and (3.25).

Thus, the transform of an (n-fold) periodic function consists of a sum of (n-dimensional) impulsive spikes. However, other functions have transforms of this kind too. In equation (4.34) we saw that the transform of the impulse $\delta_n(\mathbf{x} - \mathbf{c})$ is $e^{-i\omega \cdot \mathbf{c}}$; and, therefore, $F_n[e^{i\mathbf{x} \cdot \mathbf{c}}](\omega) = (2\pi)^n \delta(\omega - \mathbf{c})$. Suppose now that the function f has a transform consisting of a sum of spikes, of magnitude Z_r at point $\omega = \mathbf{c}_r$; that is $F_n f(\omega) = \sum_r Z_r \delta_n(\omega - \mathbf{c}_r)$. Then $f(\mathbf{x})$ has the form $f(\mathbf{x}) = \sum_r (Z_r/(2\pi)^n)\exp(i\mathbf{x} \cdot \mathbf{c}_r)$. Conversely, any function of this form has purely impulsive transform.

Weaker conditions of periodicity on a function f also yield impulsive transforms. The function $f(x,y) = \exp(-x^2)\cos y$, for instance, is periodic in y, though not in x; by separability, its transform is $F_2 f(\alpha, \beta) = \sqrt{(\tfrac{1}{2}\pi)}\exp(-\alpha^2/4)\pi(\delta(\beta + 1) + \delta(\beta - 1))$

(Fig. 4.9). This function is periodic in one variable only, so its transform is impulsive in only one dimension.

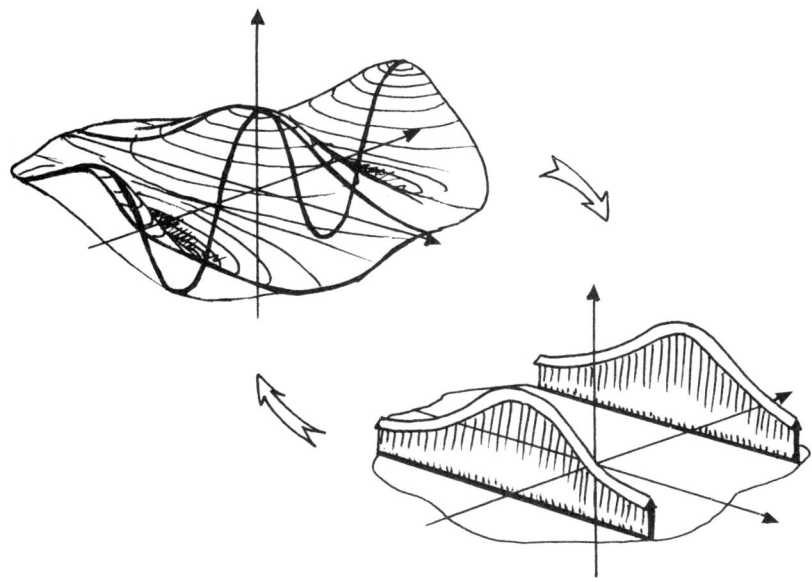

Fig. 4.9 — The function $N(x)\cos(y)$. This function is periodic in one dimension, so its transform is impulsive in one dimension.

Let us analyse functions of this type more carefully. A multidimensional function $f(\mathbf{x})$ is said to be *periodic* with *(vector) period* $\mathbf{c}$ if $f(\mathbf{x}+\mathbf{c}) = f(\mathbf{x})$ for all $\mathbf{x}$. (For example, the function $f(x,y) = e^{x+y}$ is periodic with period $(1, -1)$, since $f(x+1, y-1) = f(x,y)$ for all (x,y).) Such a function will automatically be periodic with period $n\mathbf{c}$ for all integers n.

Now if $f(\mathbf{x})$ is a continuous, nonzero periodic function which has period $\mathbf{c}$, then $|f|$ cannot have a finite total integral. (This is similar to the situation of periodic functions of one variable in section 3.9.) For values of $\boldsymbol{\omega}$ perpendicular to $\mathbf{c}$, we have $\boldsymbol{\omega} \cdot \mathbf{c} = 0$, and therefore $e^{-i\boldsymbol{\omega} \cdot \mathbf{c}} = 1$; it follows that $f(\mathbf{x})e^{-i\boldsymbol{\omega} \cdot \mathbf{x}}$ is also periodic of period $\mathbf{c}$. Thus, if the integral $\mathbf{F}_n f(\boldsymbol{\omega}) = \int f(\mathbf{x})e^{-i\boldsymbol{\omega} \cdot \mathbf{x}} d\mathbf{x}$ is nonzero, it is infinite. So the transform $\mathbf{F}_n f$ is, at points (in the $\boldsymbol{\omega}$-domain) *perpendicular* to $\mathbf{c}$, either zero or impulsive.

4.8 Lattices
A *lattice* (in n dimensions) is a subset L of real n-space satisfying the following axioms:

(1) The zero vector, $\mathbf{0} = (0,\ldots,0)$, is in L.
(2) If $\mathbf{a}$ and $\mathbf{b}$ are in L, and m and n are integers, then $m\mathbf{a} + n\mathbf{b}$ is in L.
(3) There is a positive constant c such that, for all nonzero vectors $\mathbf{a}$ in L, $\|\mathbf{a}\| \geq c$. (That is, there are not arbitrarily small vectors in L.)

A lattice in n dimensions can always be *generated* by at most n vectors. That is, there exist vectors $\mathbf{l}_1,\ldots,\mathbf{l}_r$ in L, with $r \leq n$, with the property that every point in the lattice is a sum of *integer* multiples of these $\sum_{k=1}^{\infty} m_k \mathbf{l}_k$. The number in a minimal generating set, or *basis*, vectors r is the *dimension* of the lattice; we may restrict attention to the case where the lattice 'fills' n-space, and $r = n$. Fig. 4.10(a) shows a typical two-dimensional lattice; the two points marked with light arrows form one possible basis. Two- and three-dimensional lattices are important in the study of crystal structure (see Chapter 12).

One subset of n-space that is easily seen to be a lattice is the set I of points $\mathbf{m} = (m_1,\ldots,m_n)$ whose coordinates are all integers. A basis for this lattice is $\{(1,0,0,\ldots,0), (0,1,0,\ldots,0),\ldots, (0,0,\ldots,0,1)\}$.

A *lattice function* is an impulsive function, having an impulse of unit magnitude at each point of a lattice. it takes the form

$$\delta_L(\mathbf{x}) = \sum_{\mathbf{m}} \delta_n(\mathbf{x} - \sum_{k=1}^{n} m_k \mathbf{l}_k) , \qquad (4.43)$$

the sum being over all n-tuples of integers $\mathbf{m} = (m_1,\ldots,m_n)$. Such functions are closely related to the n-dimensional sampling function, III_n. Indeed, the sampling function itself is a lattice function, based on the lattice I. Conversely, all lattice functions can be derived from the sampling function by means of a matrix transformation. If A is the $n \times n$ matrix whose columns are the vectors $\mathbf{l}_1,\ldots,\mathbf{l}_n$, then the lattice function of (4.43) satisfies the equation

$$\delta_L(A\mathbf{x}) = (1/\det A) III_n(\mathbf{x}) ,$$

so that

$$\delta_L(\mathbf{x}) = \det(A^{-1}) III_n(A^{-1}\mathbf{x}) . \qquad (4.44)$$

(The constant $1/\det A$ is necessary to normalize the impulsive spikes to unit magnitude.) It follows from equation (4.28) that the transform of δ_L is

$$F_n \delta_L(\omega) = \det(A^{-1})(1/\det(A^{-1})) F_n III_n((A^{-1})^* \omega)$$

$$= F_n III_n(A'\omega) , \qquad (4.45)$$

since A has real entries. But the transform of III_n is given in equation (4.37), so that

$$F_n \delta_L(\omega) = III_n((1/2\pi)A'\omega) . \qquad (4.46)$$

Compare this with equation (4.44). Apart from the constant, equation (4.46) represents a lattice function. If the matrix $B = 2\pi(A')^{-1}$ has columns $\mathbf{s}_1,\ldots,\mathbf{s}_n$, and S is the lattice generated by these as basis vectors, then we can write simply

$$F_n \delta_L(\omega) = (\det B) \delta_S(\omega) . \qquad (4.47)$$

So the Fourier transform of a lattice function is, apart from a constant which depends on the lattice 'size', a second lattice function. S is the *dual lattice* of L.

Lattice functions satisfy the Fourier inversion theorem. For $2\pi(B')^{-1} = A$, so that

$$F_n F_n \delta_L(x) = (\det B) F_n \delta_S(x)$$
$$= (\det A)(\det B) \delta_T(x) ,$$

where T is the lattice generated by the columns of $2\pi(B')^{-1} = A$; so $T = L$. Also $\det B = (2\pi)^n/\det A$. Hence

$$F_n F_n \delta_L(x) = (2\pi)^n \delta_L(x) .$$

By definition $\delta_L(x) = \delta_L(-x)$ for any lattice L. It follows that, in this case as in general, $F_n^{-1} \delta_L(\omega) = 1/(2\pi)^n F_n \delta_L(-\omega)$ (cf. equation (4.27)).

Fig. 4.10 shows an example of a two-dimensional lattice L, generated by (3,0) and

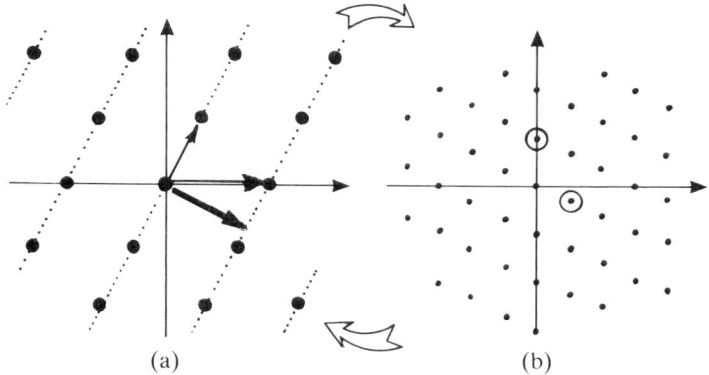

(a) (b)

Fig. 4.10 — (a) A lattice, generated by two vectors (light arrows), consists of a regular array of points. The transform of the corresponding lattice function is also a lattice function, based on the *dual lattice* (b).

Two vectors generating this lattice are circled; they are perpendicular to the generators of the lattice in (a).

Each point in (b) corresponds to a vector period of (a). The lower of the two circled points (in (b)), for instance, corresponds to the vector (in (a)) denoted by a bold arrow. The significance of this can be seen by considering the points of (a) to lie on a set of parallel lines (dotted).

(1,2), and its dual lattice S, generated by $(2\pi/3, -\pi/3)$ and $(0,\pi)$. Note the relationship

$$\begin{bmatrix} 2\pi/3 & 0 \\ -\pi/3 & \pi \end{bmatrix} = 2\pi \left\{ \begin{bmatrix} 3 & 1 \\ 0 & 2 \end{bmatrix}^t \right\}^{-1} .$$

Note also the properties observed in the previous section for periodic functions generally. L is a lattice, so δ_L is periodic; the transform of this function is impulsive (namely, $(2\pi^2/3)\delta_S$). In this transform, impulses lie at values of ω which are *perpendicular* to the periods. For a lattice function, the generators of the lattice are periods; thus $(2\pi/3, -\pi/3)$ is perpendicular to $(1,2)$, while $(0,\pi)$ is perpendicular to $(3,0)$.

The positions of the impulses in the transform lattice function δ_S can be interpreted more precisely. They indicate periodic behaviour in the original lattice δ_L. Thus, for instance, the impulse at $(2\pi/3, -\pi/3)$ corresponds to a periodicity in its direction, the direction $(2, -1)$. The length of this vector, $\pi(\sqrt{5})/3$, is the angular frequency of this periodicity. This angular frequency corresponds to a (scalar) period of $6/\sqrt{5}$; the *vector* period is therefore the vector of length $6/\sqrt{5}$ in the direction $(2, -1)$, namely $(12/5, -6/5)$. In fact, the point $(12/5, -6/5)$ does *not* lie in the lattice L; but Fig. 4.10(a) shows the significance of this vector period: it is *the spacing between the 'layers' of the lattice, measured perpendicularly, in this direction.*

EXERCISES

1. Sketch the graph of the function

$$f(x,y) = \begin{cases} 1 & (|x| \leq 1, |y| \leq 1) \\ 0 & \text{(otherwise)}, \end{cases}$$

and finds its (two-dimensional) Fourier transform.

2. Show that the function $f(x,y) = \text{Sq}(x+y)$ is doubly periodic, and find its (two-dimensional) Fourier series (i) in the constant–cos–sin form, and (ii) in the complex form.

3. More generally, let $g(x)$ be a periodic function of a single variable, and let α and β be real constants. Define $f(x,y) = g(\alpha x + \beta y)$. Show that f is doubly periodic. Can the two-dimensional series for f be deduced from the one-dimensional series for g?

4. Prove that, for a real-valued function $f(\mathbf{x})$, we have $\overline{\mathbf{F}_n f(\boldsymbol{\omega})} = \mathbf{F}_n f(-\boldsymbol{\omega})$.

5. Show how the equation (4.25) can be regarded as a special case of equation (4.28).

6. For suitably smooth functions $f(x)$ of a single variable, the derivative theorem for Fourier transforms (equation (3.35)) states that $F(df/fx)(\omega) = i\omega Ff(\omega)$. Find a corresponding relation between the transforms $\mathbf{F}_n(\partial f/\partial x_i)(\boldsymbol{\omega})$ of the *partial* derivatives of f, and the vector quantity $[i\mathbf{F}_n f(\boldsymbol{\omega})]\boldsymbol{\omega}$.

7. Show that $\delta_n(\mathbf{x})$ is the limit of the bell curves $t''(2\pi)^{-n/2}N_n(t\mathbf{x})$ as $t \to \infty$, and thereby deduce that $\mathbf{F}_n \delta_n(\boldsymbol{\omega}) = 1_n(\boldsymbol{\omega})$.

8. The lattice L is generated by $(a,0)$ and $(0,b)$. What is its dual lattice?

9. The lattice L is generated by $(2,1,1)$, $(1,2,1)$, and $(1,1,2)$. What is its dual lattice?

Project. In section 1.8 we analysed the motion of a vibrating string with fixed ends. The two-dimensional analogue of this is the stretched *membrane* attached to a fixed frame — like a drumskin on a drum. In the simplest case, the tension T in the membrane is uniform in all directions.

The case of a circular membrane is deferred to Chapter 8, but the equally important case of a square — or, with little increase in difficulty, a rectangular — membrane is rather easier. By considering the forces *in both directions* on an element of membrane, and following the argument in section 1.8, show that the equation of motion may be put into the form

$$\frac{\partial^2 z}{\partial x^2} + \frac{\partial^2 z}{\partial y^2} = \frac{1}{c^2}\frac{\partial^2 z}{\partial t^2} \tag{4.48}$$

(compare (1.32′)). This is the *two-dimensional wave equation*.

Find all the separated solutions to this equation, i.e. all the solutions of the form $z(x,y,t) = X(x)Y(y)T(t)$. Which of these 'fit' onto a fixed-edge rectangular membrane of sides a and b? What, therefore, are the natural frequencies of vibration of such a rectangular membrane? Sketch the vibration patterns of a few of the lowest-frequency modes.

How might this analysis be altered to account for the behaviour of a membrane under *non*uniform tension — for instance, if the tension in the x-direction were T_1 and the tension in the y-direction were T_2?

5
Convolution

A. DEFINITION AND PROPERTIES
5.1 Echoes

The sound produced by a performer on a theatre stage is not the same as the sound heard by the audience. It is altered by the acoustics of the theatre. This is because the sound reaches the auditorium along many routes, bouncing off the walls and ceiling. In some directions, especially where many bounces are made, the paths by which the sound travels may be very long, so the echo may take a considerable time to arrive; the sound via this path is likely to be much attenuated. Other paths are short, and yield rapid (and louder) echoes.

Of the sound (pressure) produced on stage, suppose that the fraction reaching the auditorium after a delay of between τ and $\tau + d\tau$ is $R(\tau)d\tau$. The function R is, for reasons that we shall see later, called the *impulse response function* of the echo system; it might typically look like Fig. 5.1. (In actuality, the impulse response will be different in different parts of the auditorium.)

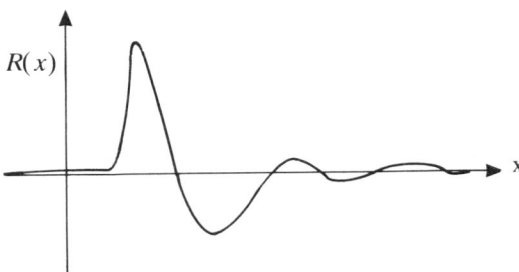

Fig. 5.1 — The acoustic impulse response of the theatre.

Now the performer produces a complex pattern of pressure changes, say $I(t)$; this is the *input* to the system. The echo in the theatre alters this, so the sound that arrives in the auditorium is different: say $O(t)$. This is the system *output*. $O(t)$ and $I(t)$ are clearly related, and related via the impulse response $R(\tau)$.

At time $t-\tau$, the input pressure is $I(t-\tau)$. Of this, $R(\tau)d\tau$ takes between τ and $\tau+d\tau$ to travel to the auditorium, so that $I(t-\tau)R(\tau)d\tau$ arrives between t and $t+d\tau$. The *total* sound arriving at time t is the *sum* of these echoes for all τ; thus

$$O(t) = \int_{-\infty}^{\infty} I(t-\tau)R(\tau)\,d\tau \ . \tag{5.1}$$

(In this case, $R(\tau)=0$ for $\tau<0$, for no echo arrives in the auditorium before the sound that yields it has been produced. For this reason, the integral in (5.1) can be restricted to the range $0<\tau<\infty$.) Typical input and output curves are shown in Fig. 5.2.

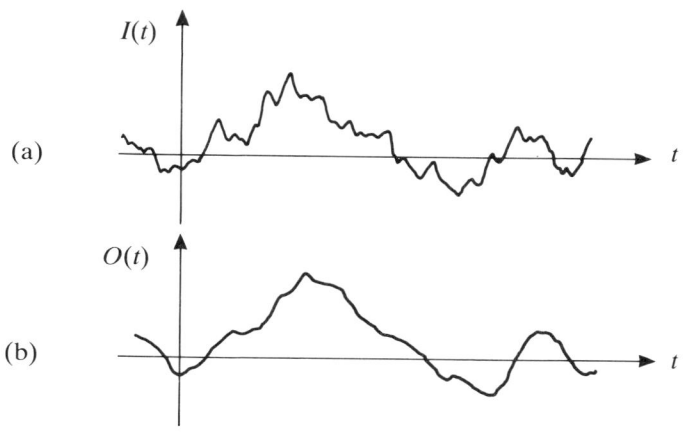

Fig. 5.2 — (a) The sound produced on stage. (b) The sound heard in the stalls.

In this analysis there are two chief assumptions: that the impulse response function $R(\tau)$ does not change with time; and that the total echo is simply the sum of all the individual echoes.

5.2 Convolution

Equation (5.1) is an example of a convolution. Convolution is an operation by which two functions $f(x)$ and $g(x)$ combine to yield a third function, denoted by $(f*g)(x)$. We say that convolution is a *binary operator*. It is defined by the integral formula

$$(f*g)(x) = \int_{-\infty}^{\infty} f(x-y)g(y)\,dy \ . \tag{5.2}$$

Convolutions typically arise in practice in situations of the following type:

(a) There is some sort of 'transmitting' object, called the *filter*.
(b) There is a signal input to the filter.

(c) There is a transmitted signal output from the filter.
(d) The filter affects the signal in a fixed way, which is not dependent on the incoming or outgoing signals, and does not change along the domain of the input.
(e) The filter has 'linear' behaviour: the output from the sum of two inputs is the sum of the outputs from the individual inputs.

In the situations described in section 5.1, (a) is the theatre building; (b) is the sound produced by the singer; (c) is the echo. Condition (d) amounts to the properties of the theatre not changing either with the sound on stage or with time; (e) we had to assume, in order that the resulting equation (5.1) be a convolution.

There are, in addition to these occurrences, situations in which convolutions arise where there is no *physical* 'transmitting object' of this kind. Some of these we shall be investigating in the next chapter.

Example Let $f(x) = e^{-ax}\chi_{(0,\infty)}(x)$ and $g(x) = e^{-bx}\chi_{(0,\infty)}(x)$. Calculate $(f*g)(x)$.

We have

$$(f*g)(x) = \int_{-\infty}^{\infty} f(x-y)g(y)\,dy$$

$$= \begin{cases} \int_0^x e^{-a(x-y)} e^{-by}\,dy & (x \geq 0) \\ 0 & (x < 0) \end{cases}$$

$$= e^{-ax}\chi_{(0,\infty)}(x) \int_0^x e^{(a-b)y}\,dy \ .$$

If $a \neq b$, then

$$(f*g)(x) = (1/(b-a))e^{-ax}\chi_{(0,\infty)}(x)(1 - e^{(a-b)x})$$
$$= \underline{(1/(b-a))(e^{-ax} - e^{-bx})\chi_{(0,\infty)}(x)} \ .$$

If $a = b$, then

$$(f*g)(x) = e^{-ax}\chi_{(0,\infty)}(x) \int_0^x 1\,dy$$
$$= \underline{xe^{-ax}\chi_{(0,\infty)}(x)} \ .$$

Convolution is a linear operator; that is to say, it has the following properties:

$$[(f+g)*h](x) = (f*h)(x) + (g*h)(x)$$
$$[(\lambda f)*g](x) = \lambda(f*g)(x)$$
$$[f*(g+h)](x) = (f*g)(x) + (f*h)(x)$$

and $\qquad [f*(\lambda g)](x) = \lambda(f*g)(x)$ \hfill (5.3)

These all follow from the linearity of integration, and are easy to check.

Less obviously, the following identities also hold:
$$[f*(g*h)](x) = [(f*g)*h](x)$$
and
$$(f*g)(x) = (g*f)(x) \ . \tag{5.4}$$

This last, the *commutativity* of the convolution operator, is particularly important (and unexpected). To prove it, we simply make the substitution $z = x - y$, $dz = -dy$ in its defining integral:

$$(f*g)(x) = \int_{-\infty}^{\infty} f(x-y)g(y)\,dy$$

$$= \int_{\infty}^{-\infty} f(z)g(x-z)(-dz)$$

$$= \int_{-\infty}^{\infty} g(x-z)f(z)\,dz$$

$$= (g*f)(x) \ .$$

It is instructive to note how the delta 'function' behaves under convolution. Convolutions involving $\delta(x)$ have a particularly simple form, in view of equation (2.17):

$$(f*\delta)(x) = \int_{-\infty}^{\infty} f(x-y)\delta(y)\,dy$$

$$= f(x) \ . \tag{5.5a}$$

Similarly, because of the symmetry of the convolution operator,
$$(\delta*f)(x) = f(x) \ . \tag{5.5b}$$

Consider what this means in terms of the theatre acoustics experiment. Our analysis there led us to conclusion that the echo $O(t)$ and performance $I(t)$ were related by the equation

$$O(t) = (I*R)(t) \tag{5.6}$$

(equation (5.1)), where $R(x)$ is a function representing the effect of the theatre hall. Now $G(t)$ and $I(t)$ are directly measurable, but $R(x)$ is not; it is an innate property of the theatre. However, if we conduct an experiment where the input is impulsive, say by clapping our hands on stage, then we will have $I(t) = \delta(t)$ and, from (5.6) and (5.5b), $O(t) = (\delta*R)(t) = R(t)$. Thus, although we cannot *directly* measure $R(x)$, we can measure it *indirectly* in this way.

5.3 Convolution and Fourier transformation
The Fourier transform of the convolution $f*g$ is surprisingly simple in form. We calculate:

$$F(f*g)(\omega) = \int_{-\infty}^{\infty} (f*g)(x)e^{-i\omega x}\,dx$$

$$= \int_{-\infty}^{\infty} \left[\int_{-\infty}^{\infty} f(x-y)g(y)\,dy\right] e^{-i\omega x}\,dx$$

$$= \int_{-\infty}^{\infty} \int_{-\infty}^{\infty} f(x-y)g(y)e^{-i\omega x}\,dy\,dx$$

$$= \int_{-\infty}^{\infty} \int_{-\infty}^{\infty} f(x-y)g(y)e^{-i\omega x}\,dx\,dy$$

$$= \int_{-\infty}^{\infty} g(y)e^{-i\omega y}\left[\int_{-\infty}^{\infty} f(x-y)e^{-i\omega(x-y)}\,dx\right]dy$$

$$= \left[\int_{-\infty}^{\infty} f(x-y)e^{-i\omega(x-y)}\,dx\right]\left[\int_{-\infty}^{\infty} g(y)e^{-i\omega y}\,dy\right]$$

(putting $z = x - y$)

$$= \left[\int_{-\infty}^{\infty} f(z)e^{-i\omega z}\,dz\right]\left[\int_{-\infty}^{\infty} g(y)e^{-i\omega y}\,dy\right]$$

$$= Ff(\omega)Fg(\omega)\ . \tag{5.7}$$

In words: the transform of a convolution is the product of the transforms. This is the *convolution theorem*.

Conversely, the convolution operation arises in the transform of a *product* of functions. From (an analogue of) equation (2.52), if fg is smooth,

$$\pi g(x) = \lim_{R\to\infty} \int_{-\infty}^{\infty} g(y)\left(\frac{\sin[R(y-x)]}{(y-x)}\right)dy\ .$$

It follows that

$$2\pi i F(fg)(\omega) = \left[\int_{-\infty}^{\infty} \left[\lim_{R\to\infty}\int_{-\infty}^{\infty} g(y)\right.\right.$$

$$\left.\left.(e^{iR(y-x)} - e^{-iR(y-x)})/(y-x)\,dy\right]f(x)e^{-i\omega x}\,dx\right.\ .$$

But

$$(e^{iR(y-x)} - e^{-iR(y-x)})/(y-x) = i\int_{-R}^{R} e^{-i\alpha(y-x)}\,d\alpha\ ,$$

so that

Convolution

$$2\pi F(fg)(\omega) = \lim_{R\to\infty} \int_{-\infty}^{\infty} \left[\int_{-\infty}^{\infty} g(y) \left(\int_{-R}^{R} e^{-i\alpha(y-x)} d\alpha\right) dy\right] f(x) e^{-i\omega x} dx$$

$$= \lim_{R\to\infty} \int_{-R}^{R} \left[\int_{-\infty}^{\infty} g(y) e^{-i\alpha y} dy\right] \left[\int_{-\infty}^{\infty} f(x) e^{-i(\omega-\alpha)x} dx\right] d\alpha$$

$$= \int_{-\infty}^{\infty} Ff(\omega - \alpha) Fg(\alpha) \, d\alpha$$

$$= (Ff * Fg)(\omega) \ . \tag{5.8}$$

(As always, the swapping of integrations, and of these with limits, needs to be justified formally.)

Where the Fourier inversion theorem (section 4.2) is applicable, (5.8) can be deduced as a corollary of the convolution theorem. For:

$$f(x)g(x) = FF^{-1}f(x) FF^{-1}g(x)$$

(by (5.7)) $\quad = F(F^{-1}f * F^{-1}g)(x) \ ,$

so

$$F^{-1}(fg)(\omega) = (F^{-1}f * F^{-1}g)(\omega) \ .$$

Using the inversion therorem in the form $F^{-1}f(\omega) = (1/2\pi)Ff(-\omega)$ (equation (3.8)), this becomes

$$F(fg)(\omega) = (1/2\pi)(Ff * Fg)(\omega) \ ,$$

as required.

It is hard to overstress the importance of the convolution theorem. We have seen how the Fourier transform is a natural way of dealing with functions associated with a linear system, and how Fourier inversion can be effected very easily. We have seen how convolutions appear, or may be introduced, quite naturally, in connection with many linear systems.

Note: the factor of $(1/2\pi)$ in equation (5.8) can be avoided if we use the Fourier frequency transform $\hat{F}$ of section 3.15 rather than the Fourier transform itself. It turns out that

$$\hat{F}(fg)(\omega) = \hat{F}f(\omega) * \hat{F}g(\omega) \ . \tag{5.9}$$

B. USES OF CONVOLUTION

5.4 Smoothing

Consider the convolution $f * g$ of two functions $f(x)$ and $g(x) = \chi_{(a,b)}(x)$. We calculate

$$(f * g)(x) = \int_{-\infty}^{\infty} f(x - y) \chi_{(a,b)}(y) \, dy$$

$$= \int_a^b f(x-y)\,dy$$

(putting $z = x - y$)

$$= \int_{x-b}^{x-a} f(z)\,dz \ . \tag{5.10}$$

If $f(z) \simeq h$ in the range $x - b \geq z \geq x - a$, then $(f*g)(x) \simeq h(b - a)$.

In addition, it follows from (5.10) that, for small δ,

$$(f*g)(x + \delta) - (f*g)(x) = \int_{x+\delta-b}^{x+\delta-a} f(z)\,dz - \int_{x-b}^{x-a} f(z)\,dz$$

$$= \int_{x-a}^{x+\delta-a} f(z)\,dz - \int_{x-b}^{x+\delta-b} f(z)\,dz \ ;$$

thus, $(f*g)(x + \delta) - (f*g)(x)$ is of the order of $2\delta h$. Comparing this with our estimate for $(f*g)(x)$, we see that the *relative* change in $(f*g)$ from x to $x + \delta$ is around $2\delta/(b - a)$. So when $\delta \ll b - a$, $(f*g)(x + \delta) \simeq (f*g)(x)$.

This is the simplest case of a more general behaviour. Precisely, suppose f and g are integrable functions; further, suppose that $g(x)$ is a positive *unimodal* function (loosely, its graph goes upwards to a unique peak, then downwards). Then the same conclusion holds: $(f*g)$ does not change too rapidly (it is *uniformly continuous*), and the rapidity of change is limited by the quantity $(\int |f(x)|\,dx)\max\{g(x)\}$.

This smoothing effect of convolution is sometimes desirable. There are circumstances under which fine detail is instrusive and/or irrelevant. Take, for example, the telescopic observation of stars. An astronomer is not usually interested in the flickering caused by atmospheric effects, still less in brief occlusions by passing birds and aircraft. Because of such effects, the direct recording of telescopic brightness ($I(t)$, say) will show all sorts of meaningless spikes. Fig. 5.3(a) shows such a trace as it might appear. In this case, we might decide that we are only interested in *slow* changes in brightness, those happening over (say) 100s or longer. Such changes may be relatively small, and be effectively obscured by all the flickering.

In order to effect the desired smoothing, we could convolve the measured light intensity with the function $\chi_{(-50,50)}(x)$. This is a mathematical process, not a physical one: this is an example of convolution with no physical filter. Better yet, since the sharp edges of the characteristic function are not really desirable in a smoothing operation, we might choose the function

$$R(x) = \exp(-x^2/2.50^2) \ ,$$

which also has an effective extent of $-50 \leq x \leq 50$.

There are two properties of the convolving function $R(x)$ which need to be considered if the smoothed input $(I*R)(t)$ is to be comparable with the input $I(t)$. First, $R(x)$ must be centred on $x = 0$. This means that $I*R$ will not be shifted in time

Ch. 5] Convolution 135

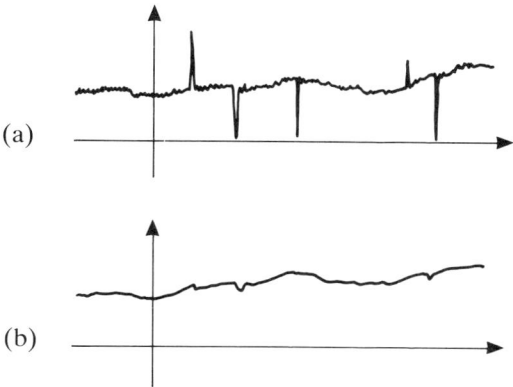

Fig. 5.3 — (a) 'Raw' light curve, showing 'blips'. (b) Smoothed curve.

compared with I. (Compare equation (5.10): g is centred on $\frac{1}{2}(a+b)$, and $(f*g)(x)$ is based in values of f around $x - \frac{1}{2}(a+b)$.)

Secondly, in order to preserve the scale of our original data, we must make sure that the convolving function has total integral $\int_{-\infty}^{\infty} R(x)\,dx = 1$. (Again, compare equation (5.10). There, $\int_{-\infty}^{\infty} g(x)\,dx = b - a$; while if $f(x) \simeq h$, then $(f*g)(x) \simeq (b-a)h$.) Thus, we normalize our chosen smoothing function, to

$$R(x) = (1/50\sqrt{(2\pi)})\exp(-x^2/2.50^2) \ . \tag{5.11}$$

Fig. 5.3(b) shows the effect of this smoothing.

The convolution theorem is useful for smoothing. In order to calculate the smoothed function $I*R$, we take Fourier transforms of the input to be smoothed I and of the chosen smoothing function R; multiply, frequency by frequency; and then apply an inverse transform. The resulting function is the required convolution, for equation (5.10) may be written as

$$(I*R)(t) = F^{-1}(FIFR)(t)$$
$$(= (1/2\pi)F(FIFR)(-t)) \ .$$

The advantage of doing convolutions by means of Fourier transforms is this. Fourier transformation is a *unary* operator: it acts on a single function. Convolution is *binary*: it acts on two functions. The computational 'effort' needed to calculate a convolution is greater because of this. Thus, for large amounts of input data, it is usually much quicker to perform three Fourier transforms, and a quick multiplication, than a direct convolution. (However, in section 9.9 we shall look at some particular convolutions that may be achieved more quickly by direct calculation.) This indirect calculation is shown diagrammatically in Fig. 5.4.

5.5 Truncation

When we measure a physical quantity, we only measure it over a limited range. A time-dependent variable will be recorded for a finite time: whether it be a meteoro-

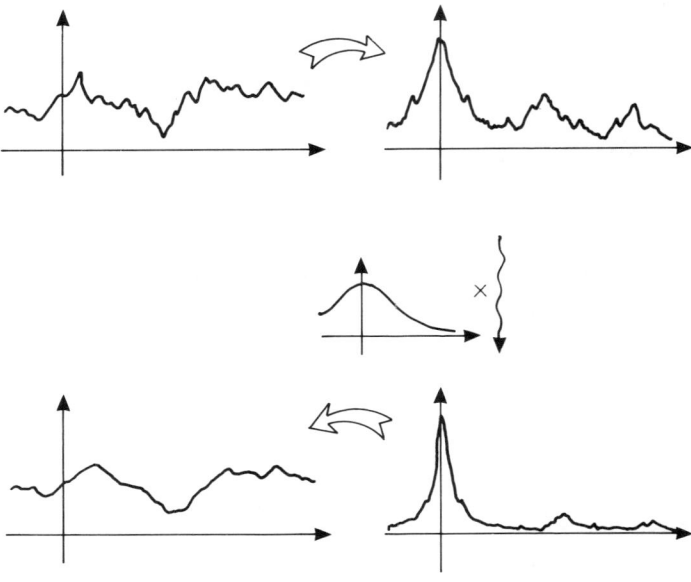

Fig. 5.4 — Convolution is achieved by a multiplication in the ω-domain.

logical temperature record that is kept for many years, or the temperature record in an experimental explosion which lasts for a fraction of a second. A spatial variable, likewise, will be recorded over a restricted region.

Abstractly, suppose we have the value of a function $f(x)$ only for a restricted range of x, say for $a<x<b$. With this limited information, we cannot evaluate the Fourier transform of f. The best we can do is to find the transform of the *truncated* function $g(x)$,

$$g(x) = \begin{cases} 0 & (x \leq a) \\ f(x) & (a<x<b) \\ 0 & (x \geq b) \ . \end{cases}$$

We can write this as $g(x) = f(x)\chi_{(a,b)}(x)$.

Now the relationship of $Fg(\omega)$ to $Ff(\omega)$ is given by the converse of the convolution theorem, equation (5.8); we have:

$$Fg(\omega) = (1/2\pi)(Ff * F\chi_{(a,b)})(\omega) \ . \tag{5.12}$$

The second term of this convolution is given by equation (3.16) as

$$F\chi_{(a,b)}(\omega) = i(e^{-i\omega b} - e^{-i\omega a})/\omega \ ;$$

in particular, if $a = -b$ (so that the interval of our information is $-b<x<b$), then we have

$$F\chi_{(-b,b)}(\omega) = 2\sin(\omega b)/\omega \;, \tag{5.13}$$

as in equation (3.17).

Now for large b, the function $2\sin(xb)/x$ behaves similarly to the impulsive function $\pi\delta(x)$, inasmuch as (for suitably smooth functions)

$$\int f(x)(\sin(xb)/x)\,dx \to \pi f(0) \quad \text{as } b\to\infty$$

(see equation (2.52)). It follows that

$$Fg(\omega) = (1/2\pi)(Ff*F\chi_{(-b,b)})(\omega)$$

$$= (1/2\pi)\int_{-\infty}^{\infty} Ff(\omega - y)(2\sin(yb)/y)\,dy$$

$$\to Ff(\omega) \quad \text{as } b\to\infty \;. \tag{5.14}$$

That is to say, if $g(x)$ represents the function $f(x)$ over a *large* range, then the transform of g approximately closely the transform of f. Because of this fact, we are justified in taking information over finite ranges.

Note, however, that the criteria for (2.52) — and hence for (5.14) — require that $f(x)$ be a smooth, absolutely integrable function. Many functions (like temperature records) are smooth, but not absolutely integrable, for they do not 'tail off' towards $\pm\infty$. In these circumstances, the justification for Fourier analysis on a restricted range is slightly different: we calculate Fg not to approximate Ff (which does not exist), but rather simply on the grounds that the range $(-b,b)$ contains a *representative sample* of the function f. (The PSDF of certain such functions is defined using precisely these truncations; see equation (3.11).)

Note too that the truncation of f leads to a convolution, and thus a smoothing, of its transform. Moreover, truncation by the function $\chi_{(-b,b)}$, with an information range of length $2b$, leads to smoothing by the function $2\sin(\omega b)/\omega$, with a 'width' scale of around $2\pi/b$. (See Fig. 5.5). This is a theme to which we shall return (section 5.9).

5.6 Amplitude modulation

Here is yet another common occurrence in which the convolution theorem can help us.

When we analysed the (slightly) damped extension oscillator in section 1.6, we concluded that the position of the bob, $x(t)$, took the form $x(t) = A\,e^{-\gamma t}\cos(\alpha t + \varphi)$ (equation (1.22)). We described this behaviour as sinusoidal oscillation, whose amplitude, $A\,e^{-\gamma t}$, diminishes steadily. The oscillation is not really sinusoidal, but the description is convenient.

The idea of sinusoidal signal whose amplitude varies is a common one. Typically, there is a system that produces a (genuine) sinusoidal output, connected to a second system by a coupling of variable strength. If the coupling strength varies as $f(t)$, then the input to the second system is something like

$$g(t) = f(t)\cos(\alpha t) \;. \tag{5.15}$$

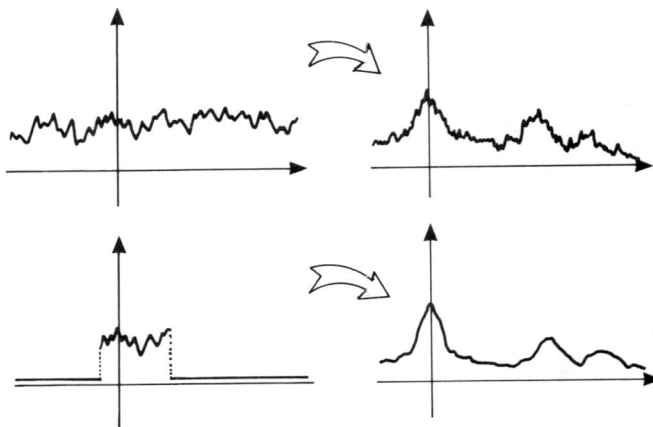

Fig. 5.5 — Truncating a signal smooths its transform.

The sinusoidal signal is said to be *amplitude-modulated*; the *modulating function* is $f(t)$.

The most familiar example of amplitude modulation, or AM, is in radio wave transmission. In this case, there is a signal $f(t)$ (representing speech, music, or whatever) to be transmitted. But it cannot be sent directly over the airwaves, because the frequencies involved are far too low for effective broadcasting. Instead, a *carrier frequency* is chosen; say, an angular frequency of α. This carrier is modulated by $f(t)$, and the modulated signal is transmitted.

Since (5.15) expresses f as a product of functions, we can invoke the convolution theorem. Let $h(x) = \cos(\alpha x)$; then $g(t) = f(t)h(t)$, and

$$Fg(\omega) = F(fh)(\omega)$$
$$= (1/2\pi)(Ff * Fh)(\omega) ,\qquad(5.16)$$

by equation (5.8). From (3.26), $Fh(\omega) = \pi(\delta(\omega + \alpha) + \delta(\omega - \alpha))$; substituting this in (5.16) and using (5.5) gives

$$Fg(\omega) = (1/2)(Ff(\omega + \alpha) + Ff(\omega - \alpha)) .\qquad(5.17)$$

Thus, the Fourier transform of the amplitude-modulated signal consists of two copies of the transform of f, one shifted to the right and one shifted to the left, by the frequency of the carrier signal. (See Fig. 5.6.)

5.7 Frequency modulation

There is a second sort of modulation commonly encountered, called *frequency modulation* (or FM). The best-known use of FM, as of AM, is in radio wave transmission.

The mathematical formulation of FM is a little more complicated than that of AM. Suppose our signal to be transmitted is $f(x)$, and our carrier is $\cos(\alpha x)$. Define $F(x)$ to be the indefinite integral of $f(x)$:

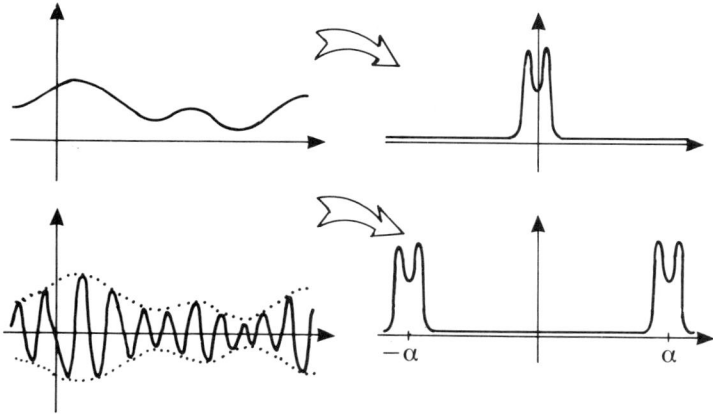

Fig. 5.6 — Amplitude modulation, and its effect in the ω-domain.

$$F(x) = \int_0^x f(t)\,dt \ . \tag{5.18}$$

Then the modulated signal is

$$g(x) = \cos[\alpha(x + F(x))] \ . \tag{5.19}$$

From Taylor's theorem, we have the estimate

$$F(x+t) \simeq F(x) + tf(x) \qquad \text{for small } t\ ; \tag{5.20}$$

substituting this into equation (5.18) gives

$$g(x+t) \simeq \cos[\alpha(x+t+F(x)+tf(x))]$$
$$= \cos[(\alpha + \alpha f(x))t + (\alpha x + \alpha F(x))] \ . \tag{5.21}$$

So, fixing x and letting t vary over small values, we see that the function g in the vicinity of x acts as a sinusoid with angular frequency $\alpha + \alpha f(x)$ (and phase $\alpha x + \alpha F(x)$).

Fig. 5.7 shows a typical FM signal.

Because FM is a nonlinear effect, Fourier analysis cannot help a great deal in its analysis. We can, however, say a few things, particularly in the case of random noise modulating a sinusoidal signal. (This sort of signal arises frequently in, for instance, the oscillations of a 'harmonic' system whose parameters are not quite constant.) Such a signal will often have constant character. If the modulating noise $f(x)$ is small, then most of the power of g will be at angular frequencies close to α. Indeed, the smaller f is — the less power in its PSDF — the more similar g is to the harmonic function, and the narrower the peak at α in *its* PSDF.

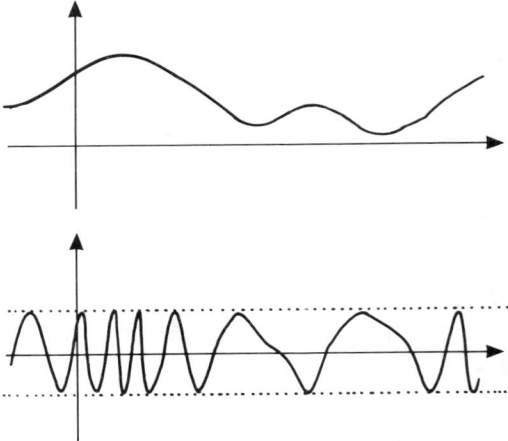

Fig. 5.7 — Frequency modulation. This is not a linear technique, so does not yield to Fourier methods.

C. CONVOLUTION AND FOURIER THEORY

5.8 The power theorem

Yet another application of the convolution theorem leads to a very useful theoretical result, usually called the *power theorem*. This is the result, for integrable functions, which corresponds to the generalized Parseval's equation for periodic functions, (2.28); it deals with the *integral of a product* of functions.

First, let $f(x)$ and $g(x)$ be two *real*-valued functions. Then:

$$\int_{-\infty}^{\infty} f(x)g(x)\,dx = \int_{-\infty}^{\infty} f(x)g(x)e^{-i0x}\,dx$$

$$= F(fg)(0)$$

$$= (1/2\pi)(Ff*Fg)(0)$$

$$= (1/2\pi)\int_{-\infty}^{\infty} Ff(\omega)Fg(0-\omega)\,d\omega\ .$$

Now g is real-valued, so $Fg(-\omega) = \overline{Fg(\omega)}$ (equation (3.32)); thus

$$\int_{-\infty}^{\infty} f(x)g(x)\,dx = (1/2\pi)\int_{-\infty}^{\infty} Ff(\omega)\overline{Fg(\omega)}\,d\omega\ . \tag{5.22a}$$

Equation (5.22a) is the power theorem for real-valued functions. The product $Ff(\omega)\overline{Fg(\omega)}$ is called the *cross-spectral power* of f and g.

Ch. 5]   **Convolution**

For *complex*-valued functions f and g, the power theorem is slightly altered:

$$\int_{-\infty}^{\infty} f(x)\overline{g(x)}\,dx = (1/2\pi) \int_{-\infty}^{\infty} Ff(\omega)\overline{Fg(\omega)}\,d\omega \quad . \tag{5.22b}$$

The special case when $f = g$ is of interest. From (5.22a) we have, for real-valued f,

$$\int_{-\infty}^{\infty} f(x)^2\,dx = (1/2\pi) \int_{-\infty}^{\infty} |Ff(\omega)|^2\,d\omega$$

$$= (1/2\pi) \int_{-\infty}^{\infty} Pf(\omega)\,d\omega \quad . \tag{5.23a}$$

When $f(x)$ is complex-valued, the equation is

$$\int_{-\infty}^{\infty} |f(x)|^2\,dx = (1/2\pi) \int_{-\infty}^{\infty} Pf(\omega)\,d\omega \quad . \tag{5.23b}$$

Equations (5.23) are called the *Rayleigh–Plancherel theorem*.

5.9 The uncertainty theorem

In the last paragraph of section 5.5, we observed that the effect of truncating a function was the smoothing of its transform; and that the amount of smoothing is inversely proportional to the amount of truncation. This was because the *effective width* of $F\chi_{(-b,b)}(\omega) = 2\sin(b\omega)/\omega$ drops as b rises. This relationship is, in fact, an example of a more general result, the *uncertainty theorem*.

The uncertainty theorem says that the effective width of a function, multiplied by the effective width of its transform, cannot be less than a certain minimum value. Thus, if f has a small effective width then $Ff(\omega)$ has a large effective width, and vice versa.

The effective width of a function f can be given a precise definition. It is denoted by $W(f)$, and defined by the formula

$$W(f)^2 = \frac{\int_{-\infty}^{\infty} |f(x)|^2 (x - M(f))^2\,dx}{\int_{-\infty}^{\infty} |f(x)|^2\,dx} \quad , \tag{5.24}$$

where $M(f)$ is the mean abscissa

$$M(f) = \frac{\int_{-\infty}^{\infty} |f(x)|^2 x\,dx}{\int_{-\infty}^{\infty} |f(x)|^2\,dx}. \tag{5.25}$$

Now let $f(x)$ be a differentiable, integrable, real-valued function, whose derivative df/dx is integrable, and for which $xf(x)^2 \to 0$ as $x \to \pm\infty$. The uncertainty theorem is then the statement that

$$W(f)W(Ff) \geq \tfrac{1}{2} . \tag{5.26}$$

We first simplify the calculation by setting

$$g(x) = \frac{f(x + M(f))}{\sqrt{\left\{\int_{-\infty}^{\infty} f(x)^2 \, dx\right\}}} . \tag{5.27}$$

Thus g is f, shifted leftwards along the x-axis by $M(f)$, and scaled by $1/\sqrt{\{\int_{-\infty}^{\infty} f(x)^2 \, dx\}}$. Then $M(g) = 0$, and because of the scaling,

$$\int_{-\infty}^{\infty} g(x)^2 \, dx = 1 . \tag{5.28}$$

In addition, g has the same effective width as f, and Fg has the same effective width as Ff. Further, the Rayleigh–Plancherel theorem (equation (5.23a)) tells us that

$$\int_{-\infty}^{\infty} |Fg(\omega)|^2 \, d\omega = 2\pi \int_{-\infty}^{\infty} g(x)^2 \, dx$$

$$= 2\pi . \tag{5.29}$$

Since g is real-valued, $Fg(-\omega) = \overline{Fg(\omega)}$ (equation (3.32)). It follows that $|Fg(\omega)|^2$ is an even function, and so $M(Fg) = 0$.

We begin the proof of (5.26) with *Schwarz's inequality*. This says that for any functions r and s, provided all the integrals exist,

$$\left[\int_{-\infty}^{\infty} r(x)s(x) \, dx\right]^2 \leq \left[\int_{-\infty}^{\infty} r(x)^2 \, dx\right]\left[\int_{-\infty}^{\infty} s(x)^2 \, dx\right] \tag{5.30}$$

(see any book on mathematical analysis). With $r(x) = xg(x)$ and $s(x) = dg/dx$, this becomes

$$\left[\int_{-\infty}^{\infty} xg(x)\frac{dg}{dx} \, dx\right]^2 \leq \left[\int_{-\infty}^{\infty} (xg(x))^2 \, dx\right]\left[\int_{-\infty}^{\infty} \left(\frac{dg}{dx}\right)^2 dx\right] . \tag{5.31}$$

But $(d/dx)(g(x)^2) = 2g(x)(dg/dx)$, so the integral on the left-hand side of (5.31) can be calculated by parts. As $x \to \infty$, $xf(x)^2 \to 0$, and so $xg(x)^2 \to 0$; therefore

$$\int_{-\infty}^{\infty} xg(x)\frac{dg}{dx} \, dx = \tfrac{1}{2}\int_{-\infty}^{\infty} x\frac{d}{dx}(g(x)^2) \, dx$$

$$= \left[\tfrac{1}{2}xg(x)^2\right]_{-\infty}^{\infty} - \tfrac{1}{2}\int_{-\infty}^{\infty} g(x)^2\,dx$$

$$= -\tfrac{1}{2}\ .\tag{5.32}$$

(5.31) then becomes

$$(1/4) \le \left[\int_{-\infty}^{\infty} (xg(x))^2\,dx\right]\left[\int_{-\infty}^{\infty} \left(\frac{dg}{dx}\right)^2 dx\right]$$

$$\le \left[\int_{-\infty}^{\infty} x^2 g(x)^2\,dx\right]\left[\int_{-\infty}^{\infty} \left(\frac{dg}{dx}\right)^2 dx\right]\ . \tag{5.33}$$

To deal with the second of the two factors on the right-hand side, we use the derivative theorem $F(dg/dx)(\omega) = i\omega Fg(\omega)$ (equation (3.35)), together with a second application of the Rayleigh–Plancherel theorem:

$$\int_{-\infty}^{\infty} \left(\frac{dg}{dx}\right)^2 dx = (1/2\pi)\int_{-\infty}^{\infty} \left|F\frac{dg}{dx}(\omega)\right|^2 d\omega$$

$$= (1/2\pi)\int_{-\infty}^{\infty} |i\omega Fg(\omega)|^2\,d\omega$$

$$= (1/2\pi)\int_{-\infty}^{\infty} |Fg(\omega)|^2 \omega^2\,d\omega\ . \tag{5.34}$$

Inserting this into equation (5.33), we get

$$(1/4) \le (1/2\pi)\left[\int_{-\infty}^{\infty} x^2 g(x)^2\,dx\right]\left[\int_{-\infty}^{\infty} |Fg(\omega)|^2 \omega^2\,d\omega\right]\ .$$

$$\le W(g)^2 W(Fg)^2$$

$$= W(f)^2 W(Ff)^2\ .$$

This proves the uncertainty theorem.

One important application of the uncertainty theorem concerns television and radio broadcasting. The broadcasting uses electromagnetic radiation, with each television or radio channel being broadcast in a narrow *range*, or *band*, of frequencies (not *at* a particular frequency, as is often considered the case). So, for example, a particular radio station may be using the frequency band from 900 to 910 kHz, corresponding to a range of ω from $5.65 \times 10^6\,\text{s}^{-1}$ to $5.71 \times 10^6\,\text{s}^{-1}$.

The broadcasting of this channel is restricted to this frequency band. The mean angular frequency, $M(Ff)$, of the broadcast signal must be about $5.68 \times 10^6\,\text{s}^{-1}$, and the spread $W(Ff)$ must be less than $0.03 \times 10^6\,\text{s}^{-1}$. From the uncertainty theorem,

$W(Ff)W(f) \geq \frac{1}{2}$; it follows that $W(f) \geq 17 \times 10^{-6}$s. In other words, the radio station cannot transmit a signal which lasts less than $17\,\mu$s.

This sounds like a very short signal. But a long signal is merely many short signals transmitted sucessively. Thus, $17\,\mu$s is a limit on the time *resolution* of the transmitted signal. The station can transmit no signal with any *detail* finer than $17\,\mu$s. Now in order to get reasonably good sound reproduction, the system must be able to transmit pitches as high as the ear can hear, which is around 20 kHz. A sound wave at this frequency has a period of $50\,\mu$s: just three times the resolution limit of the signal. There is barely room in the frequency band for a good signal.

Television systems are even more demanding. Existing European television tubes make up their pictures from 625 lines, scanned 24 times a second. For a good picture, the lines must be able to change sharply in brightness over 1/300 of their length, corresponding to a picture resolution of about 2 mm for a large screen. Thus the television signal must have a time resolution of not worse than $1/(24 \times 625 \times 300)$ seconds, or $0.2\,\mu$s. Indeed, for a colour signal, three separate lines are sent concurrently — for red, green, and blue picture elements — so the time resolution must be divided by 3. The *bandwidth* of television broadcasts must be correspondingly large.

In Chapter 11, we shall see how the uncertainty theorem relates to the celebrated uncertainty principle of quantum mechanics.

5.10 The transfer function

Let us return to the general situation described in section 5.2, where we have an input signal, transmitted through a system (the 'filter'), and an output signal. The filter has a fixed impulse response function $R(x)$, so that a general input $I(x)$ yields the output $O(x) = (R*I)(x)$. In this case, the convolution theorem tells us that $FO(\omega) = FR(\omega)FI(\omega)$.

Consider now a sinusoidal input $I(x) = ae^{i\alpha x}$. This has an impulsive transform $2\pi a\delta(\omega - \alpha)$, so that $FO(\omega) = 2\pi aFR(\alpha)\delta(\omega - \alpha)$, and the output is $O(\omega) = aFR(\alpha)e^{i\alpha x}$. Thus a sinusoidal input is merely amplified by the (complex) constant $FR(\alpha)$. But the Fourier transform of a general input $I(x)$ is a way of expressing the function as a sum of sine waves. Each of these gets scaled by the filter, and (by hypothesis) the output from the filter is the sum of these scaled sinusoids. In this sense, we may actually understand the action of the filter as being *a multiplication in the frequency domain*, rather than as being a convolution in the x-domain. The fact that $O = R*I$ is then just a result of this action in ω-space.

In this view, it is not $R(x)$ that is the fundamental innate property of the filter, but its transform $FR(\omega)$; this function is called the *transfer function* of the filter. Let us write it as $T(\omega)$. As

$$T(\omega) = FR(\omega) ,$$

so

$$R(x) = F^{-1}T(x) . \qquad (5.35)$$

In fact, we have already used the transfer function. In Chapter 1, we investigated the motion of a forced extension oscillator. The oscillator is a filter; the ceiling vibration is the input; and the motion of the bob is the output. The linearity of the system's equations of motion ensures that the output is of the convolution type.

Hence we obtained the amplitude response function, equation (1.31): a function $X(\omega)$ such that 'a harmonic input of angular frequency ω and amplitude A forces a harmonic output [of the same angular frequency and] of amplitude $AX(\omega)$'. The system's amplitude response function is no more than the absolute value of its transfer function: $X(\omega) = |T(\omega)|$.

5.11 Deconvolution

Using the convolution theorem, convolutions can be 'undone'. For, if we know that

$$FO(\omega) = FI(\omega)T(\omega)$$

and $T(\omega)$ is a *non-zero* function, then

$$FI(\omega) = FO(\omega)/T(\omega) \; , \qquad (5.36)$$

and so

$$I(t) = F^{-1}(FO/T)(t) \; . \qquad (5.37)$$

So if a filter has transfer function $T(\omega)$ and, for an unknown input $I(t)$, we measure an output $O(t)$, then we can calculate *what the input must have been*, using (5.37).

Problems arise when $T(\omega) = 0$ for one or more values of ω. What that says, effectively, is that input components of these angular frequencies ω *do not affect* the output. In this case, we cannot solve for $I(t)$ completely, because we can always add or subtract any component of any 'missing' angular frequency ω.

D. FUNCTIONS OF SEVERAL VARIABLES

5.12 Vector convolution

The convolution that we have been studying so far is an operation on functions $f(x)$ of a single variable. There is an analogue for functions of more than one variable, or functions of vectorial variables: if $f(\mathbf{x}) = f(x_1, \ldots, x_n)$ and $g(\mathbf{x}) = g(x_1, \ldots, x_n)$ are two functions on n variables, we define their *n-dimensional convolution* $f *_n g$ as

$$(f *_n g)(\mathbf{x}) = \int f(\mathbf{y}) g(\mathbf{x} - \mathbf{y}) \, d\mathbf{y} \; , \qquad (5.38)$$

where the integration is over the whole of n-space.

This operation shares many of the properties of simple convolution. It is commutative, associative, and linear: that is, equations (5.3) and (5.4) remain true if $*$ is replaced by $*_n$. Vector convolution can be used, as simple convolution can, for smoothing.

Because both the vector convolution and the vector transform treat the independent variables separately, there is an n-dimensional analogue of the convolution theorem: for functions f and g,

$$F_n(f*_n g)(\omega) = F_n f(\omega) F_n g(\omega) \ . \tag{5.39}$$

In particular, if a constant linear filter operates on an input signal $I(\mathbf{x})$ which is a function of n variables, to give an output function $O(\mathbf{x})$, then the filter has an n-dimensional transfer function $T(\omega) = T(\omega_1, \ldots, \omega_n)$, and we can write

$$F_n O(\omega) = F_n I(\omega) T(\omega) \ . \tag{5.40}$$

If $T(\omega) \neq 0$ for all ω, then we can deconvolve this equation:

$$I(\omega) = F_n^{-1}(F_n O(\omega)/T(\omega)) \ . \tag{5.41}$$

5.13 Frosted glass

As an example of multidimensional convolution and deconvolution, let us consider the effect of frosted glass. Suppose that we are looking through the glass at a *distant scene* (Fig. 5.8(a)), and that we see the scene in *black and white* only. (The addition of colour requires us to consider the chromatic properties of light and of the glass, though the analysis is basically similar.)

The surface of frosted glass consists of a large number of very small, curved features, which refract and diffract the light in peculiar ways. These features are too small for us to discern their effect individually, so that *a point source of light, seen through the glass in any position, looks like a diffuse source*. We assume that the glass is frosted *uniformly*, so that the diffusion of a point source in one position looks much like the diffusion of a point source in any other position.

Let us suppose, for example, that a point source of light is seen as a diffuse *disc* of light, of the same total brightness but spread over a disc of angular radius φ (see Fig. 5.8(b)). More precisely, suppose there is a distant point source of light, of intensity C units, at angular position $\boldsymbol{\theta} = (\theta_1, \theta_2)$ (where θ_1 is its angular position left-to-right, θ_2 the angle above the horizontal). The intensity of the source, viewed *without* the glass, is $C\delta_2(\boldsymbol{\alpha} - \boldsymbol{\theta})$ units per unit solid angle, at angular position $\boldsymbol{\alpha}$. Viewed *through* the glass, the source looks like $CR(\boldsymbol{\alpha} - \boldsymbol{\theta})$, where

$$R(\mathbf{x}) = (1/\pi\varphi^2)\chi_{(0,\varphi)}(\|\mathbf{x}\|) \tag{5.42}$$

This function, $R(\mathbf{x})$, is the *impulse response* of the system.

Now suppose we have a complex input $I(\boldsymbol{\alpha})$. This is the pattern of light intensity that would reach our eye if the glass were absent. What the eye sees through the glass is a different pattern of light intensity, say $O(\boldsymbol{\alpha})$; this represents the blurred image, and is the output of the system.

We can think of the panorama as being the 'sum' of distant point sources: $I(\boldsymbol{\alpha}) = \int I(\boldsymbol{\theta})\delta_2(\boldsymbol{\alpha} - \boldsymbol{\theta})\,d\boldsymbol{\theta}$. Each point source $I(\boldsymbol{\theta})\,d\boldsymbol{\theta}\delta_2(\boldsymbol{\alpha} - \boldsymbol{\theta})$ will appear, through the glass, to spread out into a source of angular radius φ, according to equation (5.42). What reaches the eye from this point is thus $I(\boldsymbol{\theta})\,d\boldsymbol{\theta}R(\boldsymbol{\alpha} - \boldsymbol{\theta})$; thus the eye sees the scene as

$$O(\boldsymbol{\alpha}) = \int I(\boldsymbol{\theta})R(\boldsymbol{\alpha} - \boldsymbol{\theta})\,d\boldsymbol{\theta} \ . \tag{5.43}$$

In other words, we have a convolution, $O = I *_2 R$. (In practice, the limits of integ-

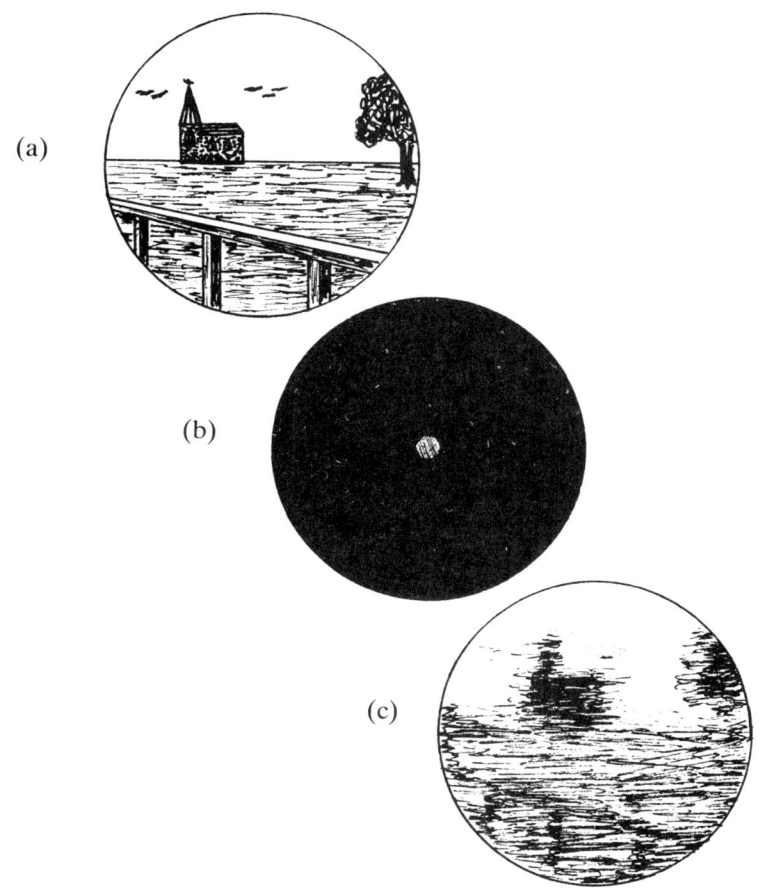

Fig. 5.8 — (a) Complex input $I(\alpha)$. (b) Impulse response function $R(\theta)$. (c) Convolved output $O(\alpha)$. The output still contains all the information of the input, however, and the input can be recovered by deconvolution.

ration in (5.43) will be determined by the breadth of the field of vision, though in theory the integral is over the whole of 2-space.)

Since frosted glass acts by convolution, its effects can be undone by deconvolution. Suppose we look through a pane of frosted glass, and see the (blurred) scene beyond as $O(\alpha)$. In order or reconstruct the scene as it *would* appear *without* the glass, we would look for the transfer function $T(\omega)$ of the glass. For example, if there were (what appeared to be) a point source of light in the scene, we could guess the impulse response $R(\theta)$ from that, and deduce $T(\omega)$ as $F_n R(\omega)$. Then we deduce $I(\alpha)$ from equation (5.41), as $I(\alpha) = F_n^{-1}(F_n O(\omega)/T(\omega))\,(\alpha)$.

Needless to say, this process of *image reconstruction* is a far from academic procedure. It has forensic value. By similar means, a blurred photograph of a car's number plate, or even a person, can be sharpened up until it is recognizable. The transfer function of a camera-plus-movement-plus-defocussing system is not always easy to find, but even trial and error will quickly yield improved picture quality; and there are techniques, based upon the fact that real scenes tend to contain many sharp edges, that can improve such pictures enormously. Unfortunately, a detailed discussion of these is beyond the scope of this book.

EXERCISES

1. Let $f(x) = \exp(-ax^2)$ and $g(x) = \exp(-bx^2)$. Calculate $(f*g)(x)$.
2. Calculate, for any integrable function f, the convolutions

    (i) $(1*f)(x)$. (ii) $(\delta*f)(x)$.

    Describe the behaviour of the convolution in the previous question

    (i) as $a \to 0$. (ii) as $a \to \infty$.

    Compare your results.

3. A sine wave is switched on for a short period of time, then switched off; the signal obtained therefore has the functional form $f(t) = \chi_{(-b,b)}(t) \cos(\alpha t)$. Treating this *both* as a truncation *and* as an amplitude modulation, find the Fourier transform of the signal.

4. How does your answer to the previous question relate to the uncertainty theorem?

5. Prove that, for any x and y,

$$\frac{1}{(1+y^2)(1+(x-y)^2)} = \frac{1}{x(4+x^2)}\left(\frac{2y+x}{1+y^2} - \frac{2y-3x}{1+(x-y)^2}\right).$$

Let $f(x) = 1/(1+x^2)$. Calculate $(f*f)(x)$.

6. Prove that convolution is *associative* (equation (5.4a)).

7. Prove that $\Lambda(x) = \frac{1}{2}(\Pi*\Pi)(2x)$ (see equations (3.54) and (3.55) for the definitions of these functions). Hence calculate the Fourier transform of $\Lambda(x)$.

8. This question investigates the smoothing effect of convolution still further. Let $f(x) = \cos(\alpha x)$. Show that $f*\Pi$ is a harmonic function of angular frequency α. What is its amplitude? What happens to the amplitude of $f*\Pi$ as (i) $\alpha \to \infty$? (ii) $\alpha \to 0$?

9. (Harder) Repeat the calculation of the previous question, using the bell-curve $N(x)$ in place of $\Pi(x)$. (This can be done without contour integration by using Fourier methods.)

10. Use the results of section 3.6 and the Rayleigh–Plancherel theorem to evaluate $\int_{-\infty}^{\infty} (\sin^2 x/x^2)\,dx$.

11. Use the result of Question 7, the fact that $F\delta(\omega) = 1(\omega)$, and the power theorem to evaluate the integral in the previous question.

12. Calculate the effective width of the function $\exp(-\alpha x^2)$ and its transform. Is their product equal to, or strictly greater than, $\frac{1}{2}$ (see equation (5.26))?

13. Rewrite Question 5, and the Project, of Chapter 3 in terms of convolutions.

14. Find a lower bound on the bandwidth of a television signal (see section 5.9). How many channels of this quality can be transmitted if the available frequencies are those between 400 and 500 MHz?

Project. The function $\Pi(x)$ is discontinuous at $x = \pm 1$. What about the function $h(x) = (\Pi*\Pi)(x)$? To what extent is h differentiable, and is its derivative continuous? Formulate and prove a conjecture about repeated convolutions of Π.

Let f and g be a pair of smooth (i.e. continuously differentiable) functions. Then $f*g$ is *twice* continuously differentiable. Is this the best possible result?

More generally, suppose $f(x)$ is *m-fold smooth* (m times continuously differentiable — that is, $d^m f/dx^m$ exists everywhere and is continuous) and $g(x)$ is *n*-fold smooth. What can you say about $f*g$?

What happens if the 'smooth' is replaced by merely 'differentiable' in the above situations?

6
Filtering

A. NOISE

6.1 Types of noise

A *signal* is a measurement of a physical quantity; it is a function. *Noise* is any extraneous and unwanted disturbance of that signal. If we have a noisy signal then it is clearly useful to be able to distinguish the signal from the noise. There are many ways of doing this, according to the type of noise present in the signal, many of which do not involve Fourier theory.

A signal can be 'noisy' in a number of ways, as the examples in Fig. 6.1 indicate.

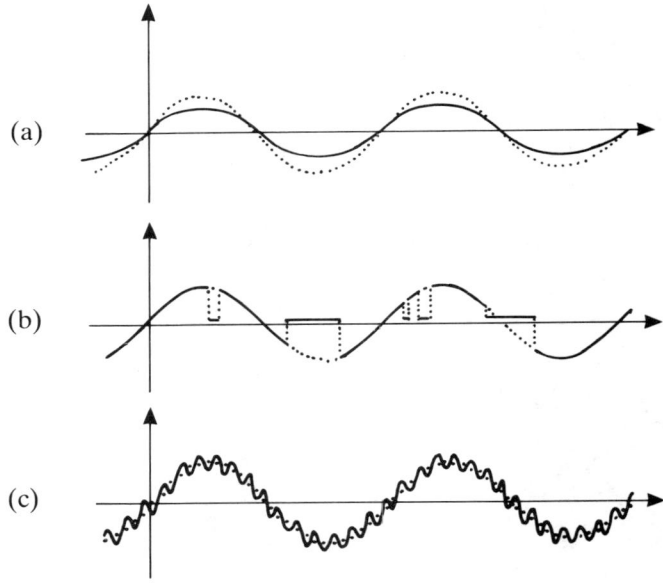

Fig. 6.1 — Three kinds of noisy sine wave: distortion (a), dead spots (b), and additive noise (c).

Each shows a 'noisy' sine wave. On (a), the signal is *distorted*, as music through a loudspeaker at too high power; in (b) the signal has *dead spots*, as an electrical signal transmitted along a frayed wire; while in (c) there is a second sinusoid superposed onto the signal, as occurs in electrical circuits which are not adequately shielded from the mains power supply (*mains hum*).

Only the last of these can be dealt with by Fourier methods: it is the case of *superposed* or *additive* noise. (The term 'noise' is, indeed, often used to refer to additive noise only.) With additive noise, the measured signal $f(x)$ is the sum of the 'clean' signal $s(x)$ and the noise $n(x)$: $f(x) = s(x) + n(x)$.

In the example shown, the added noise is periodic. More typically it has constant character; the nature of such noise is reflected in its PSDF (section 3.4).

The most important feature of constant-character noise is its *bandwidth*. (See also section 5.9.) The largest noise bandwidths of all are those of *white noise* and *pink noise*. These are kinds of noise defined by their PSDFs. The PSDF of white noise is uniform, up to some maximum frequency; that of pink noise decays as $1/\omega$, within some frequency band. Thus a function $f(x)$ is:

— white noise if $pf(\omega) = A\chi_{(-\Omega,\Omega)}(\omega)$, for some Ω;
— pink noise if $pf(\omega) = (A/\omega)\chi_{(\alpha,\beta)}(\omega)$, for some α and β. (6.1)

White noise has uniform total spectral power over each interval $(\omega, \omega + \lambda)$ within its range; pink noise has uniform total spectral power over every interval $(\omega, \lambda\omega)$ within its range. Noise of these kinds is shown in Fig. 6.2.

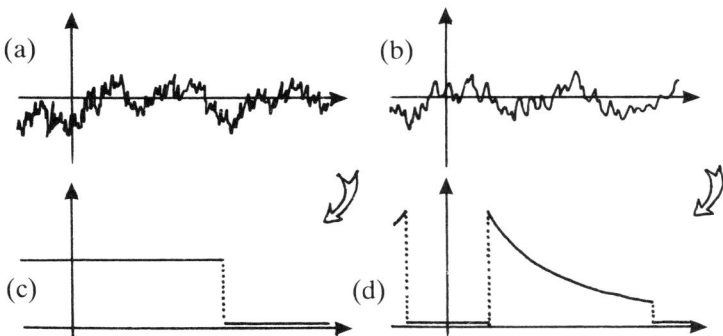

Fig. 6.2 — White noise (a) and pink noise (b). Their PSDFs are shown in (c) and (d) respectively.

Incidentally, white light, with its uniform mixture of all colours, is not merely a *model* for white noise, it *is* white noise — white noise in an electromagnetic field.

Pure white noise is sometimes defined as noise whose PSDF is uniform *throughout* the frequency range. Such a function has $pf(\omega) = A1(\omega)$. Pure white noise has infinite total spectral power over finite intervals, so can correspond to no proper function. However, impulsive 'functions' can be of this sort; for example, a function consisting of randomly spaced impulsive spikes is pure white noise. Similarly, *pure pink noise* is noise whose PSDF is $pf(\omega) = A/\omega$.

Noise of smaller bandwidth is generically termed *wideband noise* or *narrowband noise*, according to circumstances. The narrower the bandwidth of a signal, compared with its mean frequency, the more closely the signal resembles a sine wave. Conversely, the higher the bandwidth, the more closely the signal resembles white noise. The signal portrayed in Fig. 6.3(a) and (c) is narrowband noise; that in (b) and (d) is more-or-less wideband.

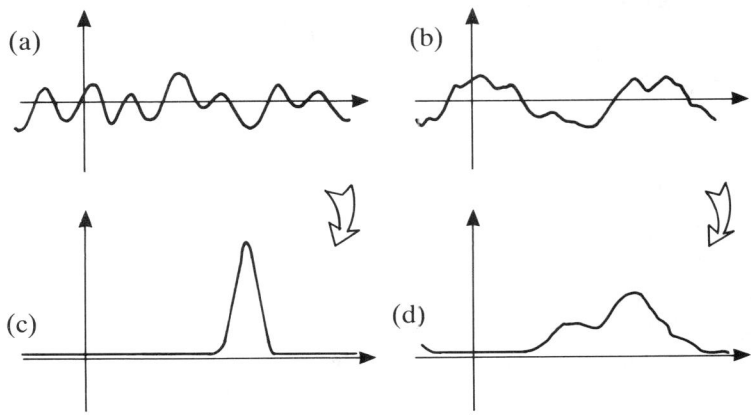

Fig. 6.3 — Narrowband (a) and wideband (b) noise, and their PSDFs ((c) and (d)).

A generalization of narrowband, almost-harmonic, noise is *almost-periodic noise*. A truly periodic function has an impulsive spectrum; an almost-periodic function has a series of finite narrow peaks in its spectrum. Noise of this type will occur where almost-harmonic narrowband noise undergoes a nonlinear mutation: Fig. 6.4 shows what happens to its spectrum. (Note that the peaks of higher 'harmonics' are, typically, proportionately broader than the 'fundamental' peak. The reader is challenged to explain this.)

6.2 Signal-to-noise ratio

The more noise there is, the harder it is to pick out the signal. The measure of 'degree of noisiness' is the *signal-to-noise ratio* (or $S:N$ *ratio*). It is defined as the ratio of the total power of the signal $s(x)$ to the power of the noise $n(x)$, over some appropriate interval. If the signal is a finite one, occurring in the range $a \leq x \leq b$, we might define the $S:N$ ratio as

$$\sigma = \frac{\int_a^b s(x)^2 \, dx}{\int_a^b n(x)^2 \, dx} . \qquad (6.2)$$

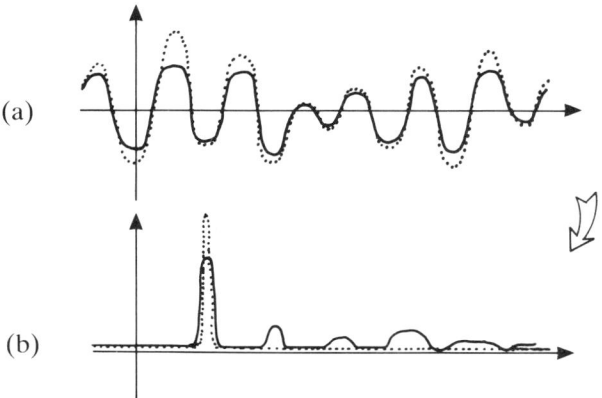

Fig. 6.4 — The effect of a nonlinear filter on narrowband (almost-harmonic) noise, and on its PSDF.

Where the frequency characteristics of the signal and of the noise are not the same, it makes sense to make this definition frequency-dependent. The (frequency-dependent) S:N ratio $\sigma(\omega)$ is defined as the ratio of the power *spectra* of s and n:

$$\sigma(\omega) = \frac{|Fs(\omega)|^2}{|F[n\chi_{(a,b)}](\omega)|^2} \ . \tag{6.3a}$$

If the signal is not finite, but instead both it and the noise have constant character, then a corresponding measure is defined using PSDFs rather than power spectra:

$$\sigma(\omega) = ps(\omega)/pn(\omega) \ . \tag{6.3b}$$

The S:N ratio of two functions has no units; it is a pure number. It is often expressed in terms of *decibels*. The decibel is a logarithmic measurement: there are 10 dB to each factor of 10. Thus, k dB corresponds to a ratio of $10^{k/10}$, while a ratio of K is $10\log_{10} K$ dB.

Note that the S:N ratio is based on *power*, not on *amplitude*. Since amplitude is squared in the calculation of power, the measurements are not interconvertible. An S:N ratio of 50 dB means a power ratio of 100 000, but an amplitude ratio of $\sqrt{100 000}$, or about 300. So if we have a noisy signal with an S:N ratio of 50 dB, the average amplitude of the noise is 1/300, or about 0.3%, of the amplitude of the signal.

Fig. 6.5 shows some waveforms having a range of S:N ratios. In each case, the 'signal' is a sinusoid, and the 'noise' is white noise.

6.3 Signal averaging in the x-domain

In order to be able to discern a signal among noise, we need first to have some idea of what kind of signal we are looking for: a sine wave at around a given frequency, maybe. The more noise there is — the lower the S:N ratio — the more signal we need to have available in order to find it.

One common case concerns a signal from a repeatable experiment. In this case,

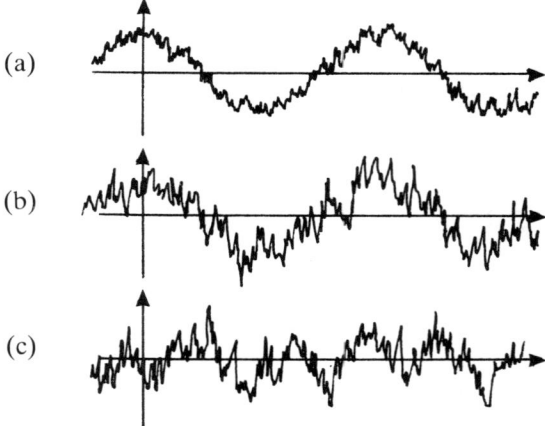

Fig. 6.5 — Sine waves with superposed white noise, with varying S:N ratios: 20 dB (a), 5 dB (b), and −10 dB (c).

we can simply average the signals from successive experiments. In order to be able to do this, the data from the experiments mnust be collected *consistently*; say, with a sharp pulse input, with data collection timed from the pulse. Then, from individual experimental results $f_1(x),\ldots,f_m(x)$, we look at the average $f(x) = (1/m)\sum_{r=1}^{m} f_r(x)$.

The advantage of this is as follows. The signal in each experiment is the same, $s(x)$, but the noise varies; so $f_r(x) = s(x) + n_r(x)$, say, with the n_r uncorrelated. The average is then $f = s + (1/m)\sum_{r=1}^{m} n_r$. As the noise is uncorrelated from experiment to experiment, if n_r has mean amplitude α, the sum $\sum_{r=1}^{m} n_r$ has mean amplitude $\alpha\sqrt{m}$; and the average $(1/m)\sum_{r=1}^{m} n_r$ has mean amplitude $\alpha/\sqrt{m}$. So if the S:N ratio of the individual signals f_r is K, then the S:N ratio of the average f is mK.

Consider the following case. We wish to investigate the waves at the side of a water-filled bowl made when a stone is dropped into the centre. At time $t = 0$, a pulse releases a stone from a fixed height. There is a water-level meter on one side of the bowl, which measures the water level at that point. Data gathering is triggered by the release of the stone.

The experiment is done a hundred separate times, and a hundred records are made. A few examples are shown in Fig. 6.6(a): there are clear indications of wave-like behaviour, but the exact response is obscured by splashing and so on, especially immediately after the stone hits the water.

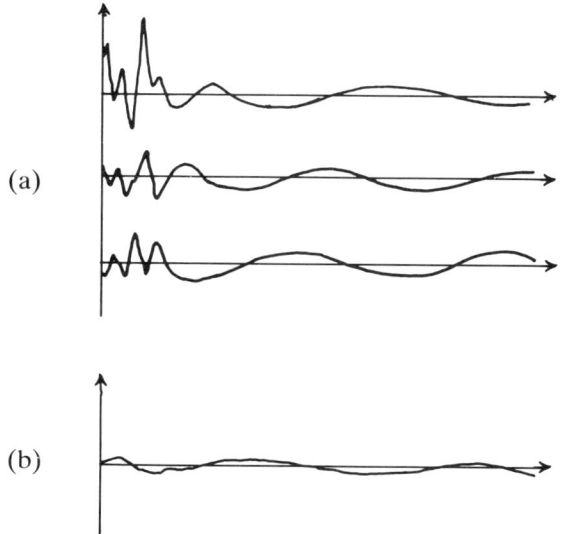

Fig. 6.6 — (a) Three separate records from individual experiments. (b) The x-domain average of many such records.

When we average these signals, point by point, the result is the function shown in Fig. 6.6(b). The wave behaviour is much clearer here. This is the 'average' pattern of water level variation, on dropping a stone.

But we have to be careful in interpreting this as the 'true' effect of dropping a stone into a bowl. First, the 'average' we obtained is not *representative*; all the data we have indicate a much less smooth response. In our extraction of signal from noise, we have omitted much that is important.

Secondly, there are likely to be small variations in the *frequency* and (more likely) the *phase* of the wave from experiment to experiment, simply because the experimental conditions are not exactly reproducible. The effect of averaging these is for the waves to begin to cancel each other out. Thus the averaged signal will show waves of slightly smaller amplitude than is actually present in the individual signals. In other words, some of the 'signal' will have been treated as 'noise'.

Where neither of these problems arises, x-domain averaging is a useful technique.

6.4 Signal averaging in the ω-domain

A second method might seem to take the point-by-point average, not of the signals themselves, but of their Fourier spectra. In fact the effect of this is exactly the same as taking the spectrum of the x-domain average, because Fourier transformation is linear. For if

$$f(x) = (1/m)[f_1(x) + \ldots + f_m(x)] \; ,$$

then

$$Ff(\omega) = F[(1/m)(f_1 + \ldots + f_m)](\omega)$$
$$= (1/m)[Ff_1(\omega) + \ldots + Ff_m(\omega)] \ . \tag{6.4}$$

So averaging the transforms suffers from the same problems as averaging in the x-domain: phase variation and small frequency variations can destroy the signal. We can regard this, in the ω-domain, as a consequence of the effect of adding complex numbers of different argument: they cancel, partially or totally.

Instead, we might average the *power* spectra of the f_r. Since power spectra have positive real values, no such cancellation can occur. What we are calculating, in doing this, is the average *strength* of each frequency component in the signal. The resulting average is the *consensus power spectrum*, or *CPS*; the CPS of the m signals $f_1(x), \ldots, f_m(x)$ is

$$\Pi f(\omega) = (1/m)[|Ff_1(\omega)|^2 + \ldots + |Ff_m(\omega)|^2]$$
$$= (1/m)[Pf_1(\omega) + \ldots + Pf_m(\omega)] \ . \tag{6.5}$$

Because the power spectrum throws away all the phase information of a signal, there is none in the CPS. It follows that no 'average signal' in the x-domain can be recovered from it.

A second feature of the CPS is that the noise in the spectra is *not* averaged out. The noise contributes power to the power spectra, and the average of this power is still present in the CPS. It may happen, then, that the signal is obscured by the noise in the CPS, just as in the x-domain. For this reason, the CPS technique is useless when the signal and the noise have *similar spectra*.

When this is not the case, the CPS is useful. For instance, if the signal has a narrow bandwidth, but the noise is broadband (a very common situation), the CPS will sharpen the spectra. An example of this sort of spectrum is given in Fig. 6.7. In

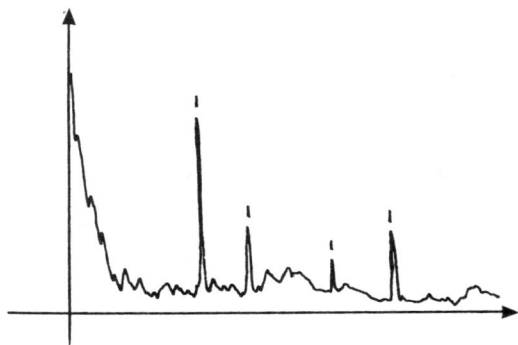

Fig. 6.7 — A typical 'diagnostic' transform, showing narrowband signals (marked spikes) against a background of broadband noise

fact, the retention of the noise in such a CPS can be very helpful, for signal and noise spectra can both be read off from a single CPS (though, admittedly, only because we have made some assumptions about the spectral nature of the two parts).

To summarize: x-domain averaging only works when the x-domain behaviour of signal and noise are sufficiently different; ω-domain averaging only works when the spectra are different.

B. FILTERING TECHNIQUES

6.5 Filtering for convenience

In section 5.2, we defined a *filter* as being some sort of device that takes in an input signal $I(x)$ and produces an output signal $O(x)$. If the filter is *linear* and *invariant* (that is, if it does not change its characteristics across the various parts of the input), then its output is related to its input by a convolution $O = I*R$; the function $R(x)$ is the impulse response function, and is an inherent property of the filter.

Some filters are physical. Such is the case with the theatre acoustics in section 5.1. Other filters are imposed by the analyst, and are purely for analytical convenience. To this category of filter belongs the smoothing of section 5.4, where we are at liberty to choose our smoothing function.

Not all operator-convenient filtering uses linear and invariant filters. The kind of filtering can be chosen to be whatever is useful at the time. Suppose, for instance, the astronomer in section 5.4 wishes to remove occasional and very brief accidental eclipses from his light intensity signal. One way of doing that is to take the recorded light-curve (Fig. 6.8(a)), identify any such 'blips' by inspection, and replace them with straight line segments (Fig. 6.8(b)). This seems a very crude filter, but it may be more appropriate in this particular case than the more sophisticated convolution method. (Filters of exactly this type are used in computerized image processing, as well as in 'cleaning up' the music of scratched old records.)

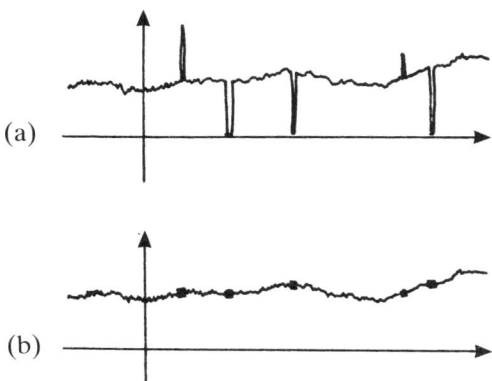

Fig. 6.8 — A useful nonlinear noise reduction filter: inspect the signal for interference, remove it, and reconnect the gaps with straight line segments.

Another, more important, example of non-linear filtering is *limiting*. This is where the signal is truncated whenever it goes below a fixed minimum value or above a fixed maximum. The effect of limiting is indicated by Fig. 6.9. It is a crude filter, and

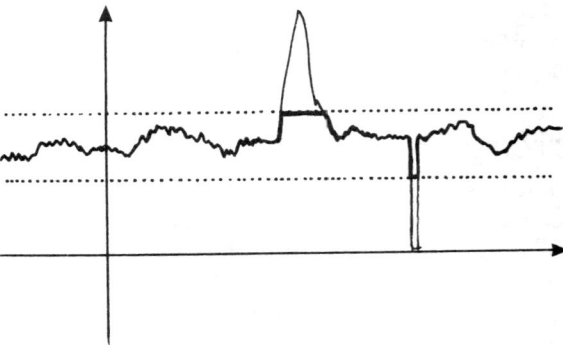

Fig. 6.9 — Limiting.

there are very few cases where limiting is used by itself. There are, however, two common sets of circumstances when it arises:

(1) The case of a signal that usually stays in a fairly narrow range, but is subject to occasional violent and undesirable bursts of noise. This happens in the output of a cracked record, for example. In this case, limiting to a fairly wide range can be a useful preparation for more sophisticated filtering. Under these circumstances, it is fair to argue that the limiting doesn't really disturb the linearity of the processing, because it hardly affects the *signal* at all — only the bursts of *noise*.
(2) The case of a signal that is actually received through a device that imposes limiting. This is a nuisance, and there is nothing that can be done about it.

Finally, there is a range of filtering techniques called *dynamic filters* that actually alter their characteristics in response to the signal. These filters need careful design to fit the expected noise behaviour. They can be very effective, but they are not invariant (by definition), and we say no more about them here.

Despite this array of techniques, it is often best to use linear and invariant filters, even if only because they are relatively easy to apply. The effect of this sort of filter in the x-domain is to convolve the input signal $I(x)$ with the impulse response $R(x)$. In the ω-domain, this becomes

$$FO(\omega) = T(\omega)FI(\omega) , \qquad (6.6)$$

where $T(\omega) = FR(\omega)$ is the transfer function of the filter (see section 5.10).

6.6 Mask filters

The most common filters are those that let through some frequencies, and stop others. These correspond to transfer functions $T(\omega)$ which take the values 1 and 0 only, for the appropriate ranges of ω. Such filters are said to be (*frequency-*)*mask filters*.

The effect of a mask filter on the spectrum of a signal is easy to compute. On the *pass bands*, the frequency ranges for which $T(\omega) = 1$, the spectrum of the output is the same as the spectrum of the input. On the *stop bands*, where $T(\omega) = 0$, the spectrum of the output is set to zero. By this means, the desired features in the input spectrum are transmitted, while the features we do not want are expunged. Fig. 6.10

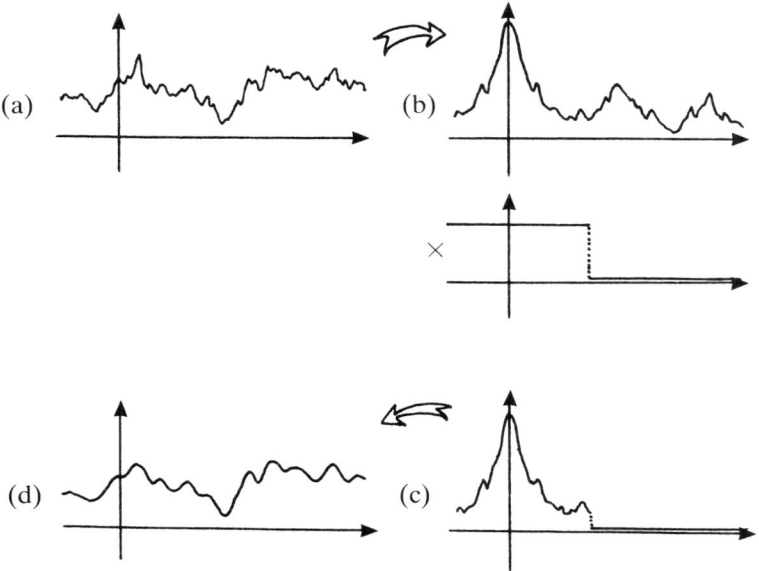

Fig. 6.10 — The operation of a mask filter. (The filter shown is an ideal low-pass filter.)

shows a typical mask filter in action.

The simplest of the mask filters are the (*ideal*) *low-pass* and the (*ideal*) *high-pass* filters. A low-pass filter lets through all frequencies *below* some critical frequency, and stops all those *above*; a high-pass filter has the opposite effect. The purposes of these are basic. If we have a signal of relatively low frequency, contaminated by high-frequency noise, then a low-pass filter of appropriate cut-off frequency will remove the noise and leave the signal. A high-pass filter deals similarly with low-frequency noise contaminating a high-frequency signal.

The transfer functions of such filters have simple forms. Remembering that we have negative frequencies to consider as well, we have

$$T(\omega) = \chi_{(-\alpha, \alpha)}(\omega) \qquad (6.7a)$$

for an ideal low-pass filter, and

$$T(\omega) = \chi_{(-\infty, -\alpha)}(\omega) + \chi_{(\alpha, \infty)}(\omega) \qquad (6.7b)$$

for an ideal high-pass filter. In either case, the parameter α is called the *critical* (or *cut-off*) angular frequency.

By combining low-pass and high-pass filters appropriately, we can construct more sophisticated mask filters. Take a low-pass filter with cut-off $\omega_c = \alpha$ and a high-pass filter with $\omega_c = \beta$; suppose $\alpha > \beta$. Applying these filters successively to a signal has the combined effect of letting through all components with angular frequency between β and α, and stopping all others. This sort of filter is called a *band-pass* filter; it is a mask filter, with pass-band $\beta < \omega < \alpha$ (see Fig. 6.11(a)).

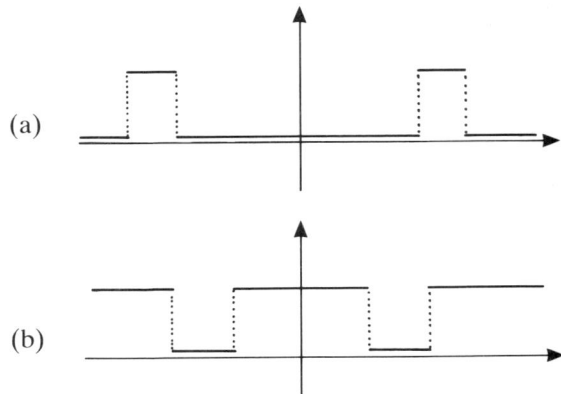

Fig. 6.11 — Ideal band-pass (a) and band-stop (b) filters.

Band-pass filters are useful for picking harmonic or narrowband signals out of wideband noise. Take the case of a radio receiver. The aerial picks up a signal covering a wide range of frequencies. If this signal were converted to sound directly, the result would be a cacophony: every station received would be reproduced simultaneously. But the different stations are transmitting in different frequency bands. Our chosen channel occupies a narrow band; all the other channels, together with any genuine noise, constitute the 'noise' to be removed, and are wideband. To select this channel, we use a narrow band-pass filter. We pick out just those frequencies associated with our channel and reproduce those. Indeed, the effect of the tuner knob on a radio is to change the pass-band of just such a filter (though not an ideal one: see section 6.8 below).

A second way of combining low- and high-pass filters is to have the cut-off angular frequency α of the low-pass filter *less* than the cut-off β of the high-pass. Now, applying the filters successively will remove *every* frequency component, and the output will be zero. So in this case, the outputs of the filters are added. That is, if $L(\omega)$ and $H(\omega)$ are the two transfer functions, then we look at the filter whose transfer function is $L(\omega) + H(\omega)$. The effect of this is precisley the opposite of a band-pass filter: it passes all angular frequencies below β or above α, but stops those components between the two. This is a *band-stop* filter; it too is a mask filter, and its mask is shown in Fig. 6.11(b).

As its properties are opposite to those of a band-pass filter, so are its uses. The band-pass filter extracts a narrowband signal from wideband noise; the band-stop purges narrowband noise from a wideband signal. For instance, a perennial

Ch. 6] **Filtering** 161

headache for electronics engineers is the *mains hum* induced in a signal by a poorly shielded mains supply. The effect of this is to superpose onto the signal an oscillating component at 50 ± 2 Hz (in the UK) — narrowband noise. To remove mains hum, we apply to the received signal a band-stop filter whose stop-band is 48–52 Hz. Of course, in doing so, we cut out those parts of the *signal* in this frequency range, as well as the noise, but that is unavoidable. (Clearly, the narrower the noise bandwidth, the easier it is, in principle, to eliminate its effects without seriously disturbing the desirable part of the signal.)

One final type of mask filtering is *comb filtering*. This is useful in dealing with nonsinusoidal but almost-periodic signals. Such signals usually reflect a nonlinearity in the system.

For example, consider again the record of air temperature θ discussed in section 3.14. There we considered the effect of solar heating. This effect is certainly almost-periodic, because of the Earth's rotation, but is not sinusoidal; as a result of this, the spectrum of θ contains, rather than a single spike at frequency 1 day^{-1}, a whole train of peaks at multiples of this frequency (see Fig. 6.12(a)). To pick out the solar

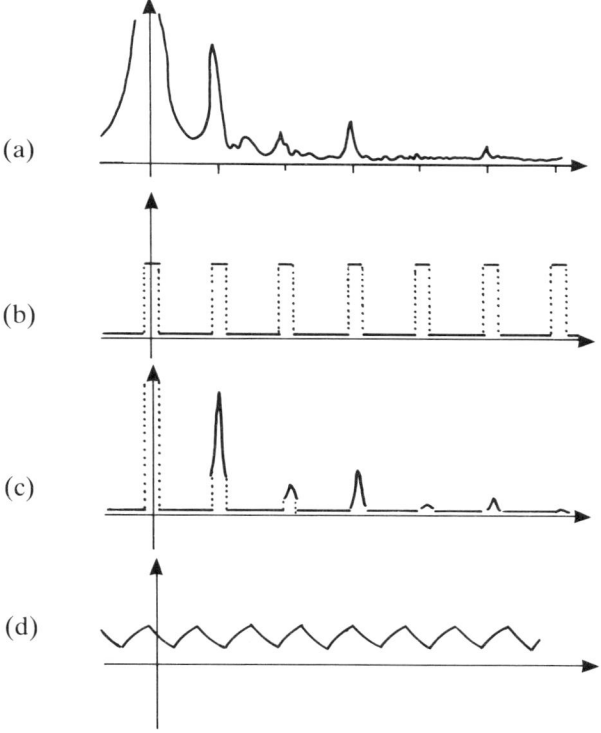

Fig. 6.12 — A comb filter may be used to extract a periodic component.

heating effect from all other effects, we filter this spectrum with a mask like 6.12(b). This mask has a narrow pass band around each multiple of the fundamental. The width of the pass bands must be decided on the basis of the likely spread of peaks. The filtered spectrum is shown in Fig. 6.12(c), and the reconstructed signal in Fig. 6.12(d). (Compare Fig. 3.9(a).)

Comb filters can be used in reverse, to *stop* almost-periodic noise from a signal. Thus, applying the filter shown in Fig. 6.13 to Fig. 6.12(a) passes all the frequencies

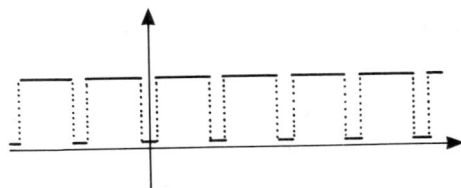

Fig. 6.13 — The 'dual' comb filter, used to *suppress* the same periodic component.

between spikes and suppresses the spikes themselves. This enables us to see more clearly what there is in 0 *apart* from solar heating.

6.7 Amplitude demodulation

In section 5.6 we saw how amplitude modulation interacts with Fourier transformation. The conclusion is that if $g(x) = f(x)\cos(\alpha x)$, then
$Fg(\omega) = \frac{1}{2}(Ff(\omega + \alpha) + Ff(\omega - \alpha))$ (equation (5.17)).

Amplitude modulation is commonly used in radio wave broadcasting. When a radio or television receiver picks up a signal, therefore, it has to *demodulate* it. The question then arises: how, given $g(x)$, can $f(x)$ be recovered? It is not as simple as dividing by $\cos(\alpha x)$, for the rapid oscillation of this carrier signal means that any slight error would be a serious distortion.

One approach is to modulate the received signal a second time, putting $h(x) = g(x)\cos(\alpha x)$; then

$$Fh(\omega) = \frac{1}{2}(Fg(\omega + \alpha) + Fg(\omega - \alpha))$$
$$= \frac{1}{2}\{\frac{1}{2}Ff(\omega + 2\alpha) + Ff(\omega) + \frac{1}{2}Ff(\omega - 2\alpha)\} \ . \tag{6.8}$$

Now the carrier wave frequency α is much higher than any significant component of f. Suppose $Ff(\omega)$ is only significant in the range $-\Omega < \omega < \Omega$, with $\Omega \ll \alpha$. We apply a low-pass filter to h, with cut-off Ω. All the components of Fh around 2α and around -2α are filtered out, and we are left with

$$Fh(\omega)\chi_{(-\Omega,\Omega)}(\omega) = \frac{1}{2}Ff(\omega) \ . \tag{6.9}$$

So we have recovered the signal f.

In practice, there is a problem with this method. The process of demodulation is only possible provided the carrier frequency α is known, since it involves multiplying by $\cos(\alpha x)$. (The phase of the carrier signal is not so important: if $h(x) =$

Ch. 6]   Filtering   163

$g(x)\cos(\alpha x + \varphi)$, then the filtered transform of h simply changes to $\frac{1}{2}\cos\varphi \, Ff(\omega)$.)
There are two ways to counter this problem. One way is for the transmitter to add to the modulated signal a strong wave actually *at* this frequency; the transmitted signal is then $g(x) = f(x)\cos(\alpha x) + A\cos(\alpha x)$. This gives rise to a large 'spikes' in the transform of g at $\omega = \pm \alpha$, which can be picked out by the receiver and used in the decoding. This is the *normal* AM method, used in most station transmission.

The second way is for the receiver simply to 'tune in'; in other words, to guess α. This is called *suppressed carrier* AM, and is used in 'citizens' band' (CB) systems. In this case the demodulated, filtered signal will often be slightly 'off-centre', so that what comes out will still be slightly displaced from the 'correct' freequency band. Instead of $Fh(\omega) = \frac{1}{2}Ff(\omega)$, we will have $Fh(\omega) = \frac{1}{4}(Ff(\omega + \delta) + Ff(\omega - \delta))$. Since the ear is most sensitive to frequencies slightly *higher* than normal speech, the result is that an ill-tuned CB set sounds squeaky, like a record played at too high a speed.

6.8 Analogue filters

From what we have said so far, the reader may have got the impression that we need actually to *perform* Fourier transforms on signals that we wish to filter. This is not true.

Remember that the term 'filter' applies to a whole range of operations, which turn an input signal into an output signal. The mask filters we have been looking at are simply examples of this behaviour. Other filters, as we have seen, are actual physical devices, like the theatre in section 5.1. These do not operate via Fourier transformation, but those that are linear and invariant may still be described in the ω-domain by their transfer functions.

One way, then, to perform a desired filtering operation on a signal is to pass it through a device whose transfer function represents that operation. Thus, to low-pass filter a signal using the mask of equation (6.7a), we need to find a device of transfer function $T(\omega) = \chi_{(-\alpha,\alpha)}(\omega)$. Ideally, we would like to find a way of building a device to have any chosen transfer function. In fact it turns out to be impossible to achieve, exactly, *every* transfer function in a physical device; the best we can do is to make some good approximations.

Take the case of electrical signals, which are probably the most common in practice. To filter these, we need a linear device. Three common linear devices are inductances, resistances, and capacitances; so any circuit that contains just these three sorts of element will be linear. Such a circuit is called an LRC circuit, after the traditional symbols for the three sorts of device.

These devices are linear inasmuch as there is a linear relationship between the voltage $V(t)$ across each device and the current $I(t)$ flowing through it. An ideal inductance of inductance L, for instance, satisfies the equation $V = L(dI/dt)$; an ideal resistance of resistance R, $V = RI$; and an ideal capacitance of capacitance C, $C(dV/dt) = I$. (Further information can be found in any electronics book.)

Fig. 6.14(a) shows a very simple filtering circuit, containing a single in-line resistance R and a single capacitance C to earth. The input to this circuit is a voltage signal $V(t)$; this represents the signal to be filtered. In order to maintain this voltage, the input has to supply a current $I(t)$. The output of the circuit is the voltage $W(t)$ between the resistance and the capacitance; this represents the filtered signal. No current is drawn from this point.

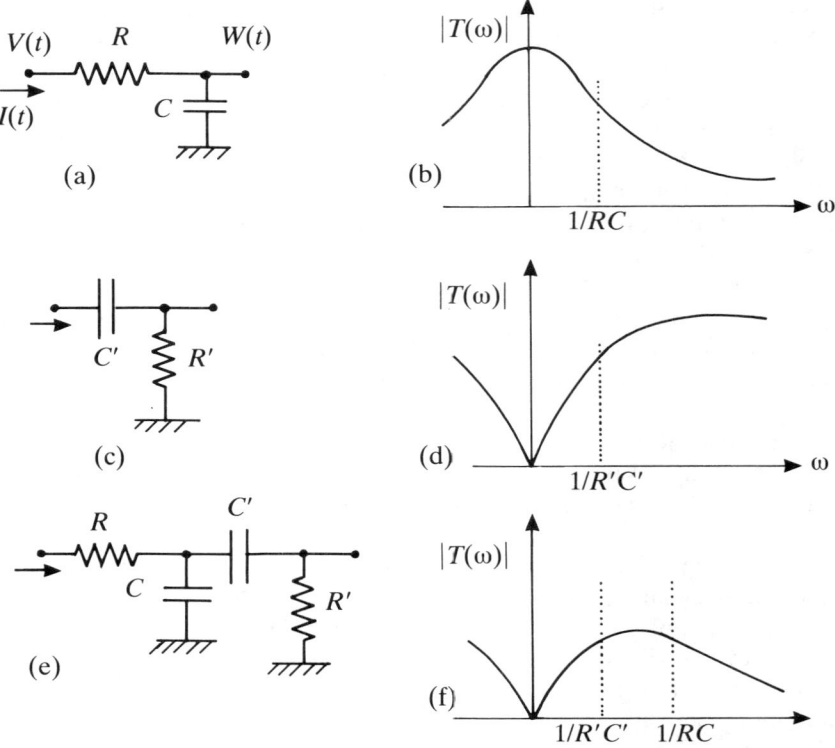

Fig. 6.14 — Elementary analogue low-pass ((a) and (b)), high-pass ((c) and (d)), and band-pass ((e) and (f)) filter circuits, and their amplitude response functions.

The behaviour of the system is governed by the equations for the two devices. In this system, they become

$$C \frac{dW}{dt} = I$$

and

$$V - W = RI \ . \tag{6.10}$$

Apply a Fourier transformation to these equations and use the derivative theorem (equation (3.30)), and they become

$$i\omega C F W(\omega) = F I(\omega)$$

and

$$FV(\omega) - FW(\omega) = RFI(\omega) \ . \tag{6.11}$$

Substituting for FI in the second equation using the first:

$$FV = (1 + i\omega RC)FW \quad ,$$

or

$$FW = FV/(1 + i\omega RC) \quad . \tag{6.12}$$

So the transfer function of this circuit is

$$T(\omega) = 1/(1 + i\omega RC) \quad . \tag{6.13}$$

Fig. 6.14(b) shows a graph of the amplitude response function $|T(\omega)|$. Note the form of this function: it is small for high frequencies (large ω), and it approaches unity for low frequencies. It is, indeed, a low-pass filter; not a very good one, though even a filter of this far-from-ideal form can be very useful for its sheer simplicity.

Because the filter is not ideal, there is no sharp cut-off frequency. For convenience, we define the *critical frequency* of a non-ideal filter like this to be the frequency at which the amplitude response function $|T(\omega)|$ falls to $1/\sqrt{2}$. For equation (6.13), this happens when $\omega RC = 1$; so the critical angular frequency of this filter is $\omega_c = 1/RC$. At this frequency, the *power* $|FW(\omega)|^2$ of the output signal falls to half the input power. This is equivalent to an input–output power ratio of $10\log_{10} 2 \simeq 3$ dB, so the critical frequency is sometimes called the *3-dB point*.

The remainder of Fig. 6.14 shows simple electronic *high*-pass and *band*-pass circuits, together with their amplitude response functions. Note that the band-pass filter consists of a low-pass filter followed by a high-pass filter. Respectively, these filters have transfer functions

$$T(\omega) = \frac{i\omega R'C'}{1 + i\omega R'C'}$$

and

$$T(\omega) = \frac{i\omega R'C'}{(1 + i\omega R'C')(1 + i\omega RC) + i\omega RC'} \quad . \tag{6.14}$$

Proving these formulae is left as an exercise for the reader.

The transfer function of the band-pass filter is *not* simply the product of its component parts, thanks to the extra term $i\omega RC'$ in the denominator. This is because, contrary to our assumptions for the low-pass filter of equation (6.13), current *is* being drawn from the low-pass section. However, it tends to this product in the limit $(R'/R) \to \infty$.

These circuits are limited by the presence of only two kinds of device, resistances and capacitances. Filters of rather better characteristics can be designed if we allow ourselves to use inductances as well. However, real inductances tend both to be bulky and to have significantly 'nonideal' behaviour; for this reason, RC filters are usually both neater and more reliable. Even better characteristics can be obtained in circuits involving other, 'active' components, like amplifiers.

There is no space here for more than this very brief survey of analogue filters. For a more detailed treatment of electronic filter design, the interested reader should consult a more specialized work (see Bibliography.)

There is an important consequence of analyses like that of electrical circuitry

Fourier theory

here, and that of mechanical systems in Chapter 1. Both low- and high-pass filtering will occur *accidentally* (albeit not in ideal form). Any equipment used to record and analyse a signal will have its own frequency response characteristics, which may be significantly distorting in the frequency range of interest. It is important to understand where and how this filtering occurs before putting faith in the output of the equipment. (A cheap microphone is useless in studying the echo-location of bats.)

EXERCISES

1. Convince yourself that any narrowband noise can be thought of as a sinusoid under the combined influence of small amounts of amplitude and frequency modulation.

2. What is the effect of an ideal low-pass filter, with $T(\omega) = \chi_{(-\alpha,\alpha)}(\omega)$, on the step function $f(x) = \chi_{(0,a)}(x)$?

(This function represents a sample of length a taken from a constant signal $I(x)$. The Fourier transform of this is $\delta(\omega)$, so that *any* low-pass filter should leave it unaffected. The finiteness of the sample, however, prevents this.)

3. Fig. 6.15 shows one method of incorporating an inductance into a filter. Is this a low-pass filter, a high-pass filter, or neither?

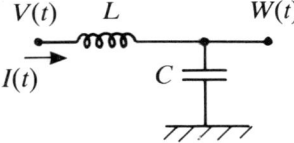

Fig. 6.15.

4. One way to improve the high-frequency suppression of the filter in Fig. 6.14(a,b) is to pass the signal through the same filter again. What is the transfer function of the overall filter? What is its critical angular frequency? (It is not $1/RC$.)

5. To achieve the 'improved', double-pass, filter of the previous question, it might be thought that a suitable circuit is that shown in Fig. 6.16. Analyse this circuit, and sketch a graph of its amplitude response function $|T(\omega)|$.

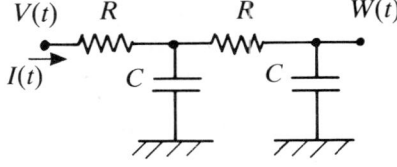

Fig. 6.16.

Project. The equations governing the behaviour of the three basic electronic linear devices (capacitances, resistances, inductances) are very similar to those governing the three basic mechanical devices (springs, dampers, masses), under the correspondence voltage↔force, current↔velocity. Show that the extension oscillator acts as a simple low-pass filter. Show that Fig. 6.17 is a low-pass filter analogous to Fig. 6.14(a). Design, and investigate the behaviour of, some mechanical filters.

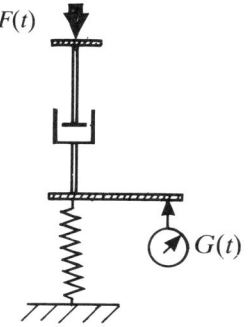

Fig. 6.17.

7

Correlation

7.1 The correlation coefficient

A statement like 'tall people are, on average, heavier than short people' describes a relationship between variables; not a perfect relationship, but a statistical one.

The statistical approach usually adopted is as follows. Suppose n people have heights $H_1, \ldots, H_n$ and weights $W_1, \ldots, W_n$; let the averages of these be $\bar{H} = (1/n) \sum_{r=1}^{n} H_r$ and $\bar{W} = (1/n) \sum_{r=1}^{n} W_r$ respectively. Now consider the product $(H_r - \bar{H})(W_r - \bar{W})$. If there is a connection between height and weight, then a greater-than-average height will usually be associated with a greater-than-average weight; for such a person, the factors $H_r - \bar{H}$ and $W_r - \bar{W}$ are both positive, and so is their product. Similarly, a less-than-average height and a less-than-average weight will tend to be associated, in which case $H_r - \bar{H}$ and $W_r - \bar{W}$ are both negative, and the product is again positive. Thus the average $(1/n) \sum_{r=1}^{n} (H_r - \bar{H})(W_r - \bar{W})$ will be positive. On the other hand, if there is *no* relationship between height and weight, a positive $H_r - \bar{H}$ will be associated with positive and negative $W_r - \bar{W}$ equally often; and the average of the $(H_r - \bar{H})(W_r - \bar{W})$ will be around zero. The quantity $(1/n) \sum_{r=1}^{n} (H_r - \bar{H})(W_r - \bar{W})$, called the *covariance* of the variables H and W and written $\text{cov}(H, W)$, is then an indicator of the relationship between height and weight. The covariance of a variable with itself is its *variance*:

$$\text{var}(H) = \text{cov}(H, H) = (1/n) \sum_{r=1}^{n} (H_r - \bar{H})^2.$$

The covariance of two real variables X and Y is at a maximum when there is a linear relationship between them: $Y_r = a + bX_r$. In this case, taking averages gives $\bar{Y} = a + b\bar{X}$, so $Y_r - \bar{Y} = b(X_r - \bar{X})$; then $(X_r - \bar{X})(Y_r - \bar{Y}) = b(X_r - \bar{X})^2 = (1/b)(Y_r - \bar{Y})^2$, so that $\text{cov}(X, Y) = b\,\text{var}(X) = (1/b)\,\text{var}(Y)$. We deduce that, in this case, $\text{cov}(X, Y)^2 = \text{var}(X)\text{var}(Y)$. Where a linear relationship does not necessarily hold, $\text{cov}(X,Y)^2 \leq \text{var}(X)\text{var}(Y)$. The *strength* of the relationship between X and Y is then indicated by the *normalized* quantity

$$r(X,Y) = \text{cov}(X,Y)/\sqrt{\{\text{var}(X)\text{var}(Y)\}} \ . \tag{7.1}$$

The quantity $r(X,Y)$ is the *correlation coefficient* of X and Y. $r(X,Y)$ always lies between ± 1. If X and Y are unrelated, $\text{cov}(X,Y) \simeq 0$, so $r(X,Y) \simeq 0$. If the relationship is a perfect linear one, then $r(H,W) = \pm 1$ (depending on whether the constant b is positive or negative).

There is a caveat to the use of the correlation coefficient: it signifies only the strength of a possible *linear* relationship. Suppose, for instance, that $n = 5$, that $X_1 = -2$, $X_2 = -1$, $X_3 = 0$, $X_4 = 1$, and $X_5 = 2$, and that $Y_r = X_r^2$ for each r; then $\text{cov}(X,Y) = 0$, so $r(X,Y) = 0$. Or, to take a more realistic case, it is not clear in our height–weight example whether W should be tested for a *linear* relationship with H, or a *cubic* one like $W = aH^3$.

Among the arithmetic properties of covariance, variance, and the correlation coefficient we may note the following:

$$\text{cov}(X,Y) = \text{var}(X), \text{ so } r(X,X) = 1;$$

$$\text{cov}(X,Y) = \text{cov}(Y,X), \text{ so } r(X,Y) = r(Y,X);$$

$$\text{cov}(\lambda X, Y) = \lambda \text{cov}(X,Y), \text{ so } r(\lambda X, Y) = r(X,Y) \text{ for } \lambda > 0;$$

$$\text{cov}(-X,Y) = -\text{cov}(X,Y), \text{ so } r(-X,Y) = -r(X,Y);$$

$$\text{cov}(X+Y,Z) = \text{cov}(X,Z) + \text{cov}(Y,Z);$$

and

$$\text{var}(X+Y) = \text{var}(X) + 2\text{cov}(X,Y) + \text{var}(Y) \ . \tag{7.2}$$

There is *no* simple connection between $r(X,Z)$, $r(Y,Z)$ and $r(X+Y,Z)$.

The nature, properties and uses of these techniques are dealt with more fully in statistics texts.

7.2 Correlation of two functions

We can define, in an analogous way, the correlation coefficient of two (real-valued) functions $f(x)$ and $g(x)$, as

$$\rho(f,g) = \frac{\int_{-\infty}^{\infty} f(x)g(x)\,dx}{\sqrt{\left\{\left(\int_{-\infty}^{\infty} f(x)^2\,dx\right)\left(\int_{-\infty}^{\infty} g(x)^2\,dx\right)\right\}}} \ . \tag{7.3}$$

For $\rho(f,g)$ to exist, each of the integrals must exist.

The properties of ρ are similar to those of r; in particular:

$$\rho(f,f) = 1;$$

$$\rho(g,f) = \rho(f,g);$$

$$\rho(\lambda f, g) = \rho(f,g) \text{ for } \lambda > 0;$$

and
$$\rho(-f,g) = -\rho(f,g). \tag{7.4}$$
Here are some examples.

Example. What is the correlation coefficient of the two functions $f(x) = \sin(x)$ and $g(x) = \cos(x)$?

It does not exist, for the functions $f(x)^2$, $f(x)g(x)$ and $g(x)^2$ are not integrable.

Example. What is the correlation coefficient between $f(x) = \exp(-x^2)$ and $g(x) = \exp(-2x^2)$?

Here the calculation goes smoothly. We have

$$\rho(f,g) = \frac{\int_{-\infty}^{\infty} f(x)g(x)\,dx}{\sqrt{\left\{\left(\int_{-\infty}^{\infty} f(x)^2\,dx\right)\left(\int_{-\infty}^{\infty} g(x)^2\,dx\right)\right\}}}$$

$$= \frac{\int_{-\infty}^{\infty} \exp(-3x^2)\,dx}{\sqrt{\left\{\left(\int_{-\infty}^{\infty} \exp(-2x^2)\,dx\right)\left(\int_{-\infty}^{\infty} \exp(-4x^2)\,dx\right)\right\}}}$$

$$= \sqrt{(\pi/3)}/\sqrt{\{\sqrt{(\pi/2)}\sqrt{(\pi/4)}\}}$$

$$= \sqrt[4]{(8/9)}$$

$$\simeq 0.97.$$

These two factors show a high positive correlation. This is because both show a single positive peak around $x = 0$. One peak, indeed, is substantially wider than the other, so that the correlation is not perfect. But ρ still reaches 0.97; this is a salutary example of the weakness of correlation coefficients.

Example. What is the correlation coefficient between $f(x) = \exp(-x^2)$ and $g(x) = \exp(-(x-2)^2)$?

The calculation is straightforward:

$$\int_{-\infty}^{\infty} f(x)g(x)\,dx = \int_{-\infty}^{\infty} \exp(-2x^2 + 4x - 4)\,dx$$

$$= e^{-2} \int_{-\infty}^{\infty} \exp(-2(x-1)^2)\, dx$$

$$= e^{-2} \sqrt{(\pi/2)} ,$$

so

$$\rho(f,g) = \frac{\int_{-\infty}^{\infty} f(x)g(x)\, dx}{\sqrt{\left\{\left(\int_{-\infty}^{\infty} f(x)^2\, dx\right)\left(\int_{-\infty}^{\infty} g(x)^2\, dx\right)\right\}}}$$

$$= \frac{e^{-2}\sqrt{(\pi/2)}}{\sqrt{\left\{\left(\int_{-\infty}^{\infty} \exp(-2x^2)\, dx\right)\left(\int_{-\infty}^{\infty} \exp(-2(x-2)^2)\, dx\right)\right\}}}$$

$$= \frac{e^{-2}\sqrt{(\pi/2)}}{\sqrt{\{\sqrt{(\pi/2)}\sqrt{(\pi/2)}\}}}$$

$$= e^{-2}$$

$$\simeq \underline{0.135} .$$

The two functions in this example have graphs of the same shape, although there is a small lateral shift between them. Yet the correlation btween them is much poorer than in the previous example, which showed *differently* shaped curves that happened both to have a peak at $x=0$.

This is important, for cases of 'delayed' correlation occur frequently in practice. For example, suppose we have a conveyor belt, carrying coal to a hopper. We put a meter at each end: one detects the rate at which coal is being poured onto the belt, the other detects its rate of discharge into the hopper. If nothing happens to disturb the coal, the two signals should be identical, apart from a time delay in the discharge signal, corresponding to the time taken for the coal to travel from one end of the belt to the other (see Fig. 7.1). The signals in this case are clearly related, and if the correlation coefficient is small, it indicates that the correlation coefficient is an inappropriate tool.

The solution to this problem is direct. Let the signal from the meter at the loading (input) end be $I(t)$, and the signal from the meter at the discharge (output) end be $O(t)$; suppose the time difference between the two signals is τ. Then $O(t) = I(t-\tau)$; so although $\rho(I(t), O(t))$ may be small, $\rho(I(t-\tau), O(t)) = 1$.

7.3 The cross-correlation function
The analysis for the conveyor belt is simple, because we know the required shift τ in advance. Typically we do not; indeed, we may be trying to *find* the shift between two

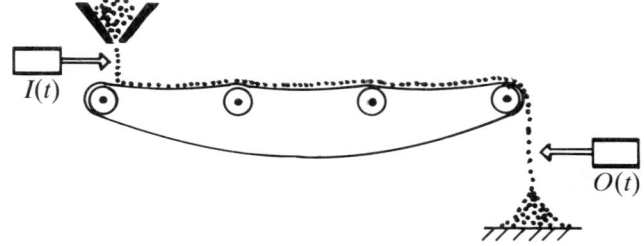

Fig. 7.1 — The input $I(t)$ to, and the output $O(t)$ from, a conveyor are monitored.

signals. The only possible recourse we have is to try out every possible shift: to calculate, for (real-valued) integrable functions f and g, the correlation $\rho(f(x+\xi),g(x))$ for every ξ. The result is a function of ξ:

$$R[f,g](\xi) = \rho(f(x+\xi),g(x))$$

$$= \frac{\int_{-\infty}^{\infty} f(x+\xi)g(x)\,dx}{\sqrt{\left(\int_{-\infty}^{\infty} f(x+\xi)^2\,dx\right)\left(\int_{-\infty}^{\infty} g(x)^2\,dx\right)}}$$

$$= \frac{\int_{-\infty}^{\infty} f(x+\xi)g(x)\,dx}{\sqrt{\left(\int_{-\infty}^{\infty} f(x)^2\,dx\right)\left(\int_{-\infty}^{\infty} g(x)^2\,dx\right)}}. \quad (7.5)$$

The function $R[f,g](\xi)$ is called the (cross-)correlation function of f and g. Note that the denominator does not depend on ξ.

Example. A conveyor is loaded with coal at rate $I(t) = A\exp(-t^2)$ and is found to discharge it from the other end at rate $O(t) = (A/\sqrt{2})\exp(-(t-10)^2/2)$. Calculate the cross-correlation function $R[I,O](\tau)$, and use it to estimate the average time taken for a lump of coal to trasverse the conveyor belt.

All the functions involved have finite integrals, so the calculation is possible. We have, from equation (7.5):

$$R[I,O](\tau) = \frac{\int_{-\infty}^{\infty} I(t+\tau)O(t)\,dt}{\sqrt{\left(\int_{-\infty}^{\infty} I(t)^2\,dt\right)\left(\int_{-\infty}^{\infty} O(t)^2\,dt\right)}}.$$

The denominator terms are easily calculated as $A^2\sqrt{\pi/2}$ and $A^2\sqrt{\pi/4}$; the numerator is

$$\int_{-\infty}^{\infty} I(t+\tau)O(t)\,dt$$

$$= (A^2/\sqrt{2}) \int_{-\infty}^{\infty} \exp(-(t+\tau)^2)\exp(-(t-10)^2/2)\,dt$$

$$= (A/^2/\sqrt{2}) \int_{-\infty}^{\infty} \exp(-(3/2)t^2 - (2\tau-10)t - (\tau^2+50))\,dt$$

$$= A^2\sqrt{(\pi/3)}\exp(-(\tau+10)^2/3) \ .$$

Thus

$$R[I,O](\tau) = \sqrt[4]{(8/9)}\exp(-(\tau+10)^2/3) \ .$$

This function has a single peak, at $\tau = -10$, at which point its value is $\sqrt[4]{(8/9)}$, or ~ 0.97 (see Fig. 7.2). Thus the best match between I and O is between $I(t-10)$ and $O(t)$, and we estimate that the particles take 10s to traverse the belt.

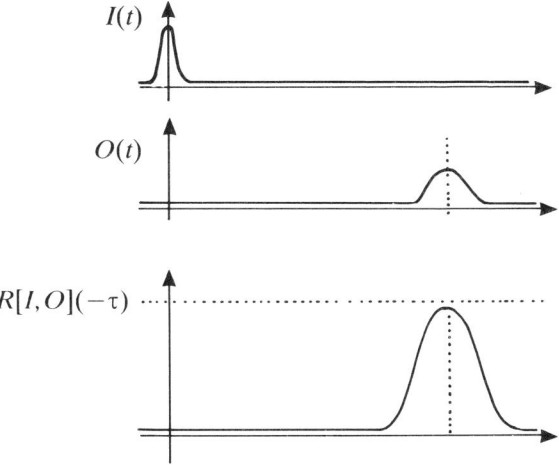

Fig. 7.2 — Their cross-correlation function.

Note, in this example, that the breadth of the peak of $R[I,O]$ does not allow us to specify precisely the time. For instance, the value of $R[I,O](-9.5)$ is still ~ 0.89.

$R[f,g]$ is derived from the correlation coefficient of two variables. There is a corresponding functional analogue of covariance. We may call this the *covariance function*:

$$T[f,g](\xi) = \int_{-\infty}^{\infty} f(x+\xi)g(x)\,dx \ . \tag{7.6}$$

$R[f,g]$ is a 'scaled' version of $T[f,g]$, in the same way that $r(X,Y)$ is a 'scaled' version of $\text{cov}(X,Y)$. It is related to $T[f,g]$ by the equation

$$R[f,g](\xi) = \frac{T[f,g](\xi)}{\sqrt{\left\{\left(\int_{-\infty}^{\infty} f(x)^2\,dx\right)\left(\int_{-\infty}^{\infty} g(x)^2\,dx\right)\right\}}} \ ;$$

the denominator is a constant, depending on the functions f and g but not on ξ. $R[f,g]$ lies, as do $r(X,Y)$ and $\rho(f,g)$, always between -1 and $+1$.

(The terminology here is unfortunately not standardised. Many authors refer to $T[f,g]$ as the cross-correlation between f and g, and call $R[f,g]$ the *normalized cross-correlation*.)

The cross-correlation and covariance functions have the following properties.

$$T[f+g,h](\xi) = T[f,h](\xi) + T[g,h](\xi) \ ;$$
$$T[\lambda f,g](\xi) = \lambda T[f,g](\xi) \ , \text{ so } R[\lambda f,g](\xi) = R[f,g](\xi) \text{ for } \lambda > 0;$$
$$R[f,f](0) = 1 \ . \tag{7.7}$$

However, they are *not* symmetric in f and g. Rather,

$$T[g,f](\xi) = \int_{-\infty}^{\infty} g(x+\xi)f(x)\,dt$$
$$= \int_{-\infty}^{\infty} g(y)f(y-\xi)\,dy \qquad \text{(putting } y = x+\xi\text{)}$$
$$= T[f,g](-\xi) \ . \tag{7.8a}$$

Hence

$$R[g,f](\xi) = R[f,g](-\xi) \ . \tag{7.8b}$$

As with the finite statistics, the cross-correlation function is limited in significance to simple linear relationships: of $f(x)$ to $g(x) = \alpha f(x-\xi)$. More complex relationships between functions will not be adequately represented; so if $g(x)$ is related to $f(\lambda x)$, $f(x^2)$, or $f(x)^2$, the cross-correlation function will not uncover this relationship.

Functions of constant character must be dealt with slightly differently. There, we define

$$T[f,g](\xi) = \lim_{t \to \infty} (1/2t) \int_{-t}^{t} f(x+\xi)g(x)\,dx$$

$$= \lim_{t \to \infty} (1/2t) T[f, g\chi_{(-t,t)}](\xi) \;.$$

and

$$R[f,g](\xi) = \lim_{t \to \infty} (1/2t) R[f, g\chi_{(-t,t)}](\xi) \;.$$

7.4 The autocorrelation function

The cross-correlation of an integrable function with itself is called its *autocorrelation function*, and denoted by $S[f](\xi)$. The formula given by (7.5) simplifies in this case:

$$S[f](\xi) = R[f,f](\xi)$$
$$= \rho(f(x+\xi), f(x))$$

$$= \frac{\int_{-\infty}^{\infty} f(x+\xi) f(x) \, dx}{\int_{-\infty}^{\infty} f(x)^2 \, dx} \;. \tag{7.9}$$

Similarly, there is a *variance function*,

$$U[f](\xi) = T[f,f](\xi) = \int_{-\infty}^{\infty} f(x+\xi) f(x) \, dx \;. \tag{7.10}$$

(Again, the terminology is not standardized; some authors call $U[f]$ the autocorrelation function, and $S[f]$ the normalized autocorrelation function.)

At the point $\xi = 0$, the comparison is between f and f unshifted, so we get

$$U[f](0) = \int_{-\infty}^{\infty} f(x)^2 \, dx$$

and

$$S[f](0) = 1 \;. \tag{7.11}$$

That is, there is perfect correlation between f and itself. Using this, we can rewrite equations (7.8) and (7.9) as

$$S[f](\xi) = \frac{U[f](\xi)}{U[f](0)}$$

and

$$R[f,g](\xi) = \frac{T[f,g](\xi)}{\sqrt{\{U[f](0) U[g](0)\}}} \;. \tag{7.12}$$

It follows from the definitions that $U[f]$ is an even function. For, using (7.8a), we have

$$U[f](\xi) = T[f,f](\xi) = T[f,f](-\xi) = U[f](-\xi) \ .$$

Similarly, $S[f](\xi) = S[f](-\xi)$, and $S[f]$ is even. (Note the irony: $U[f]$ and $S[f]$ are 'symmetric' functions because T and U are *anti*-symmetric operators.)

One useful property of the autocorrelation function is that it helps us to pick out 'almost-periodicities' in a function. (A truly periodic function will not be integrable over $(-\infty, \infty)$, so its autocorrelation function does not exist.) If f is 'almost-periodic' with period a, then $f(x + a)$ and $f(x)$ will be highly correlated, and $S[f](a)$ will be close to 1.

In fact, we can say more than this. If f is 'almost-periodic' with period a, then $f(x + \xi + a) \simeq f(x + \xi)$, and

$$\rho(f(x + \xi + a), f(x)) \simeq \rho(f(x + \xi), f(x))$$

for all ξ; that is,

$$S[f](\xi + a) \simeq S[f](\xi) \tag{7.13}$$

for all ξ. So $S[f]$ is 'almost-periodic' too, with the same period a.

Example. Let $f(x) = \chi_{(-\pi,\pi)}(x)\cos(nx)$, n being a positive integer (Fig. 7.3(a) shows this for $n = 6$). What is the autocorrelation function $S[f](\xi)$ of f?

This function is 'almost periodic' of period $2\pi/n$. The term $\chi_{(-\pi,\pi)}$ ensures that the functions have finite integrals, so we can calculate:

$$S[f](\xi) = \frac{\int_{-\infty}^{\infty} f(x + \xi) f(x) \, dx}{\int_{-\infty}^{\infty} f(x)^2 \, dx}$$

$$= \frac{\int_{-\pi}^{\pi} \chi_{(-\pi,\pi)}(x + \xi) \cos(n(x + \xi))\cos(nx) \, dx}{\int_{-\pi}^{\pi} \cos(nx)^2 \, dx}$$

$$= (1/\pi)\left(\int_{-\pi}^{\pi} \chi_{(-\pi,\pi)}(x + \xi) \tfrac{1}{2}(\cos(2nx + n\xi) + \cos(n\xi)) \, dx\right)$$

$$= \chi_{(-2\pi,2\pi)}(\xi)[(1 - |\xi|/2\pi)\cos(n\xi) - (1/2n\pi)\sin(n|\xi|)] \ .$$

Fig. 7.3(b) shows the graph of $S[f](\xi)$, when $n = 6$. If this is compared with the graph of $f(x)$ itself, Fig. 7.3(a), the inheritance of 'almost-periodicity' is apparent.

For functions of constant character, we define

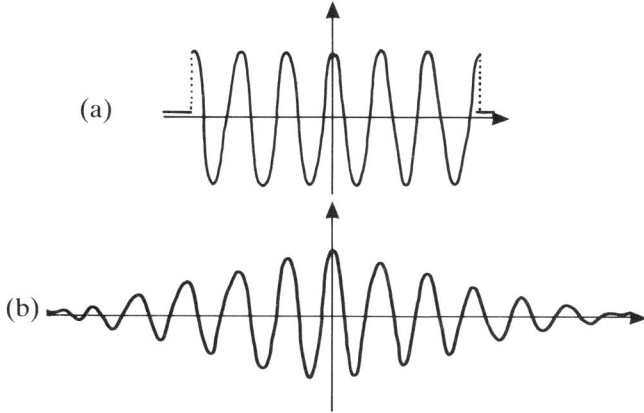

Fig. 7.3 — (a) The function $f(x) = \chi_{(-\pi,\pi)}(x)\cos(6x)$. (b) Its autocorrelation function.

$$U[f](\xi) = \lim_{t \to \infty} (1/2t) U[f\chi_{(-t,t)}](\xi)$$

and

$$S[f](\xi) = \lim_{t \to \infty} (1/2t) S[f\chi_{(-t,t)}](\xi) \ .$$

These definitions are, in fact, consistent with those of T and R, in that $U[f] = T[f,f]$ and $S[f] = R[f,f]$.

7.5 Correlation and Fourier transformation
There is a close connection between the correlation (more particularly, the covariance function) of two functions and the convolution operator. Recall the definitions:

$$(f*g)(x) = \int_{-\infty}^{\infty} f(x-y)g(v)\,dy$$

and

$$T[f,g](\xi) \int_{-\infty}^{\infty} f(x+\xi)g(x)\,dx \ .$$

Substituting $y = -x$ into this second definition gives

$$T[f,g](\xi) = \int_{-\infty}^{\infty} f(\xi - y)g(-y)\,dy \ ,$$

and so, if $h(x) = g(-x)$, then

$$T[f,g](x) = (f*h)(x) \tag{7.14}$$

Now we are supposing that $g(x)$ is real-valued. Applying the convolution theorem, equation (5.7), to equation (7.14), we get:

$$FT[f,g](\omega) = F(f*h)(\omega)$$
$$= Ff(\omega)Fh(\omega) \ . \qquad (7.15)$$

But from equations (3.32) and (3.34), $Fh(\omega) = \overline{Fg(\omega)}$; it follows that

$$FT[f,g](\omega) = Ff(\omega)\overline{Fg(\omega)} \ . \qquad (7.16)$$

This is *not* symmetric in f and g, because T is not (see equation (7.8a)).

When $f = g$, (7.16) becomes

$$FU[f](\omega) = FT[f,f](\omega)$$
$$= Ff(\omega)\overline{Ff(\omega)}$$
$$= Pf(\omega) \ . \qquad (7.18)$$

In other words, *the transform of the variance function is the power spectrum*. (This is sometimes called the *Wiener–Khintchine* theorem.)

The transforms of the (normalized) cross-correlation and autocorrelation functions do not yield such neat formulae. However, we may invoke the Rayleigh–Plancherel theorem (equation (5.23a)):

$$\int_{-\infty}^{\infty} f(x)^2 \, dx = (1/2\pi) \int_{-\infty}^{\infty} Pf(\omega) \, d\omega \ .$$

Since these integrals do not involve ξ, we have from (7.8) and (7.16) that

$$FR[f,g](\omega) = FT[f,g](\omega) \bigg/ \sqrt{\left\{\left(\int_{-\infty}^{\infty} f(x)^2 \, dx\right)\left(\int_{-\infty}^{\infty} g(x)^2 \, dx\right)\right\}}$$
$$= 2\pi Ff(\omega)\overline{Fg(\omega)} \bigg/ \sqrt{\left\{\left(\int_{-\infty}^{\infty} Pf(\alpha) \, d\alpha\right)\left(\int_{-\infty}^{\infty} Pg(\alpha) \, d\alpha\right)\right\}} \ . \qquad (7.19)$$

Again, the formula simplifies when $f = g$. We have:

$$FS[f](\omega) = FR[f,f](\omega)$$
$$= 2\pi Pf(\omega) \bigg/ \left(\int_{-\infty}^{\infty} Pf(\alpha) \, d\alpha\right) \ . \qquad (7.20)$$

We can now understand better the inheritance of periodicity by the autocorrelation function. If $f(x)$ is almost-periodic with period a, then its spectrum $Ff(\omega)$ has a large spike at about $\omega = 2\pi/a$. Then $Pf(\omega) = |Ff(\omega)|^2$ is still larger at this frequency; that is, $FS[f](\omega)$ has a spike at $\omega = 2\pi/a$. This means that $S[f](x)$ has an important component of angular frequency $2\pi/a$, and therefore $S[f](x)$ is almost periodic, of period a.

7.6 Coherence

The cross-correlation and autocorrelation functions are, notwithstanding the previous section, principally operations in the x-domain. But, as we saw in section 6.4, it is sometimes useful to perform statistics on the spectra of functions, rather than on the functions themselves. There, we sampled the power spectrum of a noisy signal $f(x)$, and averaged to get a CPS, $\Pi f(\omega)$.

Now suppose we have *two* noisy signals, $f(x)$ and $g(x)$, arising from the same system. Take a simultaneous sample record of each of these functions, say $f_1(x)$ and $g_1(x)$; repeat this n times to get records $f_1(x), \ldots, f_n(x)$ and $g_1(x), \ldots, g_n(x)$. To find out what these say about f and g separately, we would take the CPSs of these samples. But if we are looking for a possible connection between the two functions, then we can use the ideas of section 7.1. For each angular frequency ω, we treat $Ff_r(\omega)$ as the rth meaurement of a variable which represents the amplitude and phase of this frequency in f; then we define

$$K(f,g)(\omega) = \mathrm{cov}(Ff_r(\omega), \overline{Fg_r(\omega)})$$

$$= (1/n) \sum_{r=1}^{n} Ff_r(\omega)\overline{Fg_r(\omega)}$$

and

$$\gamma(f,g)(\omega) = r(Ff_r(\omega), \overline{Fg_r(\omega)})$$
$$= K(f,g)(\omega)/\sqrt{\{\Pi f(\omega)\Pi g(\omega)\}} \ . \tag{7.21}$$

$K(f,g)$ is the *consensus cross-spectral power* (or *CCSP*) of the n records; while $\gamma(f,g)$, the frequency-by-frequency correlation coefficient, is their *(complex) coherence*.

Notice some properties of these operators:

$$K(f,f)(\omega) = \Pi f(\omega);$$

$K(\alpha f,g)(\omega) = \alpha K(f,g)(\omega)$ for any complex α

$$K(g,f)(\omega) = \overline{K(f,g)(\omega)} \ ;$$

$$\gamma(f,f)(\omega) = 1 \ ;$$

and

$$|\gamma(f,g)(\omega)| \leq 1 \ . \tag{7.22}$$

The argument of $K(f,g)(\omega)$ (or, which is the same thing, of $\gamma(f,g)(\omega)$) is an estimate of the *relative phase* of f and g at this angular frequency. More important is the modulus $|\gamma(f,g)(\omega)|$ of the coherence, which shows the degree of consistency of the relationship between these components, across the various records. This modulus is sometimes called the *absolute coherence*.

For display purposes, the absolute coherence and relative phase are usually plotted.

The connection between coherence and the x-domain correlation techniques is interesting. Equation (7.16), for instance, tells us that $K(f,g)(\omega)$ is an estimate of the Fourier transform of the covariance function,

$$T[f,g](x) = \int_{-\infty}^{\infty} f(y+x)g(y)\,dy \ .$$

It follows that the CPSs $\Pi f(\omega)$ and $\Pi g(\omega)$ are estimates of the power spectra $Pf(\omega)$ and $Pg(\omega)$. They do *not* estimate the total power $\int Pf$ and $\int Pg$. Thus, the coherence $\gamma(f,g)(\omega)$ is *not* an estimate of the transform of the cross-correlation function $R[f,g](x)$ (equations (7.19) and (7.21)).

Coherence is a statistical measure, and is at its most useful when we have *sampled* data to deal with. We shall look more closely at its interpretation in Chapter 9, when we deal with finite data sets.

EXERCISES

1. Five people are measured for height and income, the results being as follows:

	Height (m)	Income (£pa)
Person 1	1.6	12 000
Person 2	1.5	15 000
Person 3	1.9	17 000
Person 4	1.7	10 000
Person 5	1.8	11 000

Calculate the covariance and the correlation coefficient between these variables. Plot a graph of height against income; what do you notice?

2. Find the correlation coefficient $\rho(f,g)$ of the two functions $f(x) = e^{-x^2}$ and $g(x) = x^2 e^{-x^2}$.

3. Find, and sketch the graph of, the cross-correlation function between the two functions f and g of the previous question.

4. Find and sketch the autocorrelation function of the g of Question 2.

5. Let $f(x) = \cos(\omega x)$, and let $g(x)$ be a real-valued function. What can you say about $R[f,g](x)$?

6. Now let $f(x) = \exp(-x^2)\cos(\omega x)$. What can you say about $R[f,g](x)$ now?

7. Find, and sketch the graph of, the autocorrelation function of function f of the previous question.

8. The functions $A(x)$ and $B(x)$ are narrowband noise of identical PSDF. What can you say about the coherence of the functions $f(x) = A(x) + B(x)$ and $g(x) = A(x) - B(x)$?

Project. Two parameters f and g of a system are simultaneously measured n times, with a view to finding a relationship between them. It turns out that $f_i(x) = a_i N(x+b)$, $g_i = \lambda a_i N(x+\alpha+b_i)$ $(i=1,\ldots,n)$, for certain constants α and λ. Calculate the CCSP of these samples of f and g. Hence calculate and sketch their absolute coherence and the relative phase spectra.

Is the same true when the bell curve N is replaced with an arbitrary integrable function h, so that $f_i = a_i h(x+b_i)$, $g_i = \lambda a_i h(x+\alpha+b_i)$? How about if the function changes, so that $f_i = a_i h_i(x)$, $g_i = \lambda a_i h_i(x+\alpha)$?

This investigates the effect on coherence of one signal being a linear multiple of another, and the effect of a delay. Often this will not be the case, of course.

Suppose n_1 and n_2 are uncorrelated noise functions of the same PSDF $R(\omega)$, and that s is a signal with PSDF $S(\omega)$. What can you say about the coherence of the functions $f(x) = s(x)$ and $g(x) = s(x) + n_1(x)$? How about the coherence of g and $h(x) = s(x) + n_2(x)$?

8

Other series and transforms

A. SERIES

8.1 Orthogonal functions

Fourier series have two important properties. First, the range of functions having Fourier series expansions is quite wide (sections 2.1 and 2.16). Secondly, the Fourier expansion of such a function can be derived in a simple manner from the function, using equations (2.3) to calculate the coefficients. This second property arises because the set of functions $\{1(x), \cos(nx), \sin(nx): n = 1, 2, \ldots\}$ satisfies the orthogonality relations of equations (2.2).

We saw in Chapter 2 that the harmonic functions themselves can be used in several different ways to produce Fourier series, of which the classical series (equation (2.1)) is but one example. There are further generalizations of Fourier series which do not involve the harmonic functions; for other families of functions that satisfy orthogonality relations can be found.

For example, one orthogonal set is the sequence of polynomials $\{p_n\}$, where $p_n(x) = (1/n!2^n)(d^n/dx^n)((x^2 - 1)^n)$. These are the *Legendre polynomials*. The first few are $p_0(x) = 1$, $p_1(x) = x$, $p_2(x) = \frac{1}{2}(3x^2 - 1)$, $p_3 = \frac{1}{2}(5x^3 - 3x) \ldots$; their graphs are shown in Fig. 8.1. It is not hard to see, in fact, that p_n is a polynomial of degree n. The

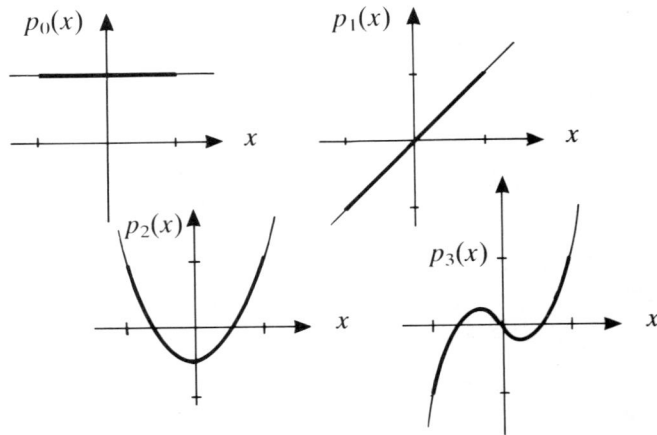

Fig. 8.1 — The first few Legendre polynomials.

polynomials p_n can be shown to satisfy the orthogonality relations

$$\int_{-1}^{1} p_m(x)p_n(x) \, dx = 0 \quad \text{if } m \neq n;$$

and

$$\int_{-1}^{1} p_n(x)^2 \, dx = 1/(n + \tfrac{1}{2}). \tag{8.1}$$

These orthogonality conditions may be used, as in the case of harmonic series, to find (what is called) the *generalized Fourier series* — or in this case, the *Fourier–Legendre series* — expansion of a function $f(x)$, where it exists. By this is meant an expression of the form $f(x) = \sum_{n=0}^{\infty} A_n p_n(x)$. If $f(x)$ has such an expansion, then multiplying by $p_m(x)$ and integrating gives

$$\int_{-1}^{1} f(x)p_m(x) \, dx = \int_{-1}^{1} p_m(x) \left\{ \sum_{n=0}^{\infty} A_n p_n(x) \right\} dx$$

$$= \sum_{n=0}^{\infty} A_n \left[\int_{-1}^{1} p_m(x)p_n(x) \, dx \right]$$

$$= A_m/(m + \tfrac{1}{2}),$$

and therefore

$$A_m = (m + \tfrac{1}{2}) \int_{-1}^{1} f(x)p_m(x) \, dx. \tag{8.2}$$

(Compare equations (2.3).)

Even if an (integrable) function $f(x)$ does not have such an expansion, it will still generate a Fourier–Legendre series $f(x) \to \sum_{n=0}^{\infty} A_n p_n(x)$, using the coefficients calculated in (8.2). For the Legendre polynomials, the series so generated is related closely to the Taylor series for f, where that exists. Indeed, since the harmonic functions $1(x)$, $\cos(nx)$, $\sin(nx)$ all have convergent Taylor series, this Fourier-type series *is* an expansion for all functions f which possess a classical Fourier expansion. In other words, the Fourier-type series based on the orthogonal sequence $\{p_n(x)\}$ is as widely applicable as the classical Fourier series.

It is true that the Legendre polynomials lack the physical significance of the harmonic functions, which are useful in the analysis of physical systems involving the harmonic equation (or a 'similar' differential equation). However, the polynomial approach is often used in cases where no such physical system exists, such as curve-fitting (see section 2.14); the polynomial approximations they produce are often more efficient, and easier so calculate with, than the rather cumbersome truncated Fourier series.

Other families of functions exist which satisfy different orthogonality relations. One such family is the set of *Hermite polynomials* $H_n(x)$, defined by the equation $H_n(x) = (-1)^n e^{x^2} (d^n/dx^n) (e^{-x^2})$. The first few Hermite polynomials are $H_0(x) = 1$, $H_1(x) = 2x$, $H_2(x) = 4x^2 - 2$, ...; again, $H_n(x)$ is a polynomial of degree n. Like the polynomials p_n, the Hermite polynomials satisfy orthogonality relations — but their orthogonality is rather different:

$$\int_{-\infty}^{\infty} H_m(x) H_n(x) e^{-x^2} dx = 0 \quad \text{if } m \neq n,$$

while

$$\int_{-\infty}^{\infty} H_n(x)^2 e^{-x^2} dx = 2^n n! \sqrt{\pi}. \tag{8.3}$$

If a function $f(x)$ has a Fourier–Hermite expansion $f(x) = \sum_{n=0}^{\infty} A_n H_n(x)$, then the coefficient A_m can be found by multiplying by $H_m(x) e^{-x^2}$, and integrating:

$$\int_{-\infty}^{\infty} f(x) H_m(x) e^{-x^2} dx = \int_{-\infty}^{\infty} H_m(x) \left\{ \sum_{n=0}^{\infty} A_n H_n(x) \right\} e^{-x^2} dx$$

$$= \sum_{n=0}^{\infty} A_n \left[\int_{-\infty}^{\infty} H_m(x) H_n(x) e^{-x^2} dx \right]$$

$$= A_m 2^m m! \sqrt{\pi},$$

and therefore

$$A_m = \left(\frac{2^{-m}}{m! \sqrt{\pi}} \right) \int_{-\infty}^{\infty} f(x) H_m(x) e^{-x^2} dx. \tag{8.4}$$

8.2 Bessel functions

Yet another family of orthogonal functions involves the *Bessel function* $J_0(x)$. The Bessel functions of order ν are the principal solutions, J_ν and Y_ν, of the differential equation

$$\frac{d^2 f}{dx^2} + \frac{1}{x} \frac{df}{dx} + \left(1 - \frac{\nu^2}{x^2}\right) f = 0. \tag{8.5}$$

For $\nu \geq 0$, the functions $J_\nu(x)$ are bounded as $x \to 0$; these are called Bessel functions 'of the first kind', and are defined by the asymptotic approximation

$J_\nu(x) \simeq (x/2)^\nu/\Gamma(\nu+1)$ as $x \to 0$ (where $\Gamma(t) = \int_0^\infty x^{t-1}e^{-x}\,dx$ is the gamma function). The functions $Y_\nu(x)$, Bessel functions 'of the second kind', are unbounded as $x \to 0$. All Bessel functions look rather like damped harmonic functions.

The Bessel equation can be solved by the series method, by looking for a solution of the form $f(x) = \sum_{n=0}^\infty \alpha_n x^{n+c}$. For the case $\nu = 0$, a series solution to (8.5) satisfies $n^2\alpha_n = -\alpha_{n-2}$. The function $J_0(x)$ is the particular solution when $c=0$, $\alpha_0 = 1$, $\alpha_1 = 0$; thus

$$J_0(x) = \sum_{n=0}^\infty (-1)^n \frac{x^{2n}}{2^{2n}(n!)^2}$$
$$= 1 - (x^2/4) + (x^4/64) - \ldots \quad (8.6)$$

This series converges for all x. (Compare the power series expansion $\cos(x) = \sum_{n=0}^\infty (-1)^n x^{2n}/(2n!)$.) Fig. 8.2 shows the graph of $J_0(x)$; note that $J_0(0) = 1$ and $J_0'(0) = 0$.

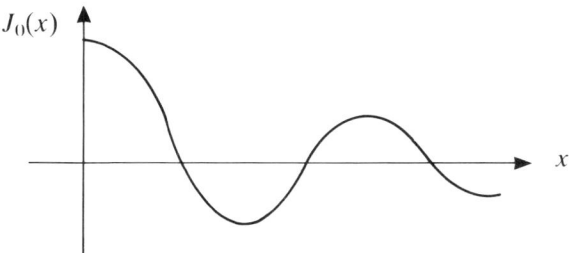

Fig. 8.2 — The Bessel function of order 0.

A second important case of the Bessel equation is the case $\nu = \frac{1}{2}$. In this case, we make the substitution $g(x) = x^{1/2}f(x)$, so that $f(x) = x^{-1/2}g(x)$. Then, with the notation $y' = dy/dx$, we have that $f' = x^{-1/2}g' - \frac{1}{2}x^{-3/2}f$ and $f'' = x^{-1/2}g'' - x^{-3/2}g' + \frac{3}{4}x^{-5/2}g$. Substituting for f, f' and f'' in (8.5) with $\nu = \frac{1}{2}$ yields

$$(x^{-1/2}g'' - x^{-3/2}g' + \tfrac{3}{4}x^{-5/2}g) + (1/x)(x^{-1/2}g' - \tfrac{1}{2}x^{-3/2}g)$$
$$+ (1 - 1/4x^2)x^{-1/2}g = 0,$$

or just

$$x^{-1/2}(g'' + g) = 0.$$

But this is the harmonic equation, with solutions $g = A\cos x + B\sin x$. Hence the general solution to (8.5) with $\nu = \frac{1}{2}$ is $x^{-1/2}(A\cos x + B\sin x)$. In fact, we have $J_{1/2}(x) = (\tfrac{1}{2}x\pi)^{-1/2}\sin x$.

The values of Bessel functions may be found from tables, in the same way as the values of $\sin(x)$ and $\cos(x)$. For small values of x, $J_0(x)$ may be calculated quite efficiently using (8.6).

8.3 Fourier–Bessel series

The Bessel function J_0 can be used to construct a sequence of orthogonal functions, analogous to the functions $\{\cos(nx)\}$. Suppose the roots of $J_0(x)$ are $\lambda_1, \lambda_2, \ldots$; then the set of functions $\{J_0(\lambda_r x)\}$ satisfies the orthogonality relations

$$\int_0^1 x J_0(\lambda_m x) J_0(\lambda_r x) \, dx = 0 \quad \text{if } m \neq n,$$

and

$$\int_0^1 x J_0(\lambda_r x)^2 \, dx = \tfrac{1}{2} J_0'(\lambda_n)^2 \ . \tag{8.7}$$

To prove the first of these, put $f(x) = J_0(\lambda_m x)$ and $g(x) = J_0(\lambda_r x)$. From equation (8.5), f and g satisfy the equations

$$(xf')' + \lambda_m^2 x f = 0,$$
$$(xg')' + \lambda_n^2 x g = 0, \tag{8.8}$$

where $y' = dy/dx$. Since $J_0'(0) = 0$, it follows that $f'(0) = g'(0) = 0$. Moreover, since λ_m and λ_n are roots of J_0, $f(1) = g(1) = 0$.

Multiply the first of equations (8.8) by g and the second by f, and subtract:

$$(xf')'g - (xg')'f + (\lambda_m^2 - \lambda_n^2) x f g = 0. \tag{8.9}$$

Now let $u = xf'g - xg'f$; then $u(0) = u(1) = 0$. Also,

$$\begin{aligned} u' &= (xf')'g + (xf')g' - (xg')'f - (xg')f' \\ &= (xf')'g - (xg')'f \\ &= (\lambda_n^2 - \lambda_m^2) x f g, \end{aligned} \tag{8.10}$$

from (8.8). Integrating this from 0 to 1 gives

$$(\lambda_n^2 - \lambda_m^2) \int_0^1 x f g = u(1) - u(0) = 0 \ . \tag{8.11}$$

Upon division by $\lambda_n^2 - \lambda_m^2$ (which is nonzero, since $m \neq n$), we are left with the first of equations (8.7).

Next, note that

$$\begin{aligned} (x^2 g^2)' &= 2x g^2 + 2x^2 g g' \\ &= 2x g^2 - (xg')(2/\lambda_n^2)(xg')', \end{aligned}$$

as $xg = (1/\lambda_n^2)(xg')'$ (from equation (8.8)). But

$$((xg')^2)' = 2(xg')(xg')';$$

hence we may write this as

$$(x^2g^2)' = 2xg^2 - (1/\lambda_n^2)((xg')^2)'. \tag{8.12}$$

Now x^2g^2 and $(xg')^2$ are both zero at $x=0$. Also, since $g(1)=0$, x^2g^2 is zero at $x=1$ too. Thus integration of (8.12) yields

$$2\int_0^1 xg^2 = (1/\lambda_n^2)(g'(1))^2.$$

Since $g'(x) = \lambda_n J_0'(\lambda_n x)$, this is the second of equations (8.7).

Now suppose $f(x)$ is a function on $0 \leq x \leq 1$. The *Fourier–Bessel* series generated by f is the series

$$f(x) \to \sum_{n=1}^{\infty} A_n J_0(\lambda_n x),$$

where

$$A_n = (2/J_0'(\lambda_n)^2)\int_0^1 xf(x)J_0(\lambda_n x)dx. \tag{8.13}$$

If f has Fourier–Bessel *expansion* (that is, if $f(x)$ can be written in the form $f(x) = \sum_{n=1}^{\infty} \alpha_n J_0(\lambda_n x)$), then the series of (8.13) gives that expansion. To see this, multiply the expansion by $xJ_0(\lambda_m x)$ and integrate from 0 to 1:

$$\int_0^1 xf(x)J_0(\lambda_m x)\,dx = \int_0^1 xJ_0(\lambda_m x)\left\{\sum_{n=1}^{\infty} \alpha_n J_0(\lambda_n x)\right\}dx$$

$$= \sum_{n=1}^{\infty} \alpha_n\left[\int_0^1 xJ_0(\lambda_m x)J_0(\lambda_n x)\,dx\right];$$

by the orthogonality relations (8.7), this simplifies to

$$\int_0^1 xf(x)J_0(\lambda_m x)\,dx = \alpha_m J_0'(\lambda_m)^2/2,$$

and therefore $\alpha_m = A_m$.

The range of functions f which have Fourier–Bessel expansions is the same as the range of functions having classical (trigonometric) Fourier expansions.

Bessel functions of orders other than zero can also be used to construct orthogonal sequences in a similar manner.

8.4 Circularly symmetric waves

Let us now examine a real, and common, situation in which the Fourier–Bessel series may usefully be applied. The situation is that of circularly symmetric waves on a circular membrane with fixed edge — perhaps the vibration of a drumskin.

The analysis of waves on membranes follows an argument similar to that used for

188　　　　　　　　　　　　　**Fourier theory**　　　　　　　　　　　　　[Pt. I]

the vibrating string (section 1.8). The resulting differential equation is the two-dimensional analogue of equation (1.32'), the *two-dimensional wave equation*:

$$\frac{\partial^2 f}{\partial x^2} + \frac{\partial f^2}{\partial y^2} = \frac{1}{c^2}\frac{\partial f^2}{\partial t^2} \tag{8.14}$$

(see the Project at the end of Chapter 4). Because of the circular symmetry of the system, it is convenient to change to a polar coordinate system, $x = r\cos\theta$, $y = r\sin\theta$. In this new coordinate system, the boundary condition is easier to state, becoming simply that $f = 0$ at $r = l$ (l being the radius of the membrane).

Using the chain rule for partial derivatives,

$$r\frac{\partial f}{\partial r} = r\left(\frac{\partial x}{\partial r}\frac{\partial f}{\partial x} + \frac{\partial y}{\partial r}\frac{\partial f}{\partial y}\right)$$

$$= r\left(\cos\theta\frac{\partial f}{\partial x} + \sin\theta\frac{\partial f}{\partial y}\right)$$

$$= x\frac{\partial f}{\partial x} + y\frac{\partial f}{\partial y} \tag{8.15}$$

and therefore

$$r\frac{\partial}{\partial r}\left(r\frac{\partial f}{\partial r}\right) = x\frac{\partial}{\partial x}\left(x\frac{\partial f}{\partial x} + y\frac{\partial f}{\partial y}\right) + y\frac{\partial}{\partial y}\left(x\frac{\partial f}{\partial x} + y\frac{\partial f}{\partial y}\right)$$

$$= x\frac{\partial f}{\partial x} + x^2\frac{\partial^2 f}{\partial x^2} + 2xy\frac{\partial^2 f}{\partial x\partial y} + y\frac{\partial f}{\partial y} + y^2\frac{\partial^2 f}{\partial y^2}\,; \tag{8.16}$$

similarly,

$$\frac{\partial f}{\partial \theta} = \frac{\partial x}{\partial \theta}\frac{\partial f}{\partial x} + \frac{\partial y}{\partial \theta}\frac{\partial f}{\partial y}$$

$$= -r\sin\theta\frac{\partial f}{\partial x} + r\cos\theta\frac{\partial f}{\partial y}$$

$$= -y\frac{\partial f}{\partial x} + x\frac{\partial f}{\partial y}, \tag{8.17}$$

and therefore

$$\frac{\partial^2 f}{\partial \theta^2} = -y\frac{\partial}{\partial x}\left(-y\frac{\partial f}{\partial x} + x\frac{\partial f}{\partial y}\right) + x\frac{\partial}{\partial y}\left(-y\frac{\partial f}{\partial x} + x\frac{\partial f}{\partial y}\right)$$

$$= y^2\frac{\partial^2 f}{\partial x^2} - y\frac{\partial f}{\partial y} - 2xy\frac{\partial^2 f}{\partial x\partial y} - x\frac{\partial f}{\partial x} + x^2\frac{\partial^2 f}{\partial y^2}\,. \tag{8.18}$$

Adding (8.16) to (8.18) gives

$$r\frac{\partial}{\partial r}\left(r\frac{\partial f}{\partial r}\right) + \frac{\partial^2 f}{\partial \theta^2} = (x^2 + y^2)\left(\frac{\partial^2 f}{\partial x^2} + \frac{\partial^2 f}{\partial y^2}\right)$$

or, since $r^2 = x^2 + y^2$,

$$\frac{\partial^2 f}{\partial x^2} + \frac{\partial^2 f}{\partial y^2} = \frac{\partial^2 f}{\partial r^2} + \frac{1}{r}\frac{\partial f}{\partial r} + \frac{1}{r^2}\frac{\partial^2 f}{\partial \theta^2} \tag{8.19}$$

The equation of motion of the system, equation (8.14), now becomes

$$\frac{\partial^2 f}{\partial r^2} + \frac{1}{r}\frac{\partial f}{\partial r} + \frac{1}{r^2}\frac{\partial^2 f}{\partial \theta^2} = \frac{1}{c^2}\frac{\partial^2 f}{\partial t^2} \tag{8.20}$$

In order to solve this, we first look for separated solutions of the form $f(r,\theta,t) = R(r)\Theta(\theta)T(t)$ (cf. equation (1.34)). Then $\partial f/\partial r = \Theta T(dR/dr)$, etc.; and (8.20) becomes

$$\Theta T \frac{d^2 R}{dr^2} + \frac{\Theta T}{r}\frac{dR}{dr} + \frac{RT}{r^2}\frac{d^2 \Theta}{d\theta^2} = \frac{R\Theta}{c^2}\frac{d^2 T}{dt^2}.$$

Dividing through by $R\Theta T$, we get

$$\frac{1}{R}\left(\frac{d^2 R}{dr^2} + \frac{1}{r}\frac{dR}{dr}\right) + \frac{1}{r^2 \Theta}\frac{d^2 \Theta}{d\theta^2} = \frac{1}{c^2 T}\frac{d^2 T}{dt^2} \tag{8.21}$$

Now the right-hand side is a function of t only, and the left-hand side does not depend on t. Hence $(1/c^2 T)(d^2 T/dt^2)$ is constant. If this constant is negative, say $-\omega^2$, then we have

$$\frac{d^2 T}{dt^2} + (\omega c)^2 T = 0. \tag{8.22}$$

This is the harmonic equation, and has the general solution $T = A\cos(\omega c t + \varphi)$. If the constant is positive, then T is exponential; this can be ruled out on physical grounds. Also from (8.21), then, we get the equation

$$\frac{1}{R}\left(\frac{d^2 R}{dr^2} + \frac{1}{r}\frac{dR}{dr}\right) + \frac{1}{r^2 \Theta}\frac{d^2 \Theta}{d\theta^2} = -\omega^2.$$

Multiplying through by r^2 and rearranging gives

$$\frac{r^2}{R}\left(\frac{d^2 R}{dr^2} + \frac{1}{r}\frac{dR}{dr}\right) + \omega^2 r^2 = \frac{-1}{\Theta}\frac{d^2 \Theta}{d\theta^2} \tag{8.23}$$

Here again, the right-hand side is a function of θ only, and the left-hand side is a function of r only. Hence both sides are constant. Since Θ must be periodic (of period 2π), it cannot be exponential. This constant must therefore be positive, say v^2; for then $d^2\Theta/d\theta^2 + v^2\Theta = 0$, which is again the harmonic equation. Now the solution of this is $\Theta = B\cos(v\theta + \psi)$; for Θ to have period 2π, v must be an integer. We can take v to be positive, for definiteness.

From the left-hand side of equation (8.23) we now have

$$\frac{r^2}{R}\left(\frac{d^2R}{dr^2} + \frac{1}{r}\frac{dR}{dr}\right) + \omega^2 r^2 = v^2;$$

multiply through by R/r^2 and rearrange to get

$$\frac{d^2R}{dr^2} + \frac{1}{r}\frac{dR}{dr} + \left(\omega^2 - \frac{v^2}{r^2}\right)R = 0. \tag{8.24}$$

Put $s = \omega r$. Then $dR/dr = \omega(dR/ds)$, $d^2R/dr^2 = \omega^2(d^2R/ds^2)$, and (8.24) becomes

$$\frac{d^2R}{ds^2} + \frac{1}{s}\frac{dR}{ds} + \left(1 - \frac{v^2}{s^2}\right)R = 0. \tag{8.25}$$

This is the Bessel equation of order v. The only solutions that are possible physically are those that are *bounded*; it follows that $R = CJ_v(s)$, and so $R = CJ_v(\omega r)$. Combining this with the solutions for Θ and T gives

$$f(r,\theta,t) = \alpha J_v(\omega r)\cos(v\theta + \psi)\cos(\omega ct + \varphi) \tag{8.26}$$

as the general *separated* solution to (8.20) (where $\alpha = ABC$).

For a solution which is *circularly symmetric*, i.e. not dependent on θ, we need to take $v = 0$. Then

$$f(r,t) = \alpha J_0(\omega r)\cos(\omega ct + \varphi). \tag{8.27}$$

Finally, we apply the boundary condition, that $f(l,t) = 0$. It follows that ωl is a root of J_0, so that $\omega = \lambda_n/l$ for some n, and (8.27) becomes

$$f(r,t) = \alpha J_0\left(\frac{\lambda_n r}{l}\right)\cos\left(\frac{\lambda_n ct}{l} + \varphi\right). \tag{8.28}$$

These separated solutions are the (circularly symmetric) normal modes of vibration of the membrane, just as the harmonic expressions of equation (1.37) represent the normal modes of the vibrating string.

A more general solution to the problem can be obtained by superposing these solutions. This is possible because equation (8.20) is linear. Thus

$$f(r,t) = \sum_{n=1}^{\infty} \alpha_n J_0\left(\frac{\lambda_n r}{l}\right)\cos\left(\frac{\lambda_n ct}{l} + \varphi_n\right) \tag{8.29}$$

is a series solution (compare (1.38)). The *initial position* of the membrane is

$$f(r,0) = \sum_{n=1}^{\infty} \alpha_n J_0\left(\frac{\lambda_n r}{l}\right)\cos(\varphi_n); \tag{8.30}$$

this is a Fourier–Bessel series.

Example. A circular drumskin of radius l is distorted conically, with its midpoint displaced by ε, and released at time $t = 0$. What is its subsequent behaviour (assuming that it follows that kinematics of equation (8.14))?

The problem exhibits circular symmetry, so we look for a solution of the form of (8.29). Differentiating,

$$\frac{\partial f}{\partial t} = \sum_{n=1}^{\infty} (-\lambda_n c\alpha_n/l) J_0(\lambda_n r/l) \sin((\lambda_n ct/l) + \varphi_n).$$

Since the drumskin is stationary at $t = 0$, we have that

$$\sum_{n=1}^{\infty} (-\lambda_n c\alpha_n/l) J_0(\lambda_n r/l) \sin(\varphi_n) = 0.$$

Now we multiply this by $rJ_0(\lambda_m r/l)$ and integrate between 0 and l. Because of the orthogonality relations (8.7), all the cross-terms cancel, and we are left with $\sin(\varphi_m) = 0$; hence we may take $\varphi_m = 0$.

Now we have also that $f(r,0) = \varepsilon(1 - (r/l))$ for $0 \leq r \leq l$; in other words, that

$$\sum_{n=1}^{\infty} \alpha_n J_0(\lambda_n r/l) = \varepsilon(1 - (r/l)).$$

Again we multiply by $rJ_0(\lambda_n r/l)$ and integrate; this time, the orthogonality relations give

$$\alpha_n \tfrac{1}{2} l^2 J_0'(\lambda_n) = \varepsilon \int_0^l (r - (r^2/l)) J_0(\lambda_n r/l) \, dr$$

and hence

$$\alpha_n = \frac{2\varepsilon \left\{ \int_0^l (r - (r^2/l)) J_0(\lambda_n r/l) \, dr \right\}}{l^2 J_0'(\lambda_n)^2}$$

(cf. equation (8.13)). Unfortunately, explicit integration of the numerator is not possible, and the α_n must be computed numerically.

B. TRANSFORMS

8.5 The Laplace transform

As Fourier series can be generalized using orthogonal functions, so can Fourier transforms. The most immediate generalization is to *Laplace transforms*.

The Fourier transform of a function $f(x)$ is defined as $Ff(\omega) = \int_{-\infty}^{\infty} f(x) e^{-i\omega x} \, dx$.
It is a complex-*valued* function, but its *argument*, the variable ω, is real. Extending to the case of complex ω is straightforward: taking the i of the exponent into the complex ω, this becomes the *two-sided Laplace transform*:

$$_2Lf(s) = \int_{-\infty}^{\infty} f(x)e^{-sx} \, dx. \tag{8.31}$$

If $s = \alpha + i\beta$, we can calculate that $_2Lf(s) = \int_{-\infty}^{\infty} f(x)e^{-\alpha x}e^{-i\beta x} \, dx$, so that

$$_2Lf(\alpha + i\beta) = F[fe^{-\alpha x}](\beta). \tag{8.32}$$

But for many integrable functions $f(x)$, the product $f(x)e^{-\alpha x}$ is not integrable. This is the case for the function $f(x) = 1/(1 + x^2)$, for example, for all $\alpha \neq 0$. Such a function cannot have a two-sided Laplace transform. Because of this *loss* of generality, two-sided Laplace transforms are not often used.

To avoid this behaviour, we define the (*one-sided*) *Laplace transform*,

$$Lf(s) = \int_0^{\infty} f(x)e^{-sx} \, dx, \tag{8.33}$$

in which the integration is only over *positive* x. In addition, we do not insist on the existence of $Lf(s)$ for *all* complex values of s.

With these limitations, any integrable function $f(x)$ has a Laplace transform — not necessarily for all s, but certainly for every s with $\text{Re}\,s \geq 0$. Moreover, many *non*integrable functions have Laplace transforms too, at least for some s — functions like $f(x) = x$ and $f(x) = e^x$:

Examples (i) If $f(x) = x$, then $Lf(s) = 1/s^2$ for all s with $\text{Re}\,s > 0$.
(ii) If $f(x) = e^x$, then $Lf(x) = 1/(s-1)$ for all s with $\text{Re}\,s > 1$.

(i) If $f(x) = x$, then integration by parts gives

$$Lf(s) = \int_0^{\infty} xe^{-sx} \, dx$$

$$= [x(-1/s)e^{-sx}]_0^{\infty} - \int_0^{\infty} (-1/s)e^{-sx} \, dx$$

$$= [x(-1/s)e^{-sx}]_0^{\infty} - [(1/s^2)e^{-sx}]_0^{\infty}.$$

When $\text{Re}\,s > 0$, $xe^{-sx} \to 0$ as $x \to \infty$, and so

$$\underline{Lf(x) = 1/s^2}.$$

(ii) If $f(x) = e^x$, then

$$Lf(x) = \int_0^{\infty} e^x e^{-sx} \, dx$$

$$= \int_0^{\infty} e^{(1-s)x} \, dx$$

$$= [e^{(1-s)x}/(1-s)]_0^{\infty}.$$

When Res > 1, $e^{(1-s)x} \to 0$ as $x \to \infty$, so that

$$Lf(s) = 1/(s-1).$$

The functions in these examples certainly do not have Fourier transforms, not even 'improper' impulsive ones. Thus a greater range of functions is amenable to Laplace transformation than to Fourier transformation.

Functions $f(x)$ which do not have a Laplace transform $Lf(s)$ for *any* s include those that 'grow' faster than exponentially with x: $f(x) = \exp(\exp(x))$ is one such. However, very few of these arise in physical problems.

There is an inversion theorem for Laplace transforms, though it lacks the elegance of the Fourier inversion theorem. It is as follows: let f be a function whose Laplace transform $Lf(s)$ exists for all s with Res $> k - \varepsilon$. Then

$$f(x) = (1/2\pi i) \int_{k-i\infty}^{k+i\infty} Lf(s) e^{sx} \, ds. \tag{8.34}$$

(Compare (3.4).)

There is also a version of the convolution theorem for the Laplace transform, but it is much less simple than the Fourier convolution theorem.

Because many functions have a Laplace transform but not a Fourier transform, the Laplace transform is more useful for many applications. In particular, it has an established rôle in the solution of certain classes of differential equations (see Chapter 10). However, the Fourier transform still has an important advantage for computational purposes over the Laplace, in the simplicity of its inversion and convolution theorems.

Table 8.1 gives some common functions, together with their Laplace transforms and their region of existence.

8.6 The Mellin transform

Another naturally arising transform is the Mellin transform. This transform is less well-known than the Fourier and Laplace transforms, for its applications are fewer.

The Mellin transform is defined as:

$$Mf(s) = \int_0^\infty x^{s-1} f(x) \, dx. \tag{8.35}$$

Thus, for a positive integer n, $Mf(n+1)$ is the *nth moment* of the function $f(x)$. Moments are useful in a wide variety of places, from mechanics (centre of gravity, moment of inertia) to statistics (mean, standard deviation, skewness), and even in pure mathematics.

If we change the variable of integration in equation (8.35) as $x = e^{-t}$, we have

$$Mf(s) = \int_0^\infty x^{s-1} f(x) \, dx$$

Table 8.1 — Some functions and their Laplace transforms

f	Lf		Exists for:
$\Pi(x)$	$(1-e^{-s})/s$		all s
$\chi_{(a,b)}(x)$	$\begin{cases} 0 \\ (1-e^{-bs})/s \\ (e^{-as}-e^{-bs})/s \end{cases}$	$(a<b\leq 0)$ $(a<0<b)$ $(0\leq a<b)$	all s all s all s
$\Lambda(x)$	$(s-1-e^{-s})/s^2$		all s
$N(x)$	$\sqrt{(2\pi)}\exp(\tfrac{1}{2}s^2)$		all s
$\delta(x)$	$\tfrac{1}{2}I(s)$		all s
$\delta(x-\alpha)$	$e^{-\alpha s}$	$(\alpha>0)$	all s
$H(x)$ or $I(x)$	$1/s$		Re$s>0$
x^n	$n!/s^{n+1}$	(n any integer)	Re$s>0$
x^α	$\Gamma(\alpha+1)/s^{\alpha+1}$		Re$s>0$
$\cos(x)$	$s/(1+s^2)$		Re$s>0$
$\sin(x)$	$1/(1+s^2)$		Re$s>0$
$J_0(x)$	$1/\sqrt{(1+s^2)}$		Re$s>0$
$J_\nu(x)$	$2^\nu\Gamma(\nu+\tfrac{1}{2})/\pi^{1/2}(1+s^2)^{\nu+1/2}$		Re$s>0$
$III(x)$	$\tfrac{1}{2}\coth s$		Re$s>0$
e^x	$1/(s-1)$		Re$s>1$
$\lambda g(x)$	$\lambda Lg(s)$		as g
$g(\lambda x)$	$(1/\lambda)Lg(s/\lambda)$	$(\lambda>0)$	Re$s>\lambda\alpha$, if Lg exists for Re$s>\alpha$
$g'(x)$	$sLg(s)-g(0)$		as g

$$= \int_0^\infty (e^{-t})^{s-1}f(e^{-t})\,d(e^{-t})$$

$$= -\int_{-\infty}^\infty e^{-(s-1)t}f(e^{-t})(-e^{-t})\,dt$$

$$= \int_{-\infty}^\infty f(e^{-t})e^{-st}\,dt.$$

Thus the Mellin transform of $f(x)$ is identical to the two-sided Laplace transform of $g(t)=f(e^{-t})$.

The uses of the Mellin transform are specialized, and we say no more about them.

8.7 Self-inversion and the Hankel transforms

The transforms that we have looked at so far — namely Fourier, one- and two-sided Laplace, and Mellin — are all integral transforms defined by formulae of the form

$$Tf(y) = \int K(x,y)f(x)\,dx. \tag{8.36}$$

The function $K(x,y)$ is called the *kernel* of the transform T. The Fourier sine and Fourier cosine transforms of equations (3.1) are also of this form; so is the operation of convolution, for if $K(x,y) = g(y-x)$, then $Tf(y) = (f*g)(y)$.

In principle, any function $K(x,y)$ of two variables gives rise to an integral transform. In practice, few kernels yield useful transforms.

One thing that makes Fourier transformation useful is that it is invertible (for a wide range of, if not all, functions). That is, a function may be recovered from its Fourier transform, by the inversion theorem. More than that, the Fourier inverse transform is almost identical to the forward transform operator. Precisely, the transform whose kernel is $K(x,y) = (1/\sqrt{(2\pi)})\,e^{-ixy}$ has the property that if

$$g(y) = \int K(x,y)f(x)\,dx$$

then

$$f(x) = \int \overline{K(x,y)} g(y)\,dy.$$

(Compare equation (3.4).)

Such a transform T (i.e. one for which $T\overline{T}f = f$) is called *conjugate-inverse*. If the formulae hold without even the complex conjugation (i.e. $TTf = f$), it is *self-inverse*. Strictly self-inverse transforms include the (normalized) *one-sided cosine* and *one-sided sine transforms*,

$$_1Cf(\omega) = \sqrt{(2/\pi)} \int_0^\infty f(x)\cos(\omega x)\,dx$$

and

$$_1Sf(\omega) = \sqrt{(2/\pi)} \int_0^\infty f(x)\sin(\omega x)\,dx,$$

and the *Hartley transform*

$$Haf(\omega) = (1/\sqrt{(2\pi)}) \int_{-\infty}^\infty f(x)\,(\cos(\omega x) + \sin(\omega x))\,dx.$$

(The similarity between this last and the Fourier transform itself is obvious.)

Of greater interest is a more general set of self-inverse transforms, constructed using the Bessel functions $J_\nu(x)$. The νth-order *Hankel transform* is defined as

$$_\nu Hf(s) = \int_0^\infty xJ_\nu(xs)f(x)\,dx. \tag{8.37}$$

Each of these transforms is self-inverse. (The proof of this fact involves some complicated analysis, and is beyond the scope of this book.)

There is a close relationship between Hankel and Fourier transforms, in several ways. First, note that if $v = \frac{1}{2}$, so that $J_v(x) = (\frac{1}{2}\pi x)^{-1/2}\sin x$, then equation (8.37) becomes

$$_{1/2}Hf(s) = \int_0^\infty x(\tfrac{1}{2}\pi xs)^{-1/2}\sin(xs)f(x)\,dx$$

$$= (\tfrac{1}{2}\pi s)^{-1/2}\int_0^\infty \sin(xs)x^{1/2}f(x)\,dx$$

$$= s^{-1/2}\,_{_1}S[x^{1/2}f](s).$$

Thus

$$_{1/2}H[_{1/2}Hf](x) = x^{-1/2}\,_{_1}S[s^{1/2}s^{-1/2}\,_{_1}S[x^{1/2}f]](x)$$
$$= x^{-1/2}\,_{_1}S\,_{_1}S[x^{1/2}f](x)$$
$$= f(x). \tag{8.38}$$

More deeply, suppose $f(\mathbf{x})$ is a function of two variables; take its two-dimensional Fourier transform $\mathbf{F}_2f(\boldsymbol{\omega})$. Write both $\mathbf{x}$ and $\boldsymbol{\omega}$ in polar coordinates, as $f(r;\theta)$ and $\mathbf{F}_2f(k;\varphi)$. With respect to the angular variables θ and φ, these functions are periodic of period 2π, so we can expand them into (complex) Fourier series

$$f(r;\theta) = \sum_{n=-\infty}^{\infty} f_n(r)e^{in\theta}$$

and

$$\mathbf{F}_2f(k;\varphi) = \sum_{n=-\infty}^{\infty} F_n(k)e^{in\varphi}, \tag{8.39}$$

where the coefficients are given by

$$f_n(r) = (1/2\pi)\int_0^{2\pi} f(r;\theta)e^{-in\theta}\,d\theta$$

and

$$F_n(k) = (1/2\pi)\int_0^{2\pi} \mathbf{F}_2f(k;\varphi)e^{-in\varphi}\,d\varphi \tag{8.40}$$

(equation (2.30)). Now, thanks to the connection between the wave equation in two dimensions and the Bessel equation (section 8.4), we have the relations

$$F_n(k) = {}_nHf_n(k)$$

and, conversely,

$$f_n(r) = {}_nHF_n(r). \tag{8.41}$$

8.8 The Radon transform

In view of this relationship, the Hankel transform of order 0 may be used instead of the two-dimensional Fourier vector transform in situations involving circular symmetry. Its properties, however, are significantly more complicated than those of the vector transform.

8.8 The Radon transform
The final transform that we shall consider in this section has been, until recently, relatively little known: the *Radon transform*. The Radon transform is used in the science of *tomography*, the art of reconstructing a function of several variables from its projections.

The best known application of this is in medicine. Suppose we wish to obtain a three-dimensional image of someone's head. To do this directly would involve taking small samples of tissue from throughout the head, which is not only extremely dangerous, but also clumsy, time-consuming, and awkward. Much easier would be to take a range of X-ray (or similar 'probing') photographs of the head from a variety of angles. But this does not give three-dimensional data, only two-dimensional *projections*. The question then arises: is it possible to reconstruct a three-dimensional map of (say) tissue density in the head from such projections? (A similar problem arises in many engineering systems, in which it is impossible, or at least unpractical, to put meters and sensors within the system.)

Before considering the full three-dimensional problem, let us analyse the two-dimensional case. Suppose we have a two-dimensional object of varying 'density' $f(x,y)$. Any line in the plane may be defined by two parameters, r and θ, in a polar coordinate system: the line $[r,\theta]$ is the line through the point $(r\cos\theta, r\sin\theta)$, and *perpendicular* to the vector $(\cos\theta, \sin\theta)$ (see Fig. 8.3). The perpendicular distance of

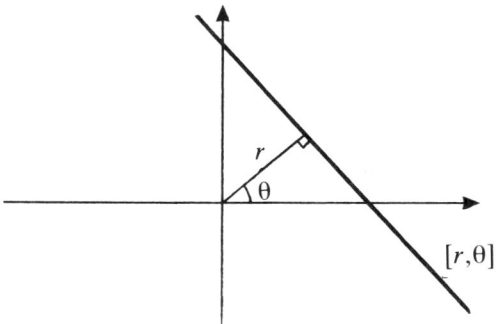

Fig. 8.3 — Representation of a line in the plane.

this line from the origin is then r. Now the cumulative density along the line $[r,\theta]$ is

$$Rf(r,\theta) = \int_{-\infty}^{\infty} f(r\cos\theta - s\sin\theta, r\sin\theta + s\cos\theta) \, ds. \tag{8.42}$$

Equation (8.42) defines the *Radon transform* of $f(x,y)$.

The tomographic problem of reconstructing f from its projections is then the problem of inverting the Radon transform, of recovering $f(x,y)$ from $Rf(r,\theta)$. Fourier analysis provides the methods.

Let us temporarily fix θ, and look at the (one-dimensional) Fourier transform of an individual projection. For convenience, define the vectors $\mathbf{u} = (\cos\theta, \sin\theta)$ and $\mathbf{v} = (-\sin\theta, \cos\theta)$; these are mutually perpendicular and of unit length. Then

$$Rf(r,\theta) = \int_{-\infty}^{\infty} f(r\mathbf{u} + s\mathbf{v}) \, ds,$$

and

$$FRf(\omega,\theta) = \int_{-\infty}^{\infty} Rf(r,\theta) e^{-i\omega r} \, dr$$

$$= \int_{-\infty}^{\infty} \left\{ \int_{-\infty}^{\infty} f(r\mathbf{u} + s\mathbf{v}) \, ds \right\} e^{-i\omega r} \, dr$$

$$= \int_{-\infty}^{\infty} \int_{-\infty}^{\infty} [f(r\mathbf{u} + s\mathbf{v}) e^{-i\omega r}] \, dr \, ds.$$

This integral is now a double integral over the whole plane. If we put $\boldsymbol{\omega} = \omega \mathbf{u}$ and $\mathbf{r} = r\mathbf{u}$, then $\omega r = \boldsymbol{\omega} \cdot \mathbf{r}$, and we therefore have

$$FRf(\omega,\theta) = \int_{-\infty}^{\infty} \int_{-\infty}^{\infty} f(x,y) e^{-i\boldsymbol{\omega} \cdot \mathbf{r}} \, dx \, dy.$$

But this is just the formula for the *two*-dimensional Fourier transform of f, at the point $\boldsymbol{\omega}$! That is,

$$FRf(\omega,\theta) = F_2 f(\omega\cos\theta, \omega\sin\theta). \tag{8.43}$$

This is for a single, fixed θ. Now, if we have all the possible projections $Rf(r,\theta)$, then we can calculate all their Fourier transforms, and therefore obtain the *complete* two-dimensional transform $F_2 f$ of f. To recover $f(x,y)$ from this, we simply apply an inverse Fourier vector transform (section 4.3).

For three-dimensional tomography, such as the reconstruction of the inside of a head, we simply (conceptually) divide the object into thin slices, and analyse each of them separately using the two-dimensional method (see Fig. 8.4). This process is called *thin-slice tomography*.

EXERCISES

1. Prove that the Legendre polynomial p_n of degree n defined in section 8.1 is a polynomial of degree exactly n, with leading term $(2n!)/(n!)^2 2^n$.

Show by repeated integration by parts that $\int_{-1}^{1} x^m p_n(x) \, dx = 0$ for all $m < n$, and that $\int_{-1}^{1} x^n p_n(x) \, dx = ((n!)^2/(2n!)) (2^{2n+1}/(2n+1))$. Hence prove the orthogonality relations between the p_n, equations (8.1).

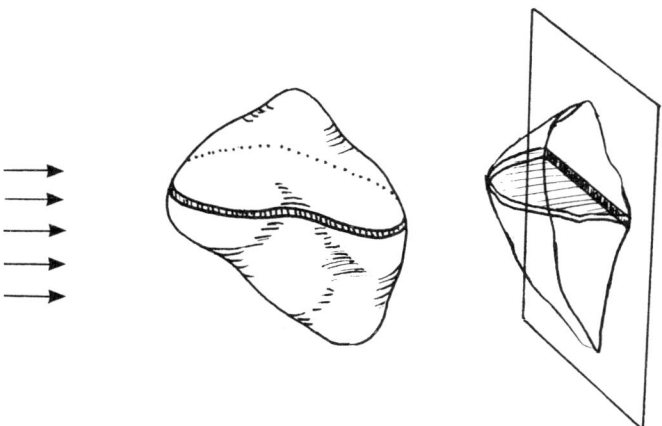

Fig. 8.4 — Two-dimensional projections of a three-dimensional object yield one-dimensional projections of its (two-dimensional) 'slices'.

2. Find, from first principles, the Laplace transforms of the functions
  (i) $f(x) = x^2$;
  (ii) $f(x) = 1/(x+a)$;
  (iii) $f(x) = e^{ax}$; and
  (iv) $f(x) = \sin x$.
For what values of s do they exist?

3. Prove the Laplace transform analogue of the derivative theorem:
$$L\frac{df}{dx}(s) = sLf(s) - f(0). \tag{8.44}$$

Use it and the results of the previous question to find the Laplace transform of the functions $1/(x+a)^n$ for all n, and of $\cos(x)$.

4. Calculate $_2LH(s)$. What happens when $s = \varepsilon + i\omega$ for small ε? How might this help define a 'Fourier transform' for $H(x)$?

5. Show that $Lf(\alpha + i\beta) = F[fHe^{-\alpha x}](\beta)$ for any function f. Hence deduce the Laplace inversion theorem, equation (8.34), from the Fourier inversion theorem, equation (3.4).

6. Find the Radon transform of the two-dimensional bell curve, $N_2(x,y)$ (equation (4.30)). (Hint: N_2 is circularly symmetric.) Hence verify that N_2 satisfies the 'Radon inversion theorem', equation (8.43).

Project. Calculate directly the first three coefficients A_0, A_1, A_2 of the Fourier–Legendre expansion of $f(x) = x^3$. Sketch the graphs of the successive approximations $f_0 = A_0 p_0$, $f_1 = A_0 p_0 + A_1 p_1$, $f_2 = A_0 p_0 + A_1 p_1 + A_2 p_2$, and compare them with the graph of $f(x)$.

Prove that the third approximation f_3 is actually equal to f.

Repeat your calculation for the function $g(x) = H(x)$; again, sketch the graphs of the first few Fourier–Legendre approximations to g. Note that the presence of the discontinuity at $x = 0$ means that *no* approximation is 'good' near the origin.

The Legendre polynomials introduced in the text have orthogonality relations involving integrals over $(-1,1)$. By considering the transformation $y = \frac{1}{2}(x+1)$, construct a sequence of polynomials that satisfy orthogonality relations over $(0,1)$. Find those relations. Calculate explicitly the first few such polynomials. Again, find the first few approximations to the function $f(x)$.

A more usual set of polynomial approximations to a function arises from the Taylor series; thus, $e^x = 1 + x + \frac{1}{2}x^2 + \ldots$ gives rise to the polynomial approximations 1, $1+x$, $1+x+\frac{1}{2}x^2$, and so on. Compare the first four terms of the truncated Taylor series with the first four Fourier–Legendre truncations (over the range $(-1,1)$) for the function $\sin(\frac{1}{2}\pi x)$.

9
Discrete Fourier transforms

A. COMPUTING FOURIER TRANSFORMS
9.1 The discrete Fourier transform

Digital computers are finite machines; any desired computation can only use a finite number of operations. No digital computer, then, can deal with real- (or complex)- valued functions of real numbers, as we have been doing in this book so far. Any such function must be *sampled* in order to be represented in a computer: instead of having $f(x)$ at all x, we have only the values of f at a finite number of points $f(x_1), f(x_2), \ldots, f(x_n)$. (Even these values are rounded off to be stored in finitely many bits, but in practice this is not often a significant limitation.)

For convenience, we usually take the sample points at regular intervals, of τ say. If we also start sampling at $x_1=0$, the sequence of sampling points becomes

$$0, \tau, 2\tau, \ldots, (n-1)\tau \tag{9.1a}$$

and the sample values are, in order,

$$f(0), f(\tau), f(2\tau), \ldots, f((n-1)\tau). \tag{9.1b}$$

This allows us to represent data functions. Next, if we are to use digital computers in Fourier analysis, we need a finite analogue of the Fourier transform too.

In the Fourier transform proper, the calculation involves the integral

$$\int_{-\infty}^{\infty} f(x) e^{-i\omega x} \, dx;$$

the integral is over all x, and is calculated for every real value of ω. In the finite analogue, the integral becomes the sum

$$\sum_{r=0}^{n-1} f(r\tau) e^{-i\omega r\tau}; \tag{9.2}$$

or, with the substitution $z = e^{i\omega\tau}$,

$$Z_{n,\tau} f(z) = \sum_{r=0}^{n-1} f(r\tau) z^{-r}. \tag{9.3}$$

Equation (9.3) defines the *z-transform*, so called because of the traditional name of the transform variable. The *z*-transform is a little like the Fourier series of a periodic function, but it is more restrictive in that it only uses a finite number of terms (compare (2.36)).

But the *z*-transform is not the desired finite analogue. It is defined, still, for all complex z of absolute value 1. We must restrict ourselves, in equation (9.2), to a finite number of values of ω.

Now the function $e^{-ir\omega\tau}$, for fixed τ, is a periodic function of ω with period $2\pi/r\tau$. It follows that the sum in (9.2) is periodic, with period $2\pi/\tau$. We may thus restrict our attention to the range $0 \leq \omega < 2\pi/\tau$. Within this range, we take as many sample points of ω as there are in the *x*-domain, spaced equidistantly. So, corresponding to the *x*-domain sample of equation (9.1a), we use the ω-domain sample:

$$0, \; 2\pi/n\tau, \; 4\pi/n\tau, \; \ldots, \; 2(n-1)\pi/n\tau. \tag{9.4}$$

Equation (9.2) then becomes the *discrete Fourier transform* (or *DFT*):

$$D_{n,\tau}f(2\pi k/n\tau) = (1/n)\sum_{r=0}^{n-1} f(r\tau)e^{-2\pi ikr/n}. \tag{9.5}$$

This transform depends, naturally, on n and τ.

As with 'continuous' Fourier transforms, we can split the complex function $D_{n,\tau}f$ up into *amplitude* and *phase* parts. Alternatively, we can split it into real and imaginary parts, corresponding to the *discrete cosine transform* and the *discrete sine transform* respectively. The former is usual for representational purposes, the latter for computational. The function $|D_{n,\tau}f(r)|^2$ is the *discrete power spectrum* of f.

9.2 Properties of the DFT

The DFT shares many properties with the 'continuous' Fourier transform of Chapter 3 (and indeed with the *z*-transform). It is linear, and there are 'similarity' and 'shift' theorems, analogous to equations (3.31). If $r(x) = f(\lambda x)$, then

$$D_{n,\tau/\lambda}r(2\pi\lambda k/n\tau) = D_{n,\tau}f(2\pi k/n\tau). \tag{9.6}$$

Also, if the sequence

$$f(\lambda\tau), \; f((\lambda+1)\tau), \; \ldots, \; f((n-1)\tau), \; f(0), \; f(\tau), \; \ldots, \; f((\lambda-1)\tau)$$

is regarded as being, *in this order*, an *n*-point sample of a function s (so that $s(0) = f(\lambda\tau)$, $s(\tau) = f((\lambda+1)\tau)$, and so on), then

$$D_{n,\tau}s(2\pi k/n\tau) = e^{2\pi ik\lambda/n}D_{n,\tau}f(2\pi k/n\tau). \tag{9.7}$$

Note the *wraparound*, the 'cyclic' behaviour required of the function s in this last result, compared with equation (3.31b). This is an important feature of the discrete transform, and is a consequence of the ω-periodicity of equation (9.2).

Most importantly, the DFT is invertible. This is where it is crucially important that we have all the information content of the sample in its transform, and why we therefore need an ω-domain sample as big as the *x*-domain sample. The *discrete inversion theorem* says that

Discrete Fourier transforms

$$f(r\tau) = \sum_{k=0}^{n-1} D_{n,\tau} f(2\pi k/n\tau) e^{2\pi i k r/n}. \qquad (9.8)$$

Compare this with equation (3.4). Clearly, the DFT and its inverse are similar; in fact, for $r = 0, 1, \ldots, n-1$

$$D_{n, 2\pi/n\tau} D_{n, \tau} f(r\tau) = \begin{cases} (1/n) \cdot f((n-r)\tau) & (r \neq 0) \\ (1/n) \cdot f(0) & (r = 0) \end{cases}$$

(cf. equation (3.7)). The term $f((n-r)\tau)$ appears here in place of the $f(-r\tau)$ we would expect from equation (3.7) because of the wraparound effect. (The value $f(-r\tau)$ is not part of the sample (9.1)).

To prove the discrete inversion theorem, we first prove the *finite orthogonality conditions*,

$$(1/n) \sum_{r=0}^{n-1} e^{-2\pi i s r/n} e^{2\pi i t r/n} = \begin{cases} 0 & \text{if } s \neq t \\ 1 & \text{if } s = t. \end{cases} \qquad (9.9)$$

The proof is straightforward. The sum is a geometric series, of initial value 1 and constant factor $e^{2\pi i (t-s)r/n}$. When $s = t$, this constant factor is 1, so each term is 1 and the sum is n. When $s \neq t$,

$$\sum_{r=0}^{n-1} e^{2\pi i r(t-s)/n} = (1 - e^{2\pi i (t-s)})/(1 - e^{2\pi i (t-s)/n})$$

$$= 0.$$

Now using these orthogonality conditions, we calculate that, for each r,

$$\sum_{k=0}^{n-1} D_{n,\tau} f(2\pi k/n\tau) e^{2\pi i k r/n} = \sum_{k=0}^{n-1} \left[(1/n) \sum_{s=0}^{n-1} f(s\tau) e^{-2\pi i k s/n} \right] e^{2\pi i k r/n}$$

$$= \sum_{s=0}^{n-1} f(s\tau) \left[(1/n) \sum_{k=0}^{n-1} e^{2\pi i k (r-s)/n} \right]$$

$$= f(r\tau),$$

as stated.

The discrete power transform obeys a finite analogue of the Rayleigh–Plancherel theorem, (5.23):

$$(1/n) \sum_{r=0}^{n-1} |f(r\tau)|^2 = \sum_{k=0}^{n-1} |D_{n,\tau} f(2\pi k/n\tau)|^2. \qquad (9.10)$$

To prove this, we simply multiply equation (9.8) by its complex conjugate and sum over r; using the finite orthogonality relations (9.9),

$$\sum_{r=0}^{n-1} |f(r\tau)|^2 = \sum_{r,s,k=0}^{n-1} D_{n,\tau} f(2\pi k/n\tau) \overline{D_{n,\tau} f(2\pi s/n\tau)} e^{2\pi i r(k-s)/n}$$

$$= \sum_{s,k=0}^{n-1} D_{n,\tau}f(2\pi k/n\tau)\overline{D_{n,\tau}f(2\pi s/n\tau)}\left(\sum_{r=0}^{n-1} e^{2\pi i r(k-s)/n}\right)$$

$$= n\sum_{k=0}^{n-1} |D_{n,\tau}f(2\pi k/n\tau)|^2$$

as required.

As we observed in the previous section,

$$\sum_{r=0}^{n-1} f(r\tau)e^{-2\pi i k r/n} = \sum_{r=0}^{n-1} f(r\tau)e^{-2\pi i[n+k]r/n}. \qquad (9.11)$$

In addition, for real-valued functions $f(x)$,

$$\sum_{r=0}^{n-1} f(r\tau)e^{2\pi i k r/n} = \sum_{r=0}^{n-1} \overline{f(r\tau)e^{-2\pi i k r/n}} \qquad (9.12)$$

(cf. equation (3.32)). Combining these, it follows that for *real*-valued f,

$$D_{n,\tau}f(2\pi[n-k]/n\tau) = \overline{D_{n,\tau}f(2\pi k/n\tau)}. \qquad (9.13)$$

We shall have more to say about this in the next section.

There are two additional properties of the DFT that have no analogue in the continuous case. These are technical properties, related to cases where the data in hand are augmented by zeros, either at the end of the sample sequence or interpolated into it:

(1) Suppose we have n data points, as in (9.1), and we extend these with n zeros; then we have the $2n$ data points

$$f(0), f(\tau), f(2\tau), \ldots, f((n-1)\tau), 0, 0, \ldots, 0.$$

We can regard this as a $2n$-point sample of a function g, sampled every τ. The *packing theorem* is then that

$$D_{2n,\tau}g(2\pi k/2n\tau) = \tfrac{1}{2}D_{n,\tau}f(2\pi\tfrac{1}{2}k/n\tau) \qquad \text{for even } k. \qquad (9.14)$$

The proof is easy, and is left to the reader. (Note that the angular frequency in both sides of (9.14) is $\pi k/n\tau$). The *odd* terms of the transform of g cannot be expressed so simply in terms of $D_{n,\tau}f(2\pi k/n\tau)$; the expressions involve *discrete convolution*.

(2) Suppose again we have n data points of f as in (9.1), and we interpolate zeros:

$$f(0), 0, f(\tau), 0, f(2\tau), 0, \ldots, f((n-1)\tau), 0.$$

This we regard as a $2n$-point sample of a function h, sampled every $\tfrac{1}{2}\tau$. The *stretch theorem* is then that

$$D_{2n,\frac{1}{2}\tau}h(2\pi k/2n\tfrac{1}{2}\tau) = \begin{cases} \tfrac{1}{2}D_{n,\tau}f(2\pi k/n\tau) & (k=0,\ldots,n-1) \\ \tfrac{1}{2}D_{n,\tau}f(2\pi[k-n]/n\tau) & (k=n,\ldots,2n-1). \end{cases} \qquad (9.15)$$

Fig. 9.1 shows the effect of the packing and stretch theorems.

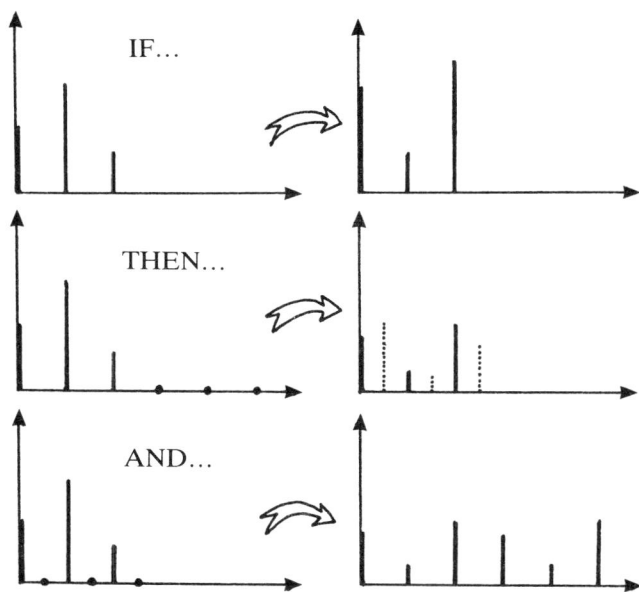

Fig. 9.1 — The 'packing' and 'stretch' theorems illustrated.

9.3 Sampling: resolution and aliasing

Equation (9.11) expresses the periodicity of the transform constructed in (9.2). The period is $2\pi/\tau$. This is why the discrete transform was constructed on an ω-domain of this extent (9.4).

However, for a *real*-valued function f, equation (9.13) says that the $(n-k)$th term of the DFT contains the same information as the kth term. This phenomenon is called *aliasing*. As a consequence, it is impossible, for such a sample, to distinguish a component of angular frequency $2\pi k/n\tau$ from one of angular frequency $2\pi(n-k)/n\tau$. More tendentiously: a component of angular frequency $2\pi(n-k)/n\tau$ will masquerade as a component of angular frequency $2\pi k/n\tau$. Because of this aliasing, the *useful* spectrum of a real-valued function consists of only half of the full spectrum, including angular frequencies up to π/τ only; conventionally, only this half is presented. (The frequency π/τ is called the (*Nyquist*) *folding frequency*.) For *complex*-valued f, this is not true: the information in the upper half of the frequency range does *not* repeat that in the lower half.

The only sure way to avoid aliasing is to know beforehand that every component of the data function has angular frequency in the range $0 \leq \omega < \pi/\tau$. Sometimes it is possible to achieve this by choosing τ appropriately. If it is not known what frequencies are involved in f, or the implied sampling rate τ is unpractical, then the data signal must be *filtered* before sampling, with a low-pass filter of cut-off $\omega_c = \pi/\tau$ (see Chapter 6 and sections 9.8–9.10 below).

In this way, the choice of the *x-resolution* τ determines the *ω-range* π/τ of the DFT. Conversely, the *x-range* — the total sample length, $n\tau$ — determines the ω-*resolution* of components. The angular frequencies in the sequence (9.5) are spaced $2\pi/n\tau$ apart. It follows that if a function f contains harmonic components whose angular frequencies are closer than $2\pi/n\tau$, the analysis will not show it. In fact, things are slightly worse than this, and for practical purposes the ω-resolution of the DFT is twice this: $4\pi/n\tau$. To see why this is so, we must examine more closely the discrete transform of a sinusoid.

9.4 Transform of a sinusoid

Let $f(x) = e^{i\alpha x}$, for some α in the range $2\pi/n\tau \ll \alpha \ll \pi/\tau$. Then $f(r\tau) = e^{i\alpha r\tau}$, and

$$\boldsymbol{D}_{n,\tau}f(2\pi k/n\tau) = (1/n)\sum_{r=0}^{n-1} e^{-ir((2\pi k/n) - \alpha\tau)}.$$

Suppose $\alpha = 2\pi m/n\tau$ for some m. When $k = m$, all the exponents in the right-hand side are zero, and $\boldsymbol{D}_{n,\tau}f(2\pi m/n\tau) = 1$. This is to be expected; $f(x)$ contains a harmonic component $e^{2\pi i m/n\tau}$, for this *is* f. At other frequencies, the sum on the right-hand side is treated as a geometric series:

$$\boldsymbol{D}_{n,\tau}f(2\pi k/n\tau) = (1/n)(1 - e^{-in((2\pi k/n) - \alpha\tau)})/(1 - e^{-i((2\pi k/n) - \alpha\tau)})$$

$$= (1 - e^{in\alpha\tau})/n(1 - e^{-i((2\pi k/n) - \alpha\tau)})$$

$$= e^{(1/2)i(n-1)\alpha\tau} e^{i\pi k/n} \sin(\tfrac{1}{2}n\alpha\tau)/n\sin(\tfrac{1}{2}\alpha\tau - (\pi k/n)). \qquad (9.16)$$

Note that if $\alpha = 2\pi m/n\tau$ and $k \neq m$, the numerator contains a factor $\sin(\tfrac{1}{2}n\alpha\tau) = \sin(\pi m) = 0$, while the denominator is nonzero. Hence for this α, the mth term of the discrete transform is 1, and all the other terms are zero.

For any α, though, (9.16) gives the discrete transform of $f(x) = e^{i\alpha x}$; the corresponding power spectrum is

$$|\boldsymbol{D}_{n,\tau}f(2\pi k/n\tau)|^2 = [\sin(\tfrac{1}{2}n\alpha\tau)/n\sin(\tfrac{1}{2}\alpha\tau - \pi k/n)]^2. \qquad (9.17)$$

From (9.16), we can deduce the discrete spectrum of any real-valued sinusoid, using the formula $\cos(\alpha x + \varphi) = \tfrac{1}{2}(e^{i(\alpha x + \varphi)} + e^{-i(\alpha x + \varphi)})$.

In equation (9.17), the kth component is small (of order $1/n^2$) unless the denominator $\sin(\tfrac{1}{2}\alpha\tau - \pi k/n)$ is small; that is, unless $\tfrac{1}{2}\alpha\tau - (\pi k/n)$ is small. Then, if $n\alpha\tau/2\pi = m + \varepsilon$, it follows that $\sin(\tfrac{1}{2}\alpha\tau - \pi m/n) = \pm\sin(\pi\varepsilon/n) \simeq \pm\pi\varepsilon/n$, and that $\sin(\tfrac{1}{2}n\alpha\tau) = \pm\sin(\pi\varepsilon)$, so that

$$|\boldsymbol{D}_{n,\tau}f(2\pi m/n\tau)|^2 \simeq (\sin(\pi\varepsilon)/\pi\varepsilon)^2.$$

Thus, if α lies between the mth and $(m+1)$th angular frequencies, the DFT of $e^{i\alpha x}$ will have large mth and $(m+1)$th components, and the remainder will be small. Further, if α is closer to $2\pi m/n\tau$ than to $2\pi(m+1)/n\tau$, the mth component will be larger than the $(m+1)$th; while if α is closer to $2\pi(m+1)/n\tau$, then this term will be the larger.

Fig. 9.2 shows the discrete power spectrum (9.17) of a sinusoidal input.

(Compare equation (9.17) with the Dirichlet kernel D_n introduced in section 2.14. In terms of $D_n(x) = \sin[(n + \tfrac{1}{2})x]/\sin[\tfrac{1}{2}x]$,

$$|\boldsymbol{D}_{n,\tau}f(2\pi k/n\tau)|^2 = [(1/n)D_{(1/2)(n-1)}(\alpha\tau - 2\pi k/n)]^2.)$$

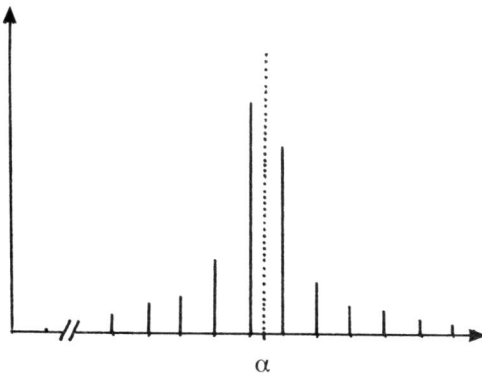

Fig. 9.2 — A sinusoid at a frequency between two 'bins' will spread power across the whole (discrete) spectrum.

Because of this possibility of a single-frequency component being 'shared' between (effectively) two terms, it is not really true to say that the transform has a ω-resolution of $2\pi/n\tau$. If we perform a DFT on a signal and find significant amplitudes in consecutive components, we *could* interpret this as two sinusoidal components whose angular frequencies differ by (exactly) $2\pi/n\tau$; but equally, it *could* reflect a single harmonic component, lying somewhere between the two. Since the latter is the simpler explanation, we are compelled to prefer it.

It follows that we cannot be sure of the existence of *two* components unless there is at least one term of the DFT between them (as in Fig. 9.3). Thus the ω-resolution of

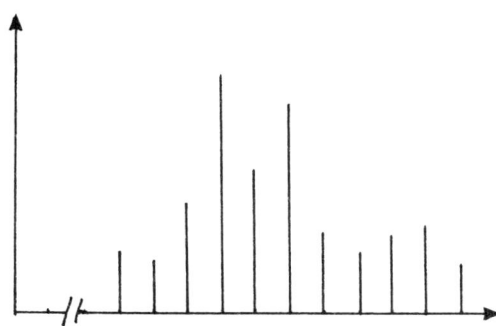

Fig. 9.3 — These two components are just barely resolvable.

the DFT is, in practice, only half as good as its theoretical value: $4\pi/n\tau$ rather than $2\pi/n\tau$.

We have seen that a sinusoid can have a complicated DFT. It is useful, then, to be able to estimate the amplitude of the original sinusoid from its DFT component amplitudes. Suppose we have a spectrum like that in Fig. 9.2, in which a real-valued function shows a peak on two successive frequencies: the mth and $(m+1)$th, say. (Because of aliasing, there will be two corresponding terms above the Nyquist folding frequency, but because only the lower half of the frequency range is shown, these do not appear in the spectrum.) We interpret this dual peak as representing a single sinusoid, of intermediate frequency. To estimate the amplitude of this putative sinusoid, we use the finite Rayleigh–Plancherel theorem, (9.10):

$$(1/n)\sum_{r=0}^{n-1} f(r\tau)^2 = \sum_{k=0}^{n-1} |D_{n,\tau}f(2\pi k/n\tau)|^2.$$

Now for $f(x) = A\cos(\alpha x + \varphi)$ with $\pi/n\tau \ll \alpha \ll \pi/\tau$, the left-hand side is

$$(1/n)\sum_{r=0}^{n-1} A^2 \cos^2(\alpha r\tau + \varphi) = (A^2/2n)\sum_{r=0}^{n-1}(1 + \cos[2(\alpha r\tau + \varphi)])$$

$$\simeq \tfrac{1}{2}A^2.$$

Since only the mth and $(m+1)$th components are significant, we may write — not forgetting the aliased peaks — that

$$\tfrac{1}{2}A^2 \simeq |D_{n,\tau}f(2\pi m/n\tau)|^2 + |D_{n,\tau}f(2\pi[m+1]/n\tau)|^2$$
$$+ |D_{n,\tau}f(2\pi[n-m-1]/n\tau)|^2 + |D_{n,\tau}f(2\pi[n-m]/n\tau)|^2;$$

so, in view of (9.13).

$$\tfrac{1}{2}A^2 \simeq 2(|D_{n,\tau}f(2\pi m/n\tau)|^2 + |D_{n,\tau}f(2\pi[m+1]/n\tau)|^2)$$

or $\qquad A \simeq 2\sqrt{(|D_{n,\tau}f(2\pi m/n\tau)|^2 + |D_{n,\tau}f(2\pi[m+1]/n\tau)|^2)}.$ \hfill (9.18)

Thus, the power in the sinusoid is the sum of the powers in the 'bins' among which it is split. Its amplitude is *not* the sum of their amplitudes.

9.5 Fourier transform algorithms

With the DFT defined as we have seen, and with the properties of transforms suitably translated into a finite context, the whole of Fourier analysis now becomes possible on digital computers. But there is a difference between *possible* and *practical*, for the *scale* of the calculation involved is substantial.

The direct way of calculating the DFT of an n-point sample is to rewrite equation (9.5) as

$$(1/n)\sum_{r=0}^{n-1} f(r\tau)\cos(2\pi kr/n) - i(1/n)\sum_{r=0}^{n-1} f(r\tau)\sin(2\pi kr/n).$$

(For the moment we shall assume that f is a real-valued data function. Complex-valued functions behave similarly.) This expression needs to be calculated for each of the n values of k. For each k, there are two sums of $2n$ terms each. Each term is a product of two factors: a value of f and a trigonometric term. The f term involves just one multiplication; the trigonometric term involves two products and a division, followed by the evaluation of a cosine or sine. Finally, there is an extra division by n on each sum.

Thus each trigonometric factor involves four operations, while each f factor involves two — calculating $r\tau$ and looking up the appropriate value of f. Multiplying these is a seventh operation. There are n terms in the sum, so each sum involves $7n + (n - 1)$ operations; a division by n makes this $8n$. Real and imaginary parts both need to be calculated, which doubles the number. Finally, for a complete transform, we need to do all this for $k = 0, 1, \ldots, [n/2]$: we need not go higher, because of aliasing. So the full calculation of $D_{n,\tau}f(2\pi k/n\tau)$ requires $8n^2$ operations.

Now, on a typical microcomputer each operation will take a few milliseconds; less for addition, substantially more for sine and cosine evaluation. Suppose for convenience each operation takes 1 ms. Then to transform a 100-point sample will take 80 s, or one and a third minutes; for a 1000-point sample, the time is over two hours. And these are average-sized spectra.

A faster computer will do better. If we have a system with compiled software and hardware multiplication, and we precalculate a sine function look-up table, we might get the operation time down to 10 μs. A 1000-point transform will still take more than a minute to perform.

Fortunately, there is a way of cutting the time down quite drastically. This is the *fast Fourier transform* (or *FFT*) technique, proposed by Cooley and Tukey in the mid-1960s. Its effects have been revolutionary, turning Fourier analysis from being merely a mathematical tool to being a practical one.

In the Cooley–Tukey algorithm, the number of sample points n must be a power of 2. (Other FFT algorithms have since been proposed that obviate this requirement, but they are less elegant. In any case, the requirement is not an onerous one.) This is necessary because the FFT method is recursive. At each stage, we divide the data in two halves, perform a (smaller) FFT on each half, and then splice the two halves back together. The splicing is quite easy; it relies on the 'shift', 'packing', and 'stretch' theorems, equations (9.7), (9.14), and (9.15).

Suppose we have a sample of $2n$ points of a function (the sampling interval is not important), whose values are $a_0, a_1, \ldots, a_{2n-1}$, and we want to calculate their $2n$-point DFT. First we split the sequence into the sum of the two sequences

$$(a_0, 0, a_2, \ldots, a_{2n-2}, 0) + (0, a_1, 0, \ldots, 0, a_{2n-1}). \qquad (9.19)$$

By linearity, the transform of our original sequence is the sum of the transforms of these two. Next, we perform two (n-point) transforms on the sequences

$$(a_0, a_2, \ldots, a_{2n-2}) \text{ and } (a_1, a_3, \ldots, a_{2n-1});$$

suppose these transforms are

$$(A_0, A_2, \ldots, A_{2n-2}) \text{ and } (A_1, A_3, \ldots, A_{2n-1})$$

respectively. Now the ($2n$-point) transforms of the two sequences in (9.19) are given by the stretch theorem and (for the second sequence) the shift theorem as, respectively,

$$(A_0, A_2, \ldots, A_{2n-2}, A_0, A_2, \ldots, A_{2n-2})$$

and

$$(A_1, A_3 e^{-\pi i/n}, \ldots, A_{2n-1} e^{-(n-1)\pi i/n}, A_1 e^{-\pi i}, A_3 e^{-(n+1)\pi i/n}, \ldots, A_{2n-1} e^{\pi i/n}). \quad (9.20)$$

Then the transform of the original sequence $(a_0, a_1, \ldots, a_{2n-1})$ is the sum of these:

$$(A_0 + A_1, A_2 + A_3 e^{-\pi i/n}, \ldots, A_{2n-2} + A_{2n-1} e^{\pi i/n}). \quad (9.21)$$

Thus we have reduced a $2n$-point transform to two n-point transforms and some elementary arithmetic.

For a transform of 2^k points, we first reduce to two 2^{k-1}-point transforms; each of these is reduced to two 2^{k-2}-point transforms; and so on. Eventually, we are left with 2^k one-point transforms. But a one-point transform is trivial; there is a single data point and a single component, for zero frequency, so the transform of (a_0) is (a_0). (From the definition, $D_{1,\tau} f(2\pi 0/\tau) = f(0) e^{-2\pi i 0} = f(0)$ for any function $f(x)$.)

In an FFT programme, then, the data sequence is first converted to a sequence of two-point transforms, then to four-point transforms, and so on. Fig. 9.4 gives such a programme. (This programme overwrites the original data array. To retain the original data, the array should be copied first.)

The number of operations required is significantly smaller than in the direct calculation. It may be shown that a 2^k-point FFT takes about $12k \times 2^k$ operations. Putting $n = 2^k$, this means about $12n\log_2 n$. So for a microcomputer capable of one operation per millisecond, a 128-point FFT will take just 10 s, while a 1024-point FFT takes about two minutes.

Nowadays, indeed, the design of digital microcircuitry has advanced to the point where single chips can be designed to perform FFTs. Because they are dedicated chips, they can be hard-wired in the most efficient way. Such a processor can perform a 1024-point FFT in a fraction of a second. With this sort of power on hand, taking Fourier transforms is no more tedious than pushing a button on a calculator.

B. USING DISCRETE TRANSFORMS

9.6 Discrete convolutions

One of the most important uses of the Fourier transform is in the analysis of convolutions, via the convolution theorem. There is a discrete analogue to this.

The convolution of two functions is given by $(f*g)(x) = \int_{-\infty}^{\infty} f(x-y)g(y)\,dy$ (equation (5.2)). In the discrete case, we would expect this to become something like

$$(f *_D g)(r\tau) = \sum_{s=0}^{n-1} f(r\tau - s\tau) g(s\tau), \quad (9.22a)$$

```
2^k→n
⟨data⟩→X(0), ..., X(n – 1)
procedure: transform
local: j,l,m,p,r,z
n/2→m
for j = 1 to k
    for l = 0 to n – 1 step 2*m
        for p = l to l + m – 1
            bitrev(p)→j
            X(p + m)*exp(2πijm/n)→z
            X(p) – z→X(p + m)
            X(p) + z→X(p)
        next p
    next l
    m/2→m
next j
for j = 0 to n – 1
    bitrev(j)→l
    if l>j then X(j)/n→z; X(l)/n→X(j); z→X(l)
next j
endprocedure

function: bitrev(t)
local: u,v
0→u
for q = 1 to k
    int(t/2)→v
    2*u + (t – 2*v)→u
next q
return u
```

Fig. 9.4 — A fast Fourier transform program. This programme is in a pseudocode that allows for X(·) and z to be complex-valued variables. For a language involving real-valued variables only, real and imaginary parts must be stored separately, and the arithmetic is more complex.

for $r = 0, 1, \ldots, n-1$. But there is a problem. The variable $r\tau - s\tau$ might be anything from $-(n-1)\tau$ to $(n-1)\tau$, depending on r and s, while the sample of f only provides $f(k\tau)$ with $k = 0, 1, \ldots, (n-1)$.

The clue to the resolution of this problem is the cyclic view of the sampling procedure. In this view, f and g are thought of as periodic functions of period $n\tau$. It follows that we treat $f(-\tau)$ as $f((n-1)\tau)$, $f(-2\tau)$ as $f((n-2)\tau)$, and so on. So equation (9.22a) is replaced by

$$(f*_\mathrm{D} g)(r\tau) = \sum_{s=0}^{n-1} f([r-s]'\tau)g(s\tau), \qquad (9.22b)$$

where $[r-s]' = (r-s) \bmod n$. For example, if $n = 4$,

$$(f*_\mathrm{D} g)(0) = f(0)g(0) + f(3\tau)g(\tau) + f(2\tau)g(2\tau) + f(\tau)g(3\tau),$$
$$(f*_\mathrm{D} g)(\tau) = f(\tau)g(0) + f(0)g(\tau) + f(3\tau)g(2\tau) + f(2\tau)g(3\tau),$$
$$(f*_\mathrm{D} g)(2\tau) = f(2\tau)g(0) + f(\tau)g(\tau) + f(0)g(2\tau) + f(3\tau)g(3\tau),$$

and $$(f*_\mathrm{D} g)(3\tau) = f(3\tau)g(0) + f(2\tau)g(\tau) + f(\tau)g(2\tau) + f(0)g(3\tau).$$

Given this definition of *discrete convolution*, it is straightforward to prove a discrete analogue of the convolution theorem; namely,

$$D_{n,\tau}(f *_D g)(2\pi k/n\tau) = n D_{n,\tau} f(2\pi k/n\tau) D_{n,\tau} g(2\pi k/n\tau). \qquad (9.23)$$

The proof is similar to the continuous case. Its converse is that, for n-point samples of two functions f and g,

$$D_{n,\tau}(fg)(2\pi k/n\tau) = D_{n,\tau} f(2\pi k/n\tau) *_D D_{n,\tau} g(2\pi k/n\tau). \qquad (9.24)$$

But the definition of $*_D$ in (9.22b) now shows *explicitly* the wraparound that is implicit in the DFT and its inverse. For many practical purposes, this is unacceptable. We can get around this problem by using the packing theorem (9.14). Suppose we have samples of f and g of n points, as usual, and we append a sequence of n zeros on to each sequence:

$$f(0), f(\tau), f(2\tau), \ldots, f((n-1)\tau), 0, 0, \ldots, 0$$

and $\quad g(0), g(\tau), g(2\tau), \ldots, g((n-1)\tau), 0, 0, \ldots, 0 \qquad (9.25)$

Treat these as $2n$-point samples of functions F and G. Then, by equation (9.22b), the discrete convolution of these sequences is

$$(F *_D G)(r\tau) = \sum_{s=0}^{r} f(s\tau)g([r-s]\tau). \qquad (9.26)$$

The packing theorem relates the DFTs of the sequences (9.25) to the original, n-point, sequences, and the discrete convolution theorem (9.23) gives the transform of (9.26).

Finally, the concept of the *transfer function of a filter* (cf. section 5.10) has an obvious discrete counterpart; and it is the DFT of the (sampled) *impulse response function* of the filter. Deconvolution, then, may be achieved as in the continuous case.

Thus, the DFT may be used to calculate discrete convolutions. Now we saw above how the FFT algorithm cuts down the amount of computation needed to produce a Fourier transform, from $\sim n^2$ operations to $\sim n\log_2 n$. There is a corresponding advantage when it is used for the indirect calculation of convolutions. This is important, for there can be no direct 'fast convolution algorithm', as the products $f(s\tau)g(r\tau - s\tau)$ are all distinct. Thus, a direct calculation of $f *_D g$ for n-point samples must take $\sim n^2$ operations. On the other hand, using (9.23) and the FFT reduces the problem to two forward DFTs ($\sim n\log_2 n$ operations each), a complex product at each frequency (totalling a few times n operations), and an inverse DFT to recover $f *_D g$ from its transform. If the *cyclic* convolution is not desired, then we can use the method of (9.25) and (9.26). As with the transforms themselves, this leads to a substantial saving in time, especially when n is large.

9.7 Windowing

One consequence of the cyclic nature of the DFT is the creation of an artificial jump at the endpoints of the data sequence. This is most evident in the discrete shift theorem, (9.7). A function that is otherwise smooth may yet have significantly different endpoints, and this 'looks like' a discontinuity in the function to the DFT. This artificial effective discontinuity is the source, for example, of the 'lobes' on the DFT of a sine wave which fails to fit a whole number of waves into the sample length (see Fig. 9.5).

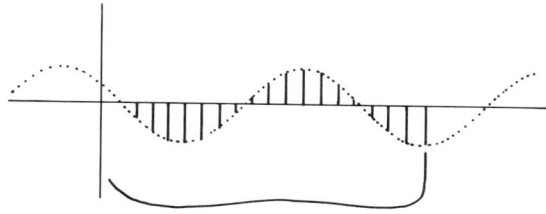

Fig. 9.5 — The apparent discontinuity in the sinusoid, caused by the 'wraparound' effect. This discontinuity is largely responsible for the spread of power in Fig. 9.2.

One way to ease these problems is to treat the data sequence before applying the DFT. The sampled function (9.1) is like a snapshot of the true function $f(x)$, with the shutter opened suddenly at $x = 0$ and shut suddenly at $x = n\tau$. Opening and closing the shutter more gently has its own distorting effect, but it at least avoids the creation of such artificial discontinuities. Thus, we multiply the sequence

$$f(0), f(\tau), \ldots, f((n-1)\tau),$$

by a *window function* $W(x)$, such as

$$W(x) = \begin{cases} 2x/n\tau & (x \leq \tfrac{1}{2}n\tau) \\ 2 - 2x/n\tau & (x > \tfrac{1}{2}n\tau) \end{cases} \tag{9.27}$$

(see Fig. 9.6), and *then* perform the DFT. (Analysing without windowing is then the

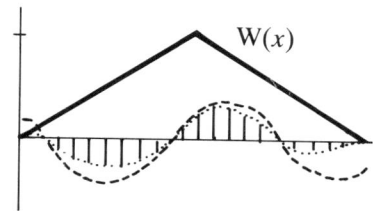

Fig. 9.6 — A typical window function, and its effect on the sinusoid of Fig. 9.5. The signal is distorted, but the discontinuity has been abolished.

same as analysing using the *uniform window function*, $W(x) = 1$.)

The effect of windowing the input to a DFT is given by (9.24) as

$$D_{n,\tau}(fW)(2\pi k/n\tau) = D_{n,\tau}f(2\pi k/n\tau) *_D D_{n,\tau}W(2\pi k/n\tau).$$

Fig. 9.7 shows the effect of applying the window of (9.27) to a sine wave.

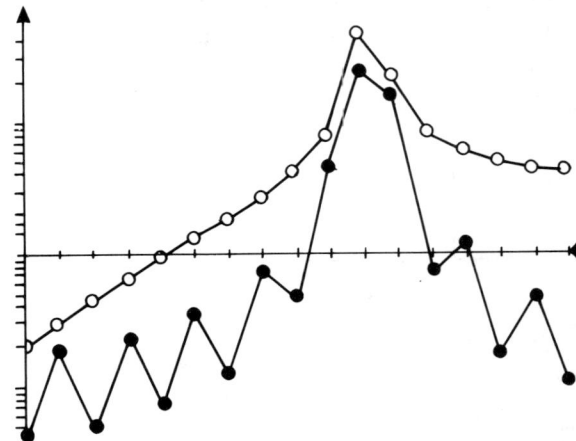

Fig. 9.7 — The effect of windowing. The figure shows the amplitude spectrum of a sinusoid, unwindowed (open circles) and windowed with the window function of Fig. 9.6 (filled circles). The vertical scale is logarithmic.

9.8 Coherence of sampled functions

Let m separate n-point samples be simultaneously taken, at intervals of τ, of two functions f and g, giving sequences $f_i(r\tau)$ and $g_i(r\tau)$, say. Then the (discrete) *coherence* of the samples is

$$\gamma[f,g](2\pi k/n\tau) = r[\boldsymbol{D}_{n,\tau} f_i(2\pi k/n\tau), \overline{\boldsymbol{D}_{n,\tau} g_i(2\pi k/n\tau)}]$$

(cf. equation (7.21)). We may use this measure for distinguishing between 'real' and 'accidental' spectral similarities.

Suppose we have discrete spectra for *single* samples of f and g. If the two spectra both have a peak at the same frequency $2\pi k/n\tau$, it *might* be that there is a connection between the two functions; that is, that these peaks reflect the oscillations of the same system. But it might simply be that f and g have components of similar frequency, and that the resolution of the spectra is simply not good enough for us to distinguish them. We might take a longer sample to improve the resolution (see section 9.4), but it might require prohibitively many data points to distinguish the peaks.

Instead, suppose we take a number of smaller samples of the two functions, spaced out along their domain. If there *is* a difference between the two frequencies, then it will show itself over this extended period. Each *individual* sample will be unable to resolve the frequencies; but because the frequencies are slightly different, the relative phase of the two components will shift slightly from sample to sample and the coherence of the two signals at this frequency will be small. If, on the other hand, there is *no* difference between the components, the relative phase will be constant and the absolute coherence will be close to 1.

9.9 Digital filters

Sampled functions can be filtered digitally. As in the continuous case, there are two ways of doing this. Either we can take the discrete Fourier transform, multiply each

Discrete Fourier transforms

component by the value of our chosen filter transfer function at that point, and perform an inverse transform; or, in the x-domain, we can take the discrete convolution of the sample with our chosen impulse response function.

Filtering via the DFT is little different from what we have seen already. *Low-pass* filters have transfer functions of the form

$$T(2\pi k/n\tau) = \begin{cases} 1 & (k \leq k_c) \\ 0 & (k > k_c); \end{cases} \quad (9.26a)$$

high-pass filters have transfer functions of the form

$$T(2\pi k/n\tau) = \begin{cases} 0 & (k \leq k_c) \\ 1 & (k > k_c); \end{cases} \quad (9.26b)$$

and *band-pass* filters have transfer functions of the form

$$T(2\pi k/n\tau) = \begin{cases} 0 & (k \leq k_l) \\ 1 & (k_l < k \leq k_u) \\ 0 & (k > k_u). \end{cases} \quad (9.26c)$$

Other types of filter can be constructed as desired.

Filtering via convolution is more interesting. It might seem that this is a poor way to filter a signal, given that the quickest way to perform a convolution in general is via FFTs. But this is not the case. We are able to choose our filters such that all but a very few of the terms of the impulse response function are zero. The calculation required is thereby much reduced, and may be less even than using the FFT.

The simplest convolution-type digital filter is the *running average*. For a sampled function

$$f(0), f(\tau), f(2\tau), \ldots, f((n-1)\tau),$$

the *three-point running average* is the sequence

$$\tfrac{1}{3}\{f(0) + f(\tau) + f(2\tau)\}, \tfrac{1}{3}\{f(\tau) + f(2\tau) + f(3\tau)\}, \tfrac{1}{3}\{f(2\tau) + f(3\tau) + f(4\tau)\}, \ldots \quad (9.27)$$

Running averages act as (nonideal) low-pass filters. Low-frequency components do not change much over a small number of successive points, so they are little affected by the averaging process; while high-frequency components oscillate rapidly, and averaging attenuates them strongly. The more points that are taken together in the averaging, the more components are filtered out, and the lower the effective 'critical frequency'. Fig 9.8 shows the effect of a three-point running average on a sampled signal.

Equation (9.27) is a convolution-type formula; its impulse response function is the sequence

$$n/3, 0, 0, \ldots, 0, n/3, n/3. \quad (9.28)$$

(Note the wraparound: the two values of $n/3$ at the end are 'really' at -2τ and $-\tau$.) Convolving the sample with (9.28) gives the sequence in (9.27).

The last two terms of the convolved sequence are

$$\ldots, \tfrac{1}{3}\{f((n-2)\tau) + f((n-1)\tau) + f(0)\}, \tfrac{1}{3}\{f((n-1)\tau) + f(0) + f(\tau)\}, \quad (9.29)$$

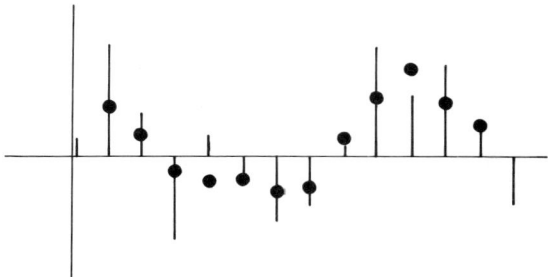

Fig. 9.8 — The effect of a three-point running average (solid circles) on a noisy data signal (lines) is that of a low-pass filter.

because of the wraparound. If this is awkward, it can be avoided by packing with zeros, but it is easier simply to ignore the last two terms.

The transfer function of the three-point running average filter is easily calculated. It is the discrete transform of (9.28):

$$T(2\pi k/n\tau) = \tfrac{1}{3}\{1 + \exp(-2\pi i k(n-2)/n) + \exp(-2\pi i k(n-1)/n)\}$$
$$= \tfrac{1}{3}\{1 + \exp(4\pi i k/n) + \exp(2\pi i k/n)\} \qquad (9.30)$$

Fig. 9.9 shows a graph of $|T(\omega)|$. Two things are clear from this graph: first, that the

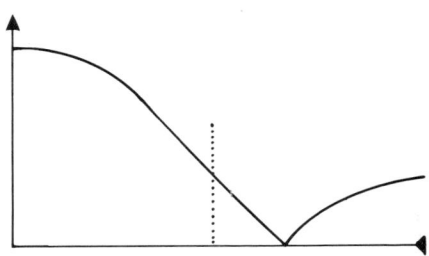

Fig. 9.9 — The amplitude response function of the three-point running average.

running average is not a very good low-pass filter; and secondly, that its attenuation at higher frequencies is very uneven.

We can improve things substantially by altering the multipliers in the impulse response, even with only three nonzero terms in the impulse response. For example, a simple low-pass filter with critical angular frequency $\omega_c = \pi/2\tau$ is given by the impulse response sequence

$$n/2,\ n/4,\ 0,\ 0,\ \ldots,\ 0,\ n/4; \qquad (9.31)$$

the corresponding transfer function is

$$T(2\pi k/n\tau) = \tfrac{1}{2}\{1 + \cos(2\pi k/n)\}, \tag{9.32}$$

and its amplitude is shown in Fig. 9.10.

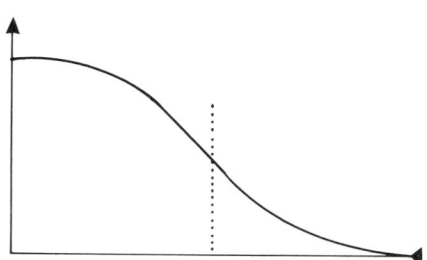

Fig. 9.10 — A better low-pass filter, obtained using a *weighted* three-point running average (equation (9.31)).

High-pass, and even band-pass, filters can all be produced in this way. Typically, the number of multipliers required will be between five and 50, depending on the user's requirements.

These are all examples of *nonrecursive* filters. There are also *recursive* filters, in which the calculation of an output value uses the output values that have already been calculated, in addition to the input values. This saves on calculation time by making maximal use of the calculations already done. Thus, suppose we have input a sequence

$$\ldots, x(r-1), x(r), x(r+1), \ldots,$$

and wish to calculate the ten-point moving average

$$y(r) = (1/10)[x(r-9) + \ldots + x(r-1) + x(r)]. \tag{9.33}$$

Each output value involves 18 sums, 10 look-ups and a division. But $y(r)$ contains a lot of information that is also in $y(r-1)$; in fact,

$$y(r) = y(r-1) + (1/10)[x(r) - x(r-10)]. \tag{9.34}$$

Thus, having calculated $y(r-1)$, we only need four sums, three look-ups and a division to get $y(r)$ if we make use of this formula.

Recursive filters (and nonrecursive filters, if they are approached via convolution rather than by transfer function) are *open-ended*, in that they can be applied to a sequence of data points of unknown length. (They may thus be interpreted as filters of the z-transform, rather than the DFT.)

As an example of what is possible with even a very simple recursive filter, consider the one defined by the formula

$$y(r) = (1/6)\{-2y(r-2) + x(r) + 3x(r-1) + 3x(r-2) + x(r-3)\}. \tag{9.35}$$

This is one of the range of so-called *Butterworth-type filters*. To calculate the transfer function of (9.35), move the y terms to the left-hand side, to get

$$6y(r) + 2y(r-2) = x(r) + 3x(r-1) + 3x(r-2) + x(r-3). \qquad (9.36)$$

Suppose $x(t) = e^{i\omega t}$, $y(t) = T(\omega)e^{i\omega t}$, and suppose the sampling interval is τ; then (9.36) becomes

$$T(\omega)e^{i\omega t}(6 + 2e^{-2i\omega\tau}) = e^{i\omega t}(1 + 3e^{-i\omega\tau} + 3e^{-2i\omega\tau} + e^{-3i\omega\tau}),$$

so that

$$T(\omega) = (1 + 3e^{-i\omega\tau} + 3e^{-2i\omega\tau} + e^{-3i\omega\tau})/(6 + 2e^{-2i\omega\tau})$$

$$= 8e^{-3i\omega\tau/2}\cos^3(\tfrac{1}{2}\omega\tau)/(6 + 2e^{-2i\omega\tau}).$$

Thus

$$|T(\omega)|^2 = 8\cos^6(\tfrac{1}{2}\omega\tau)/(5 + 3\cos(2\omega\tau))$$

$$= \cos^6(\tfrac{1}{2}\omega\tau)/(1 - 3\cos^2(\tfrac{1}{2}\omega\tau) + 3\cos^4(\tfrac{1}{2}\omega\tau))$$

$$= 1/(1 + \tan^6(\tfrac{1}{2}\omega\tau)). \qquad (9.37)$$

Fig. 9.11 shows the amplitude response function $|T(\omega)|$ of this filter.

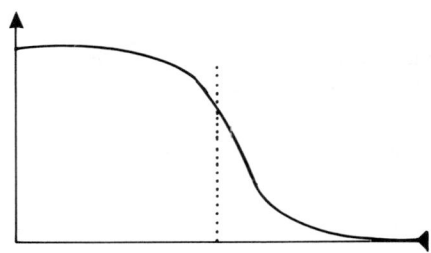

Fig. 9.11 — A still better low-pass filter, attained recursively (equation (9.35)).

The problem with recursive filters is that they may, if poorly designed, be *unstable*, in that certain (bounded) input sequences can yield exponentially growing output sequences. Such filters are unusable; thus any recursive filter must be carefully designed. Digital filter design is a substantial subject, which we cannot treat here; the interested reader is referred to the Bibliography.

9.10 Phase response of filters

When we use a filter to suppress a range of frequencies, we usually try to arrange things so that the filter transfer function $T(\omega)$ is unity on its pass-bands and zero on its stop bands. This is exactly possible if we are filtering a closed-ended data sample; take the DFT, and multiply by the ideal filter transfer function.

For open-ended data, it is not possible. Any such filter must necessarily look 'forwards' as well as 'backwards': $y(r)$ depends, not only on $x(r)$, $x(r-1)$, $x(r-2)$, ..., but also on $x(r+1)$, $x(r+2)$, ... No recursive filter can be ideal, for we would need to know $y(r+1)$, $y(r+2)$ and so on to calculate $y(r)$, and we clearly cannot.

It *is*, however, possible to design open-ended filters for which $|T(\omega)|$ is as close to ideal as desired. The problem is thus one of the *phase* of the components in the output. One compromise is to permit a phase shift from input to output *so long as* it acts simply as a time delay, consistently across the spectrum. (In filtered music, a 0.1 s delay of all the instruments would seldom be a problem, but we would find it quite disturbing if the piccolos and the bassoons were delayed by different amounts.) This means that the phase shift of the components should be *proportional to their frequency*. To see this, think of signals at 1 Hz and 2 Hz delayed by the same time, $\frac{1}{4}$ s. This corresponds to a quarter of a wave (a phase angle of $\frac{1}{2}\pi$) for the 1 Hz signal, but *half* a wave (a phase angle of π) for the 2 Hz signal.

For nonrecursive functions, a linear phase response is achieved by making the impulse response function *cyclically symmetrical*. Thus, convolving by the series

$$-\tfrac{1}{2},\ 1,\ -2,\ 1,\ -\tfrac{1}{2},\ \tfrac{1}{2},\ \tfrac{1}{2},\ 0,\ \ldots,\ 0,\ \tfrac{1}{2},\ \tfrac{1}{2}$$

will achieve a constant delay, for this sequence is cyclically symmetrical about the term -2; but convolving by the sequence

$$1,\ \tfrac{1}{2},\ \tfrac{1}{3},\ \tfrac{1}{4},\ 0,\ 0,\ \ldots,\ 0,\ 0$$

will not.

The phase response of recursive filters, similarly, is linear if the formula is symmetric. Thus, the filter

$$y(r) = 2y(r-1) - y(r-2) - \tfrac{1}{2}x(r) - \tfrac{1}{2}x(r-2)$$

is symmetric about the $r-1$ terms.

EXERCISES

1. What is the z-transform of the constant function $1(x)$? Of the function e^{-ax}? Of the delta function $\delta(x)$?

2. What is the DFT of the constant function $1(x)$?
   How would you define a 'discrete delta function'? What is its DFT?

3. What is the $2n$-point transform of the sample
   $$f(0),\ f(0),\ f(\tau),\ f(\tau),\ f(2\tau),\ \ldots,\ f((n-1)\tau)?$$
   (Treat this as sampling a function g at intervals of $\tfrac{1}{2}\tau$.)

4. Prove the packing theorem, equation (9.14).

5. (Harder) Find a convolution expression for the *odd* terms of the transform of the packed sequence $f(0), f(\tau), \ldots, f((n-1)\tau), 0, \ldots, 0$.

6. If we have an n-point sample of f, then sampling at half the interval will yield a $2n$-point sample similar to
   $$f(0),\ \tfrac{1}{2}\{f(0)+f(\tau)\},\ f(\tau),\ \tfrac{1}{2}\{f(\tau)+f(2\tau)\},\ f(2\tau),\ \ldots,$$
   in which each interpolated point is midway between its neighbours. For convenience, assume the last point is $\tfrac{1}{2}\{f((n-1)\tau)+f(0)\}$. Find an expression for the $2n$-point transform of this sample.

7. Calculate the two-point transform of $f(x) = \sin(\alpha x)$. How does this vary with α?

8. Use the FFT method to find the DFT of an arbitrary four-point sample (A_0, A_1, A_2, A_3); note that the result is indeed the DFT of that sample.

9. What is the effect of the digital (convolution) filter whose significant multipliers are, in sequence, $-\frac{1}{8}, 0, \frac{3}{8}, \frac{1}{2}, \frac{3}{8}, 0, -\frac{1}{8}$?

10. Show that the filter which turns
$$f(0), f(\tau), f(2\tau), \ldots, f((n-1)\tau)$$
into
$$-\tfrac{1}{4}f(0)+\tfrac{1}{2}f(\tau)-\tfrac{1}{4}f(2\tau),\ -\tfrac{1}{4}f(\tau)+\tfrac{1}{2}f(2\tau)-\tfrac{1}{4}f(3\tau),\ \ldots,\ -\tfrac{1}{4}f((n-1)\tau)+\tfrac{1}{2}f(0)-\tfrac{1}{4}f(\tau)$$
acts as a high-pass filter, by calculating and sketching its amplitude response function.

11. Examine the effect of the recursive filter
$$y(r)=2y(r-1)-y(r-2)+x(r).$$
using the input sequence
$$\ldots 0, 0, 0, 1, 0, 0, 0, \ldots.$$

Project. Consider the harmonic function $e^{i\alpha x}$. When $\alpha = 2\pi m/n\tau$, the mth component of its DFT is 1, and all other components vanish. Hence all the power of the original sinusoid lies in the single significant component. Is the same true of any sinusoid $A\cos(\alpha x + \varphi)$?

When α is not a multiple of $2\pi/n\tau$, the power is spread among all the frequency bins; thus equation (9.18), which takes only the two most significant components on each side of the folding frequency, *underestimates* the amplitude of a sinusoid. Establish a comparable estimate of the amplitude of the complex harmonic function $e^{i\alpha x}$; using (9.17), find how much of an underestimate this is when $\alpha = 2\pi(m + \tfrac{1}{2})/n\tau$.

Next, find how much of an underestimate equation (9.18) itself is, when $\alpha = 2\pi(m + \tfrac{1}{2})/n\tau$. What 'correction factor' does this imply should be applied to (9.18)?

A second approximation is the estimate $(1/n) \sum_{r=0}^{n-1} A^2\cos^2(\alpha x + \varphi) \simeq \tfrac{1}{2}A^2$. How accurate is this, say, when $\varphi = 0$? How justified are we, therefore, in ignoring this error, when (say) $m \simeq \tfrac{1}{4}n$?

Part II
Fourier applications

10

Mathematics

10.1 Sums of series

The range of applications of Fourier methods within pure mathematics is surprisingly wide. Many are highly technical, and therefore beyond the scope of this book. Here, we look at five comparatively straightforward, and very different, applications, beginning with the evaluation of certain sums of series.

This is an immediate 'by-product' of the Fourier series method, which of course is concerned with series of functions. Take, for example, the series

$$\sum_{n=1}^{\infty} (1/n^2) . \tag{10.1}$$

Proving *that* this series converges is straightforward; one method runs as follows. The function $1/x^2$ is decreasing on $(0, \infty)$, so $1/n^2 < \int_{n-1}^{n} (1/x^2) \, dx$ for all $n \geq 2$. It follows that

$$\sum_{n=1}^{N} (1/n^2) < 1 + \int_{1}^{N} (1/x^2) \, dx$$

$$= 2 - 1/N$$

$$< 2$$

for any N. But each $1/n^2$ is positive; the partial sums therefore form an increasing sequence, and the series converges.

Finding *to what* the series converges is not so easy. To do this, we may appeal to Fourier techniques. Let us define

$$g(x) = \sum_{n=1}^{\infty} (1/n^2)\cos(nx) ; \tag{10.2}$$

then $g(0)$ is the desired sum. If we differentiate (10.2), then

$$\frac{dg}{dx} \to \sum_{n=1}^{\infty} \frac{1}{n^2} \frac{d}{dx}(\cos(nx))$$

$$= -\sum_{n=1}^{\infty} (1/v)\sin(nx)$$

(cf. (2.24)); differentiating again gives

$$\frac{d^2g}{dx^2} \to -\sum_{n=1}^{\infty} \cos(nx) \ . \tag{10.3}$$

Now compare this with (2.19), which says that $III(x) \to 1 + \sum_{n=1}^{\infty} 2\cos(2n\pi x)$; it follows that

$$\frac{d^2g}{dx^2} = \tfrac{1}{2} - \tfrac{1}{2}III\left(\frac{x}{2\pi}\right) \ . \tag{10.4}$$

This can be integrated, for although the sampling function is an impulsive 'function', impulses are defined by an integral property. We have

$$\frac{dg}{dx} = \tfrac{1}{2}(x - (2m+1)\pi) + c \quad \text{for} \quad 2m\pi < x < 2(m+1)\pi$$

$$= \tfrac{1}{2}\pi \text{Saw}(x/\pi - 1) + c \ , \tag{10.5}$$

where c is some constant (cf. Chapter 2, Question 7). Since dg/dx has, above, a Fourier series in which the constant terms is absent, we must have $c = 0$. Integrating again gives

$$g(x) = (x - (2m+1)\pi)^2/4 + d \quad \text{for} \quad 2m\pi < x < 2(m+1)\pi \tag{10.6}$$

(see Fig. 10.1: g is a periodic function based on a section of a parabola).

Now we have a functional form for g. To find the constant d, we use the third of equations (2.3), for the constant term in the Fourier series of g is known to be zero (equation (10.2)). Thus

$$0 = \int_0^{2\pi} \left(\frac{(x-\pi)^2}{4} + d\right) dx$$

$$= \left[\frac{(x-\pi)^3}{12} + dx\right]_0^{2\pi}$$

$$= 2\pi d + \frac{\pi^3}{6} \ .$$

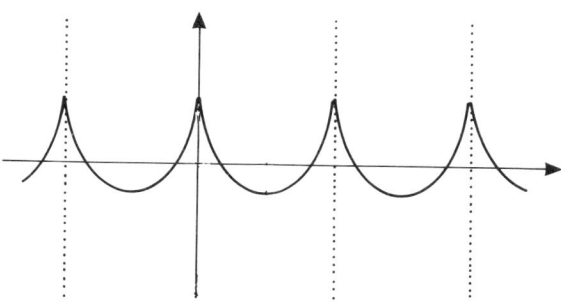

Fig. 10.1.

Thus $d = -\pi^2/12$, and

$$\sum_{n=1}^{\infty}(1/n^2)\cos(nx) = g(x)$$

$$= (x - (2m+1)\pi)^2/4 - \pi^2/12 \text{ for } 2m\pi < x < 2(m+1)\pi$$

(10.7)

(It is possible, of course, to avoid the use of impulsive functions by evaluating the Fourier coefficients of this function directly.)

From (10.7), we can make the final step

$$\sum_{n=1}^{\infty}(1/n^2) = g(0)$$

$$= \pi^2/4 - \pi^2/12$$

$$= \pi^2/6 \ .$$

(10.8)

Other series that can be evaluated in this way include $\sum_{n=1}^{\infty}(1/n^k)$ for any integer $k \geq 2$; this is left as an exercise for the reader.

10.2 The isoperimetric problem

Our second application of Fourier methods also uses Fourier series. This is the solution of the *isoperimetric problem*: what plane curve of unit perimeter encloses the largest area? The answer, as is widely known, is the circle, and Fourier methods provide a simple proof of this.

A curve in the plane may be expressed *parametrically* in the form $(x(t), y(t))$. Pick a point $O = (x_0, y_0)$ on the curve. For any other point X on the curve, let t denote the arc-length of the section between O and X measured *anticlockwise* from O; then define $x(t)$ and $y(t)$ to be the x- and y-coordinates of X. The pair of functions $x(t)$ and $y(t)$ together determine the curve.

For a *closed* curve, the functions $x(t)$ and $y(t)$ are periodic. If the perimeter of the curve is 1, then $(x(t+1), y(t+1)) = (x(t), y(t))$. Of course, $x(0) = x_0$ and $y(0) = y_0$.

Let us assume further that the curve is smooth: both dx/dt and dy/dt exist, and both are continuous. This makes the definition and calculation of the curve's perimeter straightforward. A small element dt of arc approximates to a straight line, so Pythagoras' theorem tells us that $dx^2 + dy^2 = dt^2$; hence

$$\left(\frac{dx}{dt}\right)^2 + \left(\frac{dy}{dt}\right)^2 = 1 \ . \tag{10.9}$$

In particular,

$$1 = \int_0^1 dt = \int_0^1 \left\{ \left(\frac{dx}{dt}\right)^2 + \left(\frac{dy}{dt}\right)^2 \right\} dt \tag{10.10}$$

(see Fig. 10.2).

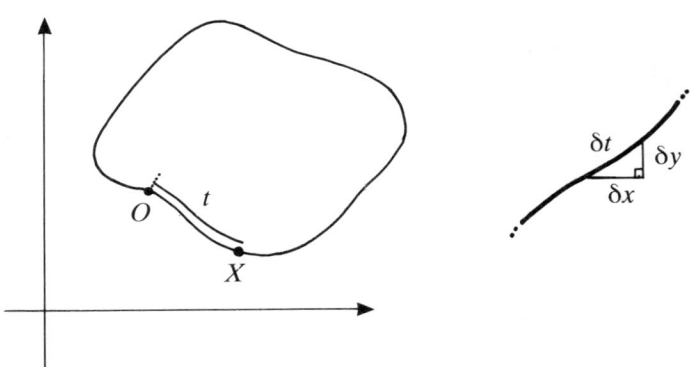

Fig. 10.2 — Perimeter of a closed curve.

For the *area* enclosed by the curve, we begin by dividing the shape into thin vertical strips, and consider the element of arc between t and $t + dt$. The area of the half-strip bounded by this element, the x-axis, and the lines $y = y(t)$ and $y = y(t + dt)$ is then

$$dA \simeq (x(t) - x(t+dt))y(t)$$
$$\simeq -\frac{dx}{dt} y(t) \, dt \tag{10.11}$$

for small dt. Thus $dA/dt = -(dx/dt)y(t)$. This is positive, for positive y, if the arc is going to the left at t, negative if to the right. Thus we cancel out areas which are 'avoided' by the curve in places where it doubles back on itself, and the total area is

$$A = -\int_0^1 y(t) \frac{dx}{dt} \, dt \tag{10.12}$$

(see Fig. 10.3).

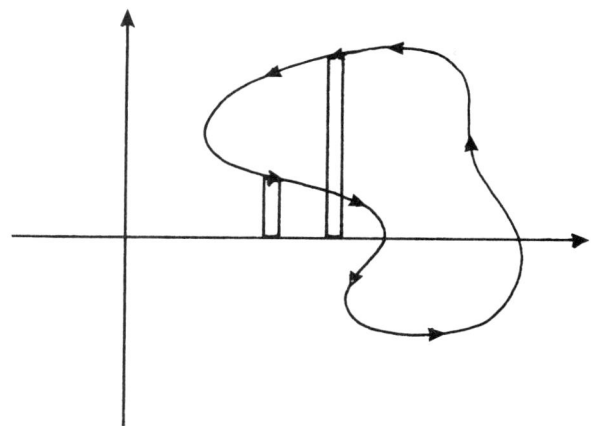

Fig. 10.3 — Area of a closed curve.

Now we are set to solve the isoperimetric problem. The functions $x(t)$ and $y(t)$ are smooth periodic functions of period 1, so that, by section 2.16, they have Fourier expansions:

$$x(t) = K + \sum_{n=1}^{\infty} A_n \cos(2n\pi t) + \sum_{n=1}^{\infty} B_n \sin(2n\pi t)$$

and

$$y(t) = L + \sum_{n=1}^{\infty} C_n \cos(2n\pi t) + \sum_{n=1}^{\infty} D_n \sin(2n\pi t) \ . \tag{10.13}$$

It follows, by differentiating, that

$$\frac{dx}{dt} \to \sum_{n=1}^{\infty}(-2n\pi A_n)\sin(2n\pi t) + \sum_{n=1}^{\infty}(2n\pi B_n)\cos(2n\pi t)$$

and

$$\frac{dy}{dt} \to \sum_{n=1}^{\infty}(-2n\pi C_n)\sin(2n\pi t) + \sum_{n=1}^{\infty}(2n\pi D_n)\cos(2n\pi t) , \qquad (10.14)$$

We may as well assume that the shape is centred on the origin, in the sense that $K = L = 0$; otherwise, we can shift it through $(-K, -L)$ to put it there.

Given these expressions for x and y, we can use Parseval's equation (2.27) to rewrite equation (10.10) as

$$1 = \int_0^1 \left\{ \left(\frac{dx}{dt}\right)^2 + \left(\frac{dy}{dt}\right)^2 \right\} dt$$

$$= \tfrac{1}{2}\sum_{n=1}^{\infty}(2n\pi A_n)^2 + \tfrac{1}{2}\sum_{n=1}^{\infty}(2n\pi B_n)^2 + \tfrac{1}{2}\sum_{n=1}^{\infty}(2n\pi C_n)^2 + \tfrac{1}{2}\sum_{n=1}^{\infty}(2n\pi D_n)^2$$

$$= 2\pi^2 \sum_{n=1}^{\infty} n^2 [A_n^2 + B_n^2 + C_n^2 + D_n^2] ; \qquad (10.15)$$

equation (2.28) applied to (10.12) yields

$$A = -\int_0^1 y(t)\frac{dx}{dt} dt$$

$$= -\sum_{n=1}^{\infty}[n\pi B_n C_n - n\pi A_n D_n]$$

$$= \pi \sum_{n=1}^{\infty} n(A_n D_n - B_n C_n) . \qquad (10.16)$$

We can combine these two equations:

$(1/\pi)(1/(2\pi) - 2A)$

$$= \sum_{n=1}^{\infty}[n^2(A_n^2 + B_n^2 + C_n^2 + D_n^2) - 2n(A_n D_n - B_n C_n)]$$

$$= \sum_{n=1}^{\infty}[(nA_n - D_n)^2 + (nB_n + C_n)^2 + (n^2 - 1)(D_n^2 + C_n^2)] . \qquad (10.17)$$

The right-hand side of (10.17) is a sum of squares, and so cannot be negative. It follows that the left-hand side is non-negative; hence the area A enclosed by the curve cannot exceed $1/4\pi$. But the area of a circle of perimeter 1 is *equal* to $1/4\pi$; hence no curve of unit perimeter encloses a greater area than a circle.

Indeed, we can say more. Suppose we have a curve enclosing an area equal to $1/4\pi$. Then the left-hand side of (10.17) is zero; it follows that the right-hand side is zero. But this is a sum of squares; it can be zero only if each of the squares is zero. In other words, each $nA_n - D_n = 0$, each $nB_n + C_n = 0$; and, for $n > 1$, each $D_n = 0$ and each $C_n = 0$. (The case $n = 1$ is not included, because $(n^2 - 1)$ is already zero in this case.) This means that $A_1 = D_1$, $B_1 = -C_1$, and $A_n = B_n = C_n = D_n = 0$ for $n > 1$. The Fourier expansions of x and y (equations (10.13)) then reduce to

$$x(t) = A_1\cos(2\pi t) + B_1\sin(2\pi t)$$

and

$$y(t) = -B_1\cos(2\pi t) + A_1\sin(2\pi t) \ . \tag{10.18}$$

But this is an equation for a circle (of radius $\sqrt{(A_1^2 + B_1^2)}$). That is, the circle is the *only* shape of unit perimeter enclosing an area as big as $1/4\pi$.

10.3 Differential equations

We have seen that there is a relationship between differentiation and Fourier transformation — namely, the derivative theorem, equation (3.35). This means that Fourier transformation is useful in the solution of certain differential equations.

The Fourier transformation is a linear operator, so it will interact simply with linear equations; but nonlinear equations will introduce convolutions. Thus we restrict attention to linear differential equations, of the form

$$c_n(x)\frac{d^n y}{dx^n} + \ldots + c_1(x)\frac{dy}{dx} + c_0(x)y = k(x) \ .$$

The coefficients c_n here are functions of x. Again, if these are anything other than constants, Fourier transformation will be complicated (though see below). We are left with equations like

$$c_n\frac{d^n y}{dx^n} + \ldots + c_1\frac{dy}{dx} + c_0 y = k(x) \ . \tag{10.19}$$

Now any solution to this is necessarily smooth. If $f(x)$ is an integrable solution, we take the Fourier transform of both sides. This gives

$$c_n F\frac{d^n f}{dx^n}(\omega) + \ldots + c_1 F\frac{df}{dx}f(\omega) + c_0 Ff(\omega) = Fk(\omega) \ ,$$

which by the derivative theorem is

$$[(i\omega)^n c_n + \ldots + i\omega c_1 + c_0]\, Ff(\omega) = Fk(\omega) \ . \tag{10.20}$$

It follows that

$$Ff(\omega) = Fk(\omega)/[(i\omega)^n c_n + \ldots + i\omega c_1 + c_0] \ ; \tag{10.21}$$

and the solution follows from taking inverse transforms.

There are problems with this approach. The most obvious is that many differential equations have nonintegrable solutions. (The harmonic equation is one such: $A\cos(x + \varphi)$ is not an integrable function for any $A \neq 0$.) We can avoid this problem to some extent by using Laplace transforms instead (see section 8.5). But using the Laplace transform has an unfortunate drawback, for negative values of x are ignored in Laplace transformation: the Laplace transform is defined (equation (8.33)) as

$$Lf(s) = \int_0^\infty f(x)e^{-sx}\,dx.$$

Consequently, the Laplace transform version of the derivative theorem is slightly different from the Fourier transform version. Integrating by parts, we have

$$L\frac{df}{dx}(s) = \int_0^\infty \frac{df}{dx} e^{-sx}\,dx$$

$$= [fe^{-sx}]_0^\infty - \int_0^\infty (-se^{-sx})f\,dx$$

$$= -f(0) + sLf(s) \,. \tag{10.22}$$

Example Solve the equation $d^2y/dx^2 + 9y = 3x$, with initial conditions $y = 0$ and $dy/dx = 1$ at $x = 0$.

Putting $y = f(x)$ and $k(x) = 3x$, and taking Laplace transforms, we obtain

$$L\frac{d^2f}{dx^2}(s) + 9Lf(s) = Lk(s) \,. \tag{10.23}$$

Two applications of (10.22) give

$$L\frac{df}{dx}(s) = sLf(s) - f(0)$$

and

$$L\frac{d^2f}{dx^2}(s) = s^2Lf(s) - sf(0) - f'(0) \,.$$

With initial conditions $f(0) = 0$, $f'(0) = 1$, this becomes

$$L\frac{d^2f}{dx^2}(s) = s^2Lf(s) - 1$$

The right-hand side of (10.23) is

$$Lk(s) = \int_0^\infty 3xe^{-sx}\,dx$$

$$= 3/s^2 \,.$$

The equation now becomes

$$(s^2 + 9)Lf(s) = (3/s^2) + 1.$$

We now have an expression for $Lf(s)$, as

$$Lf(s) = ((3/s^2) + 1)/(s^2 + 9).$$

This can be split up into partial fractions:

$$Lf(s) = (1/3)1/s^2 - (i/9)(1/(s - 3i)) + (i/9)(1/(s + 3i)). \tag{10.24}$$

To invert this transform we refer to Table 8.1. There, we find that $1/s^2$ is the transform of x and $1/(s - \alpha)$ is the transform of $e^{\alpha x}$. Thus inverting equation (10.24) gives

$$f(x) = (1/3)x - (i/9)e^{3ix} + (i/9)e^{-3ix}$$
$$= (1/3)x + (2/9)\sin(3x).$$

Thus the desired function is $\underline{(1/3)x + (2/9)\sin(3x)}$.

In this particular case, the Laplace transform method is more cumbersome than the direct: spotting $y = (1/3)x$ as a particular solution, solving the associated equation (which is in fact the harmonic equation), and finding the solution that satisfies the initial conditions. However, things will not always be so simple.

There is a second way that Fourier methods can help us in the solution of differential equation problems, which we have already seen. This uses the fact that a number of *partial* differential equations have mathematical similarities to the harmonic equation, the source of the functions used in Fourier analysis. In particular, the one-dimensional wave equation $\partial^2 y/\partial t^2 = c^2 \partial^2 y/\partial x^2$ (equation (1.32′)) submits itself to series solutions like (1.38); section 8.4 explores the two-dimensional wave equation in a similar way. These equations, and their higher-dimensional analogues, are extremely widespread in physical problems. Equally important for physicists is the fact that *Schrödinger's equation*, the governing equation for physical systems in quantum mechanics, also admits of this type of solution. We shall have more to say about this in the following chapter.

10.4 Statistics

An important problem of statistics is to find the distribution of the sum of two variables, whose individual distributions are known. For instance, a study of anthropoid fossils might include measurements of the lengths of leg bones. We might plot a bar graph of the spread of femur (thigh-bone) lengths, and a second bar graph of the spread of tibia (shin-bone) lengths. Can we deduce from these what the graph of total leg lengths would be?

A more precise statement is as follows. Suppose that X and Y are real-valued random variables. Let X have the probability density function (or p.d.f.) $f(x)$. (That is, for small δx the probability of X lying between x and $x + \delta x$ is $\text{Prob}(x < X < x + \delta x) \approx f(x)\delta x$.) Similarly, let Y have p.d.f. $g(x)$. What then is the distribution of $X + Y$?

In general, we cannot answer this question, because we know nothing about the relationship between X and Y. In the case of leg-bone lengths, it would be expected that a big hominid would have a big femur *and* a big tibia, while the bones of a small hominid would all be small. In other cases, a large value of X might be associated with a *small* value of Y. Sometimes there is *no* relationship between the variables: large values of X are associated equally often with large, small, and middling values of Y. In this last case, we say that X and Y are *independent*. Knowing X does not help you guess Y, and vice versa. (Precisely, two variables X and Y are indpendent if the 'marginal' p.d.f. of Y that arises for any fixed $X = \xi$ does not depend on ξ. As a consequence, the distribution of the correlation coefficient $r(X, Y)$, over all n-point samples of X and Y, has zero mean.)

Now the probability that X lies between x and $x + dx$ is $f(x)dx$; the probability that Y lies between y and $y + dy$ is $g(y)dy$. If X and Y are independent, the probability of *both* $x < X < x + dx$ *and* $y < Y < y + dy$ happening simultaneously is the product of these individual probabilities, or $f(x)g(y)dxdy$. Put $h(x)$ for the p.d.f. of the sum $X + Y$. Then:

$$\int_{-\infty}^{a} h(x) \, dx = \text{Prob}(X + Y < a)$$

$$= \int_{-\infty}^{\infty} \text{Prob}(x < X < x + dx \text{ and } Y < a - x) \, dx$$

$$= \int_{-\infty}^{\infty} \left\{ f(x) \int_{-\infty}^{a-x} g(y) \, dy \right\} dx$$

(putting $z = x + y$)

$$= \int_{-\infty}^{\infty} \left\{ f(x) \int_{-\infty}^{a} g(z - x) \, dz \right\} dx$$

$$= \int_{-\infty}^{a} \left\{ \int_{-\infty}^{\infty} f(x)g(z - x) \, dx \right\} dz$$

$$= \int_{-\infty}^{a} (f*g)(z) \, dz \ .$$

Differentiating this with respect to a gives

$$h(a) = (f*g)(a) \ . \tag{10.25}$$

In other words, the distribution of a *sum* of independent variables is the *convolution* of their separate distributions.

Now suppose we define the *characteristic function* of a random variable to be the Fourier transform of its p.d.f. In view of the convolution theorem (5.7), equation (10.25) can be recast into the following statement: if X and Y are independent random variables with characteristic functions $\varphi(\omega)$ and $\psi(\omega)$ respectively, then the characteristic function of $X + Y$ is the product $\varphi(\omega)\psi(\omega)$.

This fact can be developed to prove what is perhaps the most important single result in statistics, the *central limit theorem*. Becasue of it, the normal distribution is the most important single statistical distribution. Because of *that*, much of statistics can be, and is, based on normally distributed random variables. This is useful, because the normal distribution also turns out to be relatively easy to calculate with.

Suppose the continuous variable X has probability density function $f(x)$. Suppose also that X has finite *mean* and *variance*; in other words, the integrals

$$\mu = \int_{-\infty}^{\infty} xf(x)\, dx$$

and

$$\sigma^2 = \int_{-\infty}^{\infty} (x-\mu)^2 f(x)\, dx \tag{10.26}$$

exist and are finite. Let $X_1, X_2, \ldots, X_n$ be a sample of n independent determinations of the variable X, and let $\bar{X} = (1/n)(X_1 + \ldots + X_n)$ be their sample mean. Then the claim of the central limit theorem is that, for large n, the distribution of $\bar{X}$ is approximately normal, of mean μ and variance σ^2/n.

To prove this, we first deal with the 'normalized' case $\mu = 0$, $\sigma^2 = 1$. Let $F_n(x)$ be the probability density function of the sum $S_n = X_1 + \ldots + X_n$ of a sample of size n. From (10.25), F_n is given by the repeated convolution

$$F_n = f*f*\ldots*f \quad (n \text{ times}) . \tag{10.27}$$

Next, we apply the Fourier transform operator F to both sides of this equation. By the convolution theorem,

$$FF_n(\omega) = (Ff(\omega))^n \tag{10.28}$$

Thus, the characteristic function of S_n is just the nth power of the characteristic function for X.

The p.d.f. $f(x)$ has total integral 1. It follows that $Ff(0) = 1$ (for $Ff(0) = \int_{-\infty}^{\infty} f(x)\, dx$). Moreover, $Ff(\omega)$ is twice differentiable at $\omega = 0$; using the converse of the derivative theorem, equation (3.36), we have

$$\frac{dFf}{d\omega}(0) = -iF(xf)(0)$$

$$= -i\int_0^{\infty} xf(x)\, dx$$

$$= -i\mu$$

$$= 0 ,$$

and similarly

$$\frac{d^2 Ff}{d\omega^2}(0) = -F(x^2 f)(0)$$

$$= -\int_0^\infty x^2 f(x)\,dx$$

$$= -\sigma^2$$

$$= -1.$$

Further, $|Ff(\omega)|$ is *globally* maximum at $\omega = 0$, for

$$|Ff(\omega)| = \left|\int_{-\infty}^\infty f(x)e^{-i\omega x}\,dx\right|$$

$$\leq \int_{-\infty}^\infty |f(x)e^{-i\omega x}|\,dx$$

$$= \int_{-\infty}^\infty f(x)\,dx,$$

as $f(x)$ is a nonnegative function. In fact it can be shown that, if f is a continuous real-valued function, then $|Ff(\omega)| = 1$ happens *only* at $\omega = 0$.

Now we can use Taylor's theorem to write

$$Ff(\omega) \simeq 1 - \tfrac{1}{2}\omega^2$$

$$\simeq \exp(-\omega^2/2) \tag{10.29}$$

for small values of ω. It follows that $Ff(\omega)^n \simeq \exp(-\omega^2/2)^n = \exp(-n\omega^2/2)$. For values of ω away from zero, $|Ff(\omega)| < 1$, so that $Ff(\omega)^n \to 0$ as $n \to \infty$. Thus, from (10.28), we can say that

$$Ff_n(\omega/\sqrt{n}) = Ff(\omega/\sqrt{n})^n$$

$$\to \exp(-\omega^2/2) \qquad \text{as } n \to \infty. \tag{10.30}$$

For large n, then, $Ff_n(\omega) \simeq \exp(-n\omega^2/2)$. But from equations (3.13) and (3.30), this is the Fourier transform of $\sqrt{(n/2\pi)}\exp(x^2/2n)$. Hence, for large n, $F_n \simeq \sqrt{(n/2\pi)}\exp(-x^2/2n)$.

This is the distribution of the *sum* $S_n = X_1 + \ldots + X_n$. It is (approximately) normal, of mean 0 and variance n. The distribution of the sample mean $\bar{X} = S_n/n$ is therefore (approximately)

$$(1/n)\sqrt{(n/2\pi)}\exp(-(ny)^2/2n)$$

or simply

$$1/\sqrt{(2\pi n)}\exp(-nx^2/2),$$

which is the normal distribution of mean zero and variance $1/n$. Thus the central limit theorem is proved in this case.

For the more general case, where X has mean μ and variance σ^2, we put $Z = (X - \mu)/\sigma$. Then Z has mean 0 and variance 1; its sample mean $\bar{Z}$ has mean 0 and variance $1/n$, so the sample mean $\bar{X}$ has mean μ and variance σ^2/n. Fig. 10.4 shows the distribution of the mean of n samples from a uniformly distributed variable, with p.d.f. $f(x) = \chi_{(0,1)}(x)$, for small n.

The proof of the central limit theorem as I have presented it here is not rigorous, and skates over many pitfalls. One is this: we can conclude, from the fact that $Ff_n(\omega)$ converges to a certain function, that $F_n(x)$ necessarily converges to its inverse transform? Another is that there is no indication about just how 'close' to normal the function F_n is — even if the sequence does converge appropriately, it might still happen that *every* F_n is highly nonnormal. There are ways of overcoming these problems, but they require a deal of additional theory, which is beyond the scope of this book.

Let us consider the significance of the central limit theorem. We might think of the variable X as an *estimator* of the quantity μ, with σ as the 'average error' in the estimate. The theorem tells us that the mean of n independent estimates is also an estimate of μ, and that the error of this estimate is smaller than that of the individual determinations by a factor of $\sqrt{n}$.

Partly for this reason, most statistics is done on 'large' samples. The more information you have about a system, the more accurate your knowledge of it. What a statistician does with these samples depends on the system, but there is a fair

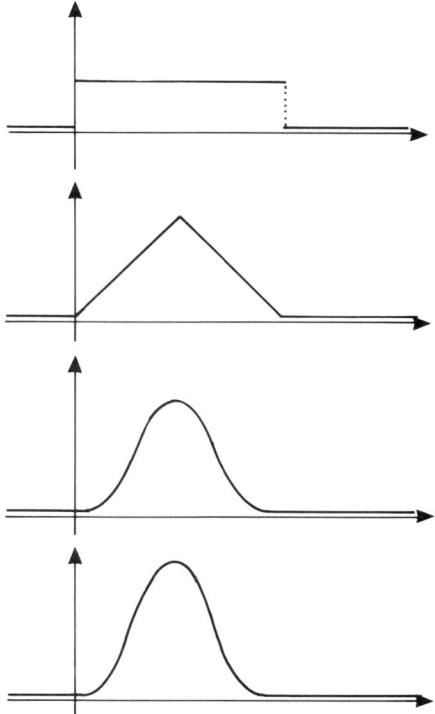

Fig. 10.4 — The p.d.f.s of sample means of (from top) one-, two-, three-, and four-point samples of a uniformly distributed variable.

arsenal of standard techniques. Of these, many of the most widely used — regression, variance analysis, χ^2 testing and so on — are based on a theory developed for *normally distributed* variables. This is important, because in many applications the variables are *not* normally distributed. It would seem, then, that the techniques are often inappropriate to the data. In fact, for many purposes (including those listed) the central limit theorem may be invoked. For large samples, many standard tests may be applied to nonnormally distributed variables *as if they were* normally distributed, and because of the central limit theorem, the conclusions will be reasonably accurate. However, the number of data points needed for meaningful analysis is likely to be higher if the underlying distributions depart significantly from the normal.

10.5 The prime number theorem

Our final example of the Fourier transform in action in mathematics is both the deepest and in many ways the most interesting. It is a salutary example of the indivisibility of mathematics that a technique is blatantly 'physical' as Fourier transformation should have, as a consequence, one of the best-known results of pure number theory; the prime number theorem. As such, this proof deserves a place in this book, and I make no excuses for including (what is) a hard piece of mathematics in its entirety.

The prime number theorem provides an estimate for the density of primes. It says that the first n natural numbers contain approximately $n/\log n$ primes, for large n (here, and throughout this section, logarithms are to base e). More precisely, let $\pi(n)$ be the number of primes in the range $1,\ldots,n$; then the theorem says that

$$\pi(n)\log n/n \to 1 \qquad \text{as } n \to \infty \ . \tag{10.31}$$

This formula does not say how *quickly* $\pi(n)$ begins to approximate $n/\log n$ with any accuracy. Indeed, it does so quite slowly. If $\tau(n) = \pi(n)\log n/n$, then we can calculate that

$$\tau(10) = 4.\log 10/10 \simeq 0.9210$$

$$\tau(100) \simeq 1.1513$$

$$\tau(1000) \simeq 1.1605$$

$$\tau(10^4) \simeq 1.1320$$

$$\tau(10^5) \simeq 1.1043$$

$$\tau(10^6) \simeq 1.0845$$

and so on. Much more sophisticated calculation is required to show that, in fact, $\tau(n) \simeq 1 + 1/\log n$ for large n.

The prime number theorem was first formulated by Gauss in the early 19th century, on the evidence of trial and error; but the first formal proof was not provided until the end of that century. The proof we give here is an adaptation from a proof given by Ikehara in 1931.

The proof is divided into five steps. Fourier methods come in only in the fifth step, but they enter in an essential manner:

Step 1. A function $Q(x)$ is introduced; we prove that the prime number theorem is equivalent to proving that $Q(x)/x \to 1$ as $x \to \infty$.

Step 2. The Riemann zeta-function $\zeta(s)$ is introduced; we prove the equation

$$-(1/\zeta)(d\zeta/ds) = s \int_1^\infty Q(x) x^{-s-1} \, dx.$$

Step 3. We find two separate expressions for $\zeta(s)$, and a technical arithmetic property.

Step 4. We define an 'error function', $f_\delta(b)$; we prove that there is a constant k such that $f_\delta(b) \leq k(1 + b^4)$ for all δ, and that the limit function $f(b) = \lim_{\delta \to 0} f_\delta(b)$ exists.

Step 5. Finally, we take the Fourier transform of the product $f(b) k_{\alpha, \beta}(b)$ for a range of functions $k_{\alpha, \beta}$. An application of the convolution theorem and some limiting arguments finish the proof.

Step 1.
Define the function $Q(x)$ by the expression

$$Q(x) = \sum_{p \text{ prime}} [\log x / \log p] \log p, \qquad (10.32)$$

where $[z]$ denotes the largest integer less than or equal to z. Notice that the sum is really only over primes $p \leq x$, because if $p > x$ then $[\log x / \log p] = 0$.

In steps 2 to 5, we prove that $Q(x)/x \to 1$ as $x \to \infty$. The prime number theorem follows from this fact. On the one hand, we have that

$$Q(x) = \sum_{p \leq x} [\log x / \log p] \log p$$

$$\leq \sum_{p \leq x} \log x$$

$$= \pi(x) \log,$$

so that

$$Q(x)/x \leq \pi(x) \log x / x. \qquad (10.33)$$

On the other hand, let δ be a small positive real number, between 0 and $\tfrac{1}{2}$, and write $y = x^{1-\delta}$. Then $y > \sqrt{x}$, so if $y < p \leq x$ then $p^2 > x$, and therefore $[\log x / \log p] = 1$. Thus we have

$$\pi(x) = \pi(y) + \sum_{y<p\leqslant x} 1$$

$$\leqslant y + \sum_{y<p\leqslant x} \log p/\log y$$

$$= y + (1/\log y) \sum_{y<p\leqslant x} [\log x/\log p]\log p$$

$$\leqslant y + \frac{Q(x)}{\log y}$$

$$= x^{1-\delta} + \frac{Q(x)}{(1-\delta)\log x}$$

It follows that

$$(1-\delta)\pi(x)\frac{\log x}{x} \leqslant (1-\delta)x^{-\delta}\log x + \frac{Q(x)}{x} \qquad (10.34)$$

The first term on the right-hand side tends to zero as $x \to \infty$, for any δ. Thus, if $Q(x)/x \to 1$ as $x \to \infty$, then (10.33) and (10.34) together imply that $\pi(x)\log x/x \to 1$.

Step 2.

The Rieman zeta-function $\zeta(s)$ is defined, for complex s, by the equation

$$\zeta(s) = \sum_{n=1}^{\infty} (1/n^s) . \qquad (10.35)$$

This sum converges for all s with $\operatorname{Re} s > 1$. Indeed, it may be shown that $\zeta(s)$ is *analytic* on the punctured half-plane $\{a+ib: a>\tfrac{1}{2}\}\setminus\{a:a\leqslant 1\}$.

Further, on any closed bounded region in the half-plane $\{a+ib: a>1\}$ equation (10.35) is *absolutely* and *uniformly* convergent, so that we can differentiate the series term by term:

$$\zeta' = \frac{d\zeta}{ds} = \sum_{n=1}^{\infty} \frac{d}{ds}(1/n^s)$$

$$= -\sum_{n=1}^{\infty} (1/n^s)\log n . \qquad (10.36)$$

This sum is over the integers 2 and above, since $\log 1 = 0$, and each such integer is a product of primes. Now the function log converts products into sums, so that if p divides n exactly m_p times, then $\log n = \sum_p m_p \log p$; this can be written as

$$\log n = \sum_p \sum_{k=1}^{\infty} \begin{cases} \log p & (p^k \text{ divides } n) \\ 0 & (\text{otherwise}) \end{cases} \qquad (10.37)$$

Thus equation (10.36) can be rewritten as

$$-\zeta'(s) = \sum_{n=1}^{\infty} (1/n^s) \log n$$

$$= \sum_{n=1}^{\infty} \sum_p \sum_{k=1}^{\infty} \begin{cases} (1/n^s) \log p & (p^k \text{ divides } n) \\ 0 & (\text{otherwise}) \end{cases}$$

$$= \sum_p \sum_{k=1}^{\infty} \sum_{n=1}^{\infty} \begin{cases} (1/n^s) \log p & (n \text{ a multiple of } p^k) \\ 0 & (\text{otherwise}) \end{cases}$$

$$= \sum_p \sum_{k=1}^{\infty} \sum_{m=1}^{\infty} (1/(mp^k)^s) \log p$$

$$= \zeta(s) \sum_p \log p \left\{ \sum_{k=1}^{\infty} (1/p^{ks}) \right\}$$

$$= \zeta(s) \sum_p \frac{\log p}{(p^s - 1)} . \qquad (10.38)$$

That is, ζ' is a multiple of ζ. Moreover, the multiplicand is also an analytic function in the half-plane $\operatorname{Re} s > 1$.

Now $\zeta(s)$ is never zero. For if z is a root of ζ, then from (10.38) its derivatives of all orders are also zero at $s = z$. But an analytic function has a Taylor expansion. Expanding $\zeta(s)$ around z gives $\zeta(s)$ as a sum of terms which are all zero; this implies that $\zeta(s)$ is identically zero, which is false.

Because $\zeta(s)$ is nonzero, which can divide (10.38) through by it. If we then 'undo' some of the steps of the previous argument, we get

$$-(1/\zeta)\zeta' = \sum_p \log p \left\{ \sum_{k=1}^{\infty} (1/p^{ks}) \right\}$$

$$= \sum_p \sum_{n=2}^{\infty} \begin{cases} (1/n^s) \log p & (\text{if } n \text{ is a power of } p) \\ 0 & (\text{otherwise}) \end{cases} \qquad (10.39)$$

We may use this expression to relate ζ to the function Q of Step 1. It is the case that $Q(n) = Q(n-1)$ for most n; they differ only when n is a power of a prime, and then $Q(p^k) = Q(p^k - 1) + \log p$. (These facts follow directly from the definition of Q, (10.32).) Thus, (10.39) may be rewritten as

$$-(1/\zeta)\zeta' = \sum_{n=2}^{\infty} (1/n^s)\{Q(n) - Q(n-1)\}$$

$$= \sum_{n=1}^{\infty} Q(n)\{(1/n^s) - 1/(n+1)^s\}$$

$$= \sum_{n=1}^{\infty} Q(n) \, s \int_{n}^{n+1} (1/x^{s+1}) \, dx \ .$$

Since $Q(x)$ is clearly constant in every interval $n \leq x < n+1$, this becomes

$$-(1/\zeta)\zeta' = s \sum_{n=1}^{\infty} \left\{ \int_{n}^{n+1} Q(x)(1/x^{s+1}) \, dx \right\}$$

$$= s \int_{1}^{\infty} Q(x) x^{-s-1} \, dx \ . \tag{10.40}$$

Thus, $Q(x)$ is related to $\zeta(s)$ by an integral equation.

Step 3.
The Riemann zeta-function $\zeta(s)$ is a well-known function, and its properties have been investigated quite closely. Equation (10.40) does, therefore, represent something of an advance. The zeta function has a number of expansions, two of which are useful here.

The first of these arises as a consequence of calculating the integral of the function $(n + \frac{1}{2} - x)x^{-s-1}$. We have:

$$\int_{n}^{n+1} (n + \tfrac{1}{2})x^{-s-1} \, dx = (1/s)(n + \tfrac{1}{2})\{(1/n^s) - 1/(n+1)^s\} \ .$$

Now, if we sum this over the positive integers $n = 1, 2, \ldots$, we get

$$s \int_{1}^{\infty} ([x] + \tfrac{1}{2}) x^{-s-1} \, dx = \sum_{n=1}^{\infty} (n + \tfrac{1}{2})\{(1/n^s) - 1/(n+1)^s\}$$

$$= \sum_{n=1}^{\infty} (n + \tfrac{1}{2})(1/n^s) - \sum_{n=2}^{\infty} (n - \tfrac{1}{2})(1/n^s)$$

$$= (3/2) + \sum_{n=2}^{\infty} (1/n^s)$$

$$= \tfrac{1}{2} + \zeta(s) \ . \tag{10.41}$$

Also, we have that $s\int_1^\infty x^{-s}\,dx = s/(s-1) = 1 + (1/(s-1))$. Subtracting this from equation (10.41) and rearranging gives

$$\zeta(s) = s\left(\int_1^\infty ([x] - x + \tfrac{1}{2})x^{-s-1}\,dx\right) + \tfrac{1}{2} + \frac{1}{(s-1)} \ . \tag{10.42}$$

In particular, the first factor of the integrand, $[x] - x + \tfrac{1}{2}$, always lies between $\pm\tfrac{1}{2}$, so that the integral exists for all s with $\mathrm{Re}\,s > 0$. Indeed the function

$$U(s) = s(s-1)\left(\int_1^\infty ([x] - x + \tfrac{1}{2})x^{-s-1}\,dx\right) + \tfrac{1}{2}(s+1)$$

$$= \begin{cases} (s-1)\zeta(s) & (s \neq 1) \\ 1 & (s = 1) \end{cases} \tag{10.43}$$

is analytic at $s = 1$; differentiating gives

$$U'(s) = (2s-1)\left(\int_1^\infty ([x] - x + \tfrac{1}{2})x^{-s-1}\,dx\right)\left[\,-s(s^2-1)\left(\int_1^\infty ([x] - x + \tfrac{1}{2})x^{-s-2}\,dx\right) + \tfrac{1}{2}\,,\right.$$

so that, in partcular,

$$U'(1) = \int_1^\infty ([x] - x + \tfrac{1}{2})x^{-2}\,dx + \tfrac{1}{2} \ .$$

Because ζ is nonzero on $\mathrm{Re}\,s > 1$, so is U.

The second rewriting of $\zeta(s)$ is known as 'Euler's product':

$$E(s) = \prod_{p\ \text{prime}} (1 - (1/p^s))^{-1} \tag{10.44}$$

Here, we proceed by a process of successive reduction. First, note that

$$\sum_{n\ \text{divisible by 2}} (1/n^s) = \sum_{n=1}^\infty (1/(2n)^s)$$

$$= (1/2^s)\sum_{n=1}^\infty (1/n^s)$$

so that

$$\sum_{n \text{ not divisible by } 2} (1/n^s) = (1 - (1/2^s))\zeta(s) \ . \tag{10.45}$$

Next, we do the same with divisibility by 3, to get

$$\sum_{n \text{ not divisible by } 2 \text{ or } 3} (1/n^s) = (1 - (1/3^s))(1 - (1/2^s))\zeta(s) \ .$$

Repeating this process, we have

$$\sum_{n \text{ not divisible by } 2,\ldots,p} (1/n^s) = (1 - (1/p^s))\ldots(1 - (1/2^s))\zeta(s) \ . \tag{10.46}$$

As more and more primes are removed, the left-hand sum includes fewer and fewer terms. In the limit just $n = 1$ is left, and we have

$$1 = \prod_p (1 - (1/p^s))\zeta(s) \ .$$

That is, $\zeta(s) = E(s)$.

This expression for $\zeta(s)$ has a useful consequence. Applying the series expansion $\log(1 - x) = -\sum_{n=1}^{\infty} (1/n)x^n$ to the Euler product formula, we obtain

$$\zeta(s) = \exp\left[\sum_p \{-\log(1 - (1/p^s))\}\right]$$

$$= \exp\left[\sum_p \sum_{n=1}^{\infty} (1/n)(1/p^{ns})\right] \ . \tag{10.47}$$

Since the absolute value of $\exp(a + ib)$ is $\exp(a)$, we have

$$|\zeta(a + ib)| = \exp\left[\sum_p \sum_{n=1}^{\infty} (1/n)\operatorname{Re}(1/p^{n(a+ib)})\right]$$

$$= \exp\left[\sum_p \sum_{n=1}^{\infty} (1/n)(1/p^{na})\cos(nb \log p)\right] \ . \tag{10.48}$$

Now $0 \leq 2(1 + \cos\theta)^2 = 3 + 4\cos\theta + \cos 2\theta$ for all θ. It follows from equation (10.48) that

$$\zeta(a)^3 |\zeta(a + ib)|^4 |\zeta(a + 2ib)| \geq \exp(0) = 1 \ . \tag{10.49}$$

Step 4.
Define the 'error function', $f_\delta(b)$, as follows:

$$f_\delta(b) = -\frac{1}{s\zeta(s)}\zeta'(s) - \frac{1}{(s-1)}, \qquad (10.50)$$

where

$$s = 1 + \delta + ib .$$

Using equation (10.43), we may rewrite this as

$$f_\delta(b) = -U'(s)/sU(s) .$$

But U is nonzero and analytic for $\mathrm{Re}\, s > 0$. Hence the limit

$$f(b) = \lim_{\delta \to 0} f_\delta(b) = -\frac{U'(1+ib)}{(1+ib)U(1+ib)}$$

exists for all b; indeed, it is smooth.

To investigate the behaviour of $f(x)$ for *large* x we use equation (10.49). Rearranging gives

$$|\zeta(a+ib)| \geq \zeta(a)^{-3/4}|\zeta(a+2ib)|^{-1/4} . \qquad (10.52)$$

But notice that the integral in equation (10.42) is bounded for, say, $\mathrm{Re}\, s > \tfrac{1}{2}$. It follows that $|\zeta(a+ib)| \leq k_1|b|$ for large b (and some constant k_1). Also, since $\zeta(s) \simeq 1/(s-1)$ near $s = 1$, we can write

$$|\zeta(a+ib)| \geq k_2 \varepsilon^{3/4}|b|^{-1/4} \qquad (10.53)$$

for large b and small $\varepsilon = a - 1$. Finally, differentiating (10.42) with respect to s gives the similar bound $|\zeta'(a+ib)| \leq k_3|b|$. Thus for any δ with $0 < \delta \leq \varepsilon$

$$|\zeta(1+\delta+ib)| \geq |\zeta(a+ib)| - \int_{1+\delta}^{a} \left|\frac{\partial \zeta(y+ib)}{\partial y}\right| dy$$

$$\geq k_2 \varepsilon^{3/4}|b|^{-1/4} - k_3 \varepsilon|b|$$

$$= k_2 \varepsilon^{3/4}|b|^{-1/4}(1 - k_4 \varepsilon^{1/4}|b|^{5/4}) . \qquad (10.54)$$

Now if we take $\varepsilon = c|b|^{-5}$, for small c, then equation (10.54) becomes

$$|\zeta(1+\delta+ib)| \geq k_2|b|^{-15/4}|b|^{-1/4}(1 - k_4 c)$$

$$= k_5|b|^{-4} ,$$

and therefore

$$|f_\delta(b)| = \left|\frac{1}{s\zeta(s)}\zeta'(s) + \frac{1}{(s-1)}\right|$$

$$\leq \left|\frac{1}{s\zeta(s)}\zeta'(s)\right| + \left|\frac{1}{(s-1)}\right|$$

$$= \frac{|\zeta'|}{|1+\delta+ib||\zeta|} + \frac{1}{|\delta+ib|}$$

$$\leqslant \frac{k_3|b|}{|b|k_5|b|^{-4}}+1$$

$$\leqslant k_6 b^4 . \tag{10.56}$$

Equation (10.56) holds for all sufficiently small δ and all large b. It follows that $f_\delta(b) \leqslant k_6 b^4$ for all large b. For small b, we saw that f was finite and continuous, and therefore bounded. Thus

$$f_\delta(x) \leqslant A + Bx^4 \tag{10.57}$$

for all x.

Step 5.
In terms of $f_\delta(b)$, equation (10.40) becomes

$$f_\delta(b) = \int_1^\infty Q(x) x^{-\delta - ib} \, dx + 1/(\delta + ib) .$$

Putting $y = \log x$ in the integral, and noting that $\int_0^\infty e^{-sx} \, dx = 1/s$ if $\operatorname{Re} s > 0$,

$$f_\delta(b) = \int_0^\infty \{e^{-y} Q(e^y) - 1\} e^{-\delta y} e^{-iby} \, dy . \tag{10.58}$$

Taking limits of this expression is not possible directly. This is where Fourier theory comes in; for what equation (10.58) says is that $f_\delta(\omega) = Fg_\delta(\omega)$, where

$$g_\delta(x) = \{e^{-x} Q(e^x) - 1\} e^{-\delta x} \chi_{(0, \infty)}(x) . \tag{10.59}$$

From the convolution theorem (5.7), $f_\delta(\omega) Fh(\omega) = F(g_\delta * h)(\omega)$ for any integrable function $h(x)$. If h is smooth, then so is $g_\delta * h$, and we may invert this transform, giving

$$F^{-1}(f_\delta Fh)(x) = (g_\delta * h)(x) \tag{10.60}$$

In particular, let us define, for $0 < \alpha, \beta < 1$,

$$h_{\alpha, \beta}(x) = \begin{cases} 0 & (\text{if } |x| > \alpha + \beta) \\ \left(1 - \dfrac{(|x| - \alpha)^7}{\beta^7}\right)^7 & (\text{if } \alpha \leqslant |x| \leqslant \alpha + \beta) \\ 1 & (\text{if } |x| < \alpha) . \end{cases}$$

The relevant features of this function are these:

(1) It is six times continuously differentiable; each of its derivatives is integrable.

(2) It takes the value 1 between $\pm \alpha$, and takes the value zero everywhere outside $\pm(\alpha+\beta)$.
(3) For any x, $h_{\alpha,\beta}(x)$ lies between 0 and 1.

From the results of section 3.12, property (1) means that $\omega^6 Fh(\omega)\to 0$ as $\omega\to\infty$. But $f_\delta(\omega)$ is dominated by $A+B\omega^4$; thus $f_\delta(\omega)Fh(\omega)$ tends to zero faster than $1/\omega^2$ for any $h=h_{\alpha,\beta}$. This means that the family $\{f_\delta Fh\}$ is dominated by an integrable function. Taking limits as $\delta\to 0$, we have $f_\delta Fh\to fFh$, and equation (10.60) becomes

$$F^{-1}(fFh)(x) = \int_0^\infty h(x-y)\{e^{-y}Q(e^y)-1\}\, dy$$

by the dominated convergence theorem. Using (3.8), this is

$$(1/2\pi)F(fFh)(-x) = \int_0^\infty h(x-y)\{e^{-y}Q(e^y)-1\}\, dy \ . \tag{10.61}$$

Both f and Fh are smooth, so the product is smooth. By the Riemann–Lebesgue lemma (3.38), the left-hand side tends to zero as $x\to\infty$; therefore, so does the right-hand side. Thus we have that

$$\lim_{x\to\infty}\int_0^\infty h(x-y)e^{-y}Q(e^y)\, dy = \lim_{x\to\infty}\int_0^\infty h(x-y)\, dy$$

$$= \int_{-\infty}^\infty h(y)\, dy \ . \tag{10.62}$$

Now we use properties (2) and (3) of $h_{\alpha,\beta}$, which say that $\chi_{(-\alpha,\alpha)}(x)\leq h_{\alpha,\beta}(x)\leq \chi_{(-\alpha-\beta,\alpha+\beta)}(x)$. Hence $2\alpha<\int_{-\infty}^\infty h_{\alpha,\beta}(y)\, dy<2(\alpha+\beta)$. It follows from equation (10.62) that

$$2\alpha < \lim_{x\to\infty}\int_0^\infty h_{\alpha,\beta}(x-y)e^{-y}Q(e^y)\, dy < 2(\alpha+\beta) \ .$$

Thus, for large enough x,

$$2\alpha < \int_0^\infty h(x-y)e^{-y}Q(e^y)\, dy < 2(\alpha+\beta) \ . \tag{10.63}$$

Also, the function $e^{-y}Q(e^y)$ is positive. Again using the inequalities $\chi_{(-\alpha,\alpha)}\leq h_{\alpha,\beta}\leq \chi_{(-\alpha-\beta,\alpha+\beta)}$, we obtain from (10.63):

$$\int_{x-\alpha-\beta}^{x+\alpha+\beta} e^{-y}Q(e^y)\, dy > 2\alpha$$

and

$$\int_{x-\alpha}^{x+\alpha} e^{-y}Q(e^y)\, dy < 2(\alpha+\beta) \ . \tag{10.64}$$

But e^{-y} is a *decreasing* function of y, and $Q(e^y)$ is an *increasing* function. Hence

$$\int_{x-\alpha-\beta}^{x+\alpha+\beta} e^{-y}Q(e^y)\,dy < 2(\alpha+\beta)e^{-x+\alpha+\beta}Q(e^{x+\alpha+\beta})$$

and

$$\int_{x-\alpha}^{x+\alpha} e^{-y}Q(e^y)\,dy > 2\alpha e^{-x-\alpha}Q(e^{x-\alpha}) \ .$$

It follows from (10.64) that

$$2(\alpha+\beta)e^{-x+\alpha+\beta}Q(e^{x+\alpha+\beta}) > 2\alpha$$

and

$$2\alpha e^{-x-\alpha}Q(e^{x-\alpha}) < 2(\alpha+\beta) \ .$$

Substituting $y = x + \alpha + \beta$ in the first of these and $y = x - \alpha$ in the second, we get

$$2(\alpha+\beta)e^{-y+2\alpha+2\beta}Q(e^y) > 2\alpha$$

and

$$2\alpha e^{-y-2\alpha}Q(e^y) < 2(\alpha+\beta)$$

for large enough y. Putting these inequalities together yields

$$(\alpha/(\alpha+\beta))e^{-2\alpha-2\beta} < e^{-y}Q(e^y) < ((\alpha+\beta)/\alpha)e^{2\alpha} \ . \tag{10.65}$$

But these inequalities hold for *arbitrary* (small) α and β. In particular, we can make both α and β/α as small desired. As $\alpha \to 0$ and $\beta/\alpha \to 0$, both the upper limit and the lower limit of equation (10.65) tend to 1. It follows that $e^{-y}Q(e^y) \to 1$ as $y \to \infty$. One final substitution of $z = e^y$ gives the result that

$$Q(z)/z \to 1 \text{ as } z \to \infty \ .$$

From Step 1, the prime number theorem $\pi(x)\log x/x \to 1$ follows.

11
Physics

11.1 Quantum theory

If a beam of electrons is aimed at a narrow slit, it does not simply pass through, as one would expect of a beam of particles. Instead, it is *diffracted*. Moreover, it does so in just the way one would expect of a wave phenomenon, like light or sound. For a wave, the pattern of diffraction depends on its wavelength λ; for electron beams, the *effective* (or *Compton*) *wavelength* is

$$\lambda_c = h/p , \qquad (11.1)$$

where h is Planck's constant and p is the momentum of the individual electrons. Notice that λ_c does not depend on the intensity of the beam. Equivalently, the (spatial) angular frequency $\omega_c = 2\pi/\lambda_c$ of the electron wave is related to the electrons' momentum by

$$p = \hbar\omega_c , \qquad (11.2)$$

where $\hbar = h/2\pi$. (In SI units, $h \simeq 6.626 \times 10^{-34}$ J s, while $\hbar \simeq 1.054 \times 10^{-34}$ J s.)

Again, if a monochromatic beam of X-rays is directed at atoms, the radiation scattered off the electrons shows a significant increase of wavelength. Further, the wavelength of the scattered radiation depends only on the angle of scatter. This is just what one would expect of a collision of particles; the X-ray 'particle' gives up some of its momentum to the electron, causing it to recoil. We can calculate the effective momentum p_e of an X-ray 'particle'; it is related to the angular frequency ω of the X-rays by the equation $p_e = \hbar\omega$. (These two phenomena, and others like them, are collectively termed *wave–particle duality*.)

Quantum theory proposes to resolve these problems by treating both 'waves' and 'particles' as neither pure waves nor pure particles, but as *wave packets*: wave-like structures, confined to a small region (see Fig. 11.1). The localization enables them to act as particles; the wave-like substructure enables them to act as waves. A wave packet is represented by a *wavefunction* ψ which is (it is proposed) a solution of the *Schrödinger equation* for the system. In the absence of external forces, the Schrödinger equation is

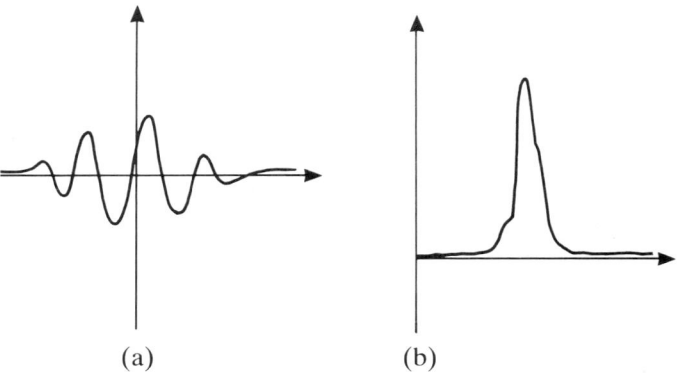

Fig. 11.1 — A wave packet, in the x-domain (a) and in the ω-domain (b).

$$\frac{\partial \psi}{\partial t} = i\alpha \nabla^2 \psi \ . \tag{11.3}$$

For a (non-relativistic) 'particle' of mass m, the constant α is $\hbar/2m$; for an electromagnetic 'wave' of spatial angular frequency ω, $\alpha = c/\omega$, where c is the speed of light ($c \simeq 299792$ km s^{-1} in a vacuum). The Schrödinger equation thus replaces, in a sense, Newton's second law. (In the absence of external forces, Newton's law says simply that $\partial^2 x/\partial t^2 = 0$.) Note that the Schrödinger equation is linear. Note too, however, the explicit appearance of i: every solution of (11.3) is necessarily complex-valued.

Unlike a particle, a wave packet does not have a precise position; it extends over a finite region. Unlike a wave, it does not have a precise wavelength either; it has a finite bandwidth. Now for a wavefunction ψ, the uncertainty theorem (equation (5.26)) tells us that $W(\psi)W(F\psi) \geq \tfrac{1}{2}$. We can think of $\Delta x = W(\psi)$ as being the accuracy to which the position of the wave packet may be measured; similarly, $\Delta \omega = W(F\psi)$ is the uncertainty in a determination of its angular frequency. Thus $\Delta x \Delta \omega \geq \tfrac{1}{2}$. Further, since its momentum is $p = \hbar \omega$, we have $\Delta p = \hbar \Delta \omega$, and

$$\Delta x \Delta p \geq \tfrac{1}{2}\hbar \ . \tag{11.4}$$

Equation (11.4) is the famous *Heisenberg uncertainty principle*. We say that position x and momentum p are *conjugate* variables; they cannot both be measured simultaneously to arbitrary precision. Other conjugate pairs of variables, also arising from a consideration of the properties of wave packets, include time (duration) and energy; and angular position and angular momentum.

Example Consider an electron in an atom. According to the classical model, the electron is a point particle, orbiting the nucleus like a planet around its sun. But if it obeyed Maxwell's equations, then, as an accelerating charge, it would radiate energy, and the atom would collapse in a tiny fraction of a second.

In the Heisenberg picture, the electron is not a point particle, but a wave

Ch. 11] Physics 249

packet ψ. The electron's mass is known: about 10^{-30} kg. The energy it takes to knock it out of the atom is also known, experimentally: about 10^{-18} J, typically. If it remains within the atom, its kinetic energy is then known to this accuracy; so, using $E = \frac{1}{2}mv^2 = p^2/2m$, its momentum is known to within $\Delta p \simeq 10^{-24}$ kg m s^{-1}. It follows from the uncertainty principle that $\Delta x \geq 10^{-10}$ m. But this is the size of a typical atom. What we have, then, is not an orbiting point particle but an electron *cloud*, smeared over the region surrounding the nucleus.

Another way of coming to the same conclusion is this. We think of the orbiting electron as spiralling in, as closely as it can. Suppose the outer electron of an atom is 'orbiting' the nucleus at a distance r. If the position of this electron were determinable to $\Delta x \ll r$, then the electron would behave classically and spiral in. Since it does not, $\Delta x \simeq r$. Now, let the charge on the electon be $-e$; the effective charge on the atom is the $+e$, owing to the shielding effect of the other electrons. The electron's electrostatic potential (energy) is then $e^2/4\pi\varepsilon r$, where ε is the local *permittivity* (for a vacuum, $\varepsilon \simeq 8.854 \times 10^{-12}$ F m^{-1}). If its mass is m, then its escape velocity from the atom is $v = \sqrt{[e^2/2\pi\varepsilon m r]}$. Thus $\Delta p \simeq mv = \sqrt{[me^2/2\pi\varepsilon r]}$; it follows that $\frac{1}{2}\hbar \simeq \Delta p \Delta x \simeq \sqrt{[mre^2/2\pi\varepsilon]}$, so that

$$r \simeq \tfrac{1}{2}\pi\varepsilon\hbar^2/me^2$$
$$\simeq 10^{-10}\,\text{m}. \tag{11.5}$$

This is, if you will, why atoms are about 10^{-10} m in radius: they can collapse no further.

Let us see how we might solve the Schrödinger equation properly for the hydrogen atom. With the addition of an electrostatic potential term, the equation becomes

$$i\hbar \frac{\partial \psi}{\partial t} = \frac{\hbar^2}{2m} \nabla^2 \psi - \frac{e^2 \psi}{\|\mathbf{x}\|} . \tag{11.6}$$

Here m is the mass, and e is the charge, on the electron; the nucleus, being much heavier, is assumed to be a point mass fixed at the origin.

As usual, we look for separated solutions. In this case, the natural coordinates are spherical polars. In particular, we look for time-independent solutions, or *stationary states*; then $\partial\psi/\partial t = 0$, we may write $\psi(\mathbf{x}) = R(r)\Theta(\theta)\Phi(\varphi)$, and (11.6) becomes

$$0 = -\frac{\hbar^2}{2m}\left\{\frac{\Theta\Phi}{r^2}\frac{d}{dr}\left(r^2\frac{dR}{dr}\right) + \frac{R\Phi}{r^2\sin\theta}\frac{d}{d\theta}\left(\sin\theta\frac{d\Theta}{d\theta}\right) + \frac{R\Theta}{r^2\sin^2\theta}\frac{d^2\Phi}{d\varphi^2}\right\} - \frac{e^2 R\Theta\Phi}{r}.$$

Division by $R\Theta\Phi/r^2\sin^2\theta$ gives

$$0 = -\frac{\hbar^2}{2m}\left\{\frac{\sin^2\theta}{R}\frac{d}{dr}\left(r^2\frac{dR}{dr}\right) + \frac{\sin\theta}{\Theta}\frac{d}{d\theta}\left(\sin\theta\frac{d\Theta}{d\theta}\right) + \frac{1}{\Phi}\frac{d^2\Phi}{d\varphi^2}\right\} - e^2 r\sin^2\theta ,$$

which allows the term in φ to be separated. The φ equation is just the harmonic equation $d^2\Phi/d\varphi^2 = k\Phi$; since Φ must be periodic of period 2π, $k = -n^2$ and $\Phi(\varphi) = A\cos(n\varphi + \kappa)$. Similarly (though not quite so simply), the r and θ equations may be separated. In fact, the function Θ turns out to be related to the Legendre polynomials of section 8.1, while R is connected with yet another family of orthogonal functions, the *Laguerre polynomials*. (These separated solutions are called *spherical harmonic functions*.)

It is fairly easy to calculate the electrostatic energy in each mode, or *state*, of the electron wave packet. The state with lowest energy is the *ground state*, and a 'cold', unexcited atom will tend to be in this state. For the hydrogen atom, the ground state wavefunction takes the form

$$\psi(\mathbf{x}) = (1/\pi a^3)^{1/2} e^{-\|\mathbf{x}\|/a} , \qquad (11.7)$$

where $a = \varepsilon \hbar^2 / m e^2$ is the *Bohr radius* of the atom.

If the electron is given energy, for instance by absorbing a photon, it may jump to a state of higher energy, with a more complex wavefunction form. It may then spontaneously *decay* to a lower-energy state, re-emitting a photon.

11.2 Lasers

The emission of photons by excited electrons is responsible for all our light, from candle flames to stars. In few situations is the quantum nature of light really important; one case where it is, however, is in lasers.

At the heart of a laser is the *laser cavity*, an 'optical resonator': an object, usually cylindrical, with mirrors at both ends (see Fig. 11.2). In practice, this object might be

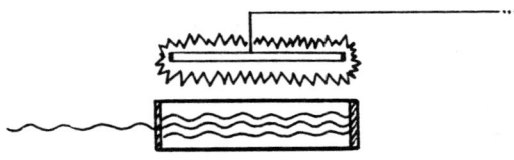

Fig. 11.2 — Laser (schematic).

a crystal, or a bottle of gas, or even a plasma. Around this resonator is an energy source: usually a bright light of some sort.

The operation of a laser can be explained, semi-classically, as follows. When the laser is switched on, the source flashes. The photons emitted are unordered, and come out in a range of energies, depending on the process that produces them. Some of these photons are absorbed by the atoms in the laser cavity, lifting the electrons in them from one state to another. Ideally, the source is tuned to lift electrons predominantly to a particular excited state. Eventually, a large fraction of the electrons in the resonator are in the higher-energy state.

Sooner or later, one of these electrons will spontaneously decay to the lower state, emitting a photon. The energy of this photon is precisely the difference in energy between the two states. This photon may strike another 'excited' electron; this has the effect of stimulating it to release *its* energy, in the same way. Interestingly, the resulting pair of photons will have not only the same energy, but also the same *phase* and the same *direction*. (The photons are said to be *coherent*). In this way, a 'chain reaction' of such *stimulated emission* sets in, the train of photons increasing exponentially in numbers. (The word 'laser', indeed, is an acronym for 'light amplification by the stimulated emission of radiation'.)

The distance a photon must travel before it stimulates another emission is quite long. Thus, photons not travelling parallel to the sides of the laser cavity will escape out of the side of the resonator before they can do anything. Only those travelling almost horizontally are reflected back and forth, between the mirrors, enough times to be able to trigger the chain reaction. Light travels very fast, of course, so the process takes very little time. Eventually, one of the mirrors is too weak to contain the pulse, and it bursts out of that end. (One mirror is therefore made deliberately weaker than the other, so that the pulses all come out of the same end.)

This explanation is simple and straightforward, but it fails on a number of counts. For example, in the uncertainty principle, position and momentum (or wavelength) are conjugate variables. Now the laser has a resonating cavity of finite length l; so any photon contained within that cavity has $\Delta x \leq l$. The cavity is typically 10^{-1} m long, so that Δp is variable by 10^{-33} kg ms^{-1} or so. In a visible-light laser, this implies a wavelength variability of a few parts per million. Thus, the photons do *not* have a 'precise' energy.

Again, the electrons stay in their excited state only for a finite time, τ say. But energy E (or temporal frequency) and time duration are conjugate variables, and $\Delta E \Delta \tau \geq \frac{1}{2}\hbar$. Therefore the energy of the photons *even in their emission* is not precisely defined; rather, it is subject to an uncertainty which is inversely proportional to the *mean lifetime* of the excited electrons. A mean lifetime of one microsecond, for instance, implies a frequency spread of the order of one megahertz; for visible light at 4×10^{14} Hz, this corresponds to one part per 10^9. (This spreading of energies is called *natural broadening*, to distinguish it from the various other processes at work in real lasers — like the recoiling of the atoms in the laser when struck by photons, which gives rise to *Doppler broadening*.)

We may avoid these problems by using the wave packet model of light. Rather than speaking about the photons as particles, we imagine them as electromagnetic *standing waves* in the laser cavity, just like the waves on a finite string (section 1.8). A closed resonator with perfect mirrors would restrict the photons that can exist within the cavity to those for which a whole number of half-waves fits into the cavity length. These are, indeed, the natural modes of an electromagnetic oscillator. More generally, the light in the laser cavity will be a superposition of such modes.

Now, the operation of stimulating the emission of the laser, at a 'precise' wavelength, becomes the excitation of exactly one of the resonant modes in the cavity. As the excitation builds, the wave will begin to emerge from the end with the thinner mirror. (This emergence can be understood in the string analogy too. Imagine an *infinite* string, for space. Attach two heavy masses, distance l apart, for the mirrors. The length of string between the masses is effectively finite, so will vibrate as a string with fixed ends. But as the vibration energy mounts, the masses will begin to move, and the wave will pass out into the 'outside world'.)

Pulsed lasers show an interesting twist. Typically, a laser pulse will last a few microseconds or milliseconds. This is much longer than the time it takes the light to traverse the length of the cavity, and the evolution of the laser output is therefore a slow process of energy being drained from the (photon) standing wave. But there exist lasers that produce ultra-short pulses: often substantially less than a picosecond. Light travels only about 0.3 mm in a picosecond, so these pulses are of very small spatial extent. Here, it would seem, the naïve quantum picture of photons bouncing backwards and forwards would seem to be a more sensible view.

In such cases, our understanding of Fourier series can help us out. The ideal laser pulse would be an impulse, travelling at the speed of light c, corresponding to a wavefunction like $\psi(x,t) = A\delta(x - ct)$. However, the laser pulse inside the cavity undergoes reflection, so that we have a *train* of impulses: something like $\psi(x,t) = A\text{III}((x - ct)/2l) - A\text{III}((x + ct)/2l)$. In terms of Fourier series, we express ψ as a sum of the various possible harmonic modes of oscillation (equation (2.19)): each mode is present with equal amplitude. Thus, in order to produce an impulsive output, we must excite all the possible photon modes in the cavity, to equal amplitude (with appropriately matched phase).

In practice, this is not possible. Even a single quantum (of energy $\hbar\omega$) in each mode would require an infinite total amount of energy to be input to the resonator. In addition, various other physical effects arise at high energies that prevent an arbitrarily good approximation to impulses. Experimentally, by attaching 'pulse compression' devices to lasers (and so effectively altering the properties of the weaker mirror), pulses as short as 10^{-14} s have been produced. This corresponds to just six waves of visible (green) light. From the uncertainty principle, since Δt is sharply constrained, ΔE must be quite large; for a 10 fs pulse of visible light, the photon energy is indeterminate by several per cent.

11.3 Scattering

In section 11.1 we saw how the picture of an electron as a point particle, orbiting around a nucleus, had to be revised in the light of its quantum properties; it 'smears out' over the atom. But the nucleus itself has quantum properties too. It is true that the mass of the nucleus is thousands of times the mass of the electron, so that it 'smears' over a much smaller volume; from the point of view of the 'cold', low-energy electron, it is reasonable to regard it as a point. But when collisions at higher energies, and higher momenta, are considered, the quantum structure of the nucleus becomes significant.

The archetype of collision events is the scattering experiment, in which a beam of particles — electrons, say — is directed at a target — a nucleus, say. To analyse this system, we begin by making a number of simplifications. We place the target nucleus at the origin, and treat it as an immobile point mass (though, as we said, these assumptions are invalid at high energies). Incident to it is a beam of electrons of momentum $\hbar\omega$, which are duly scattered. We assume that the only force involved is electrostatic, and that the nucleus, being much heavier than the electrons, does not recoil in any collision. The electrons have mass m and charge $-e$, the nucleus has charge $+Ze$.

Classical mechanics would lead us to expect the incoming electrons to adopt a *hyperbolic* orbit under the attraction of the nuclear charge (Fig. 11.3). In quantum mechanics, the electrons are not particulate, but instead are associated with a

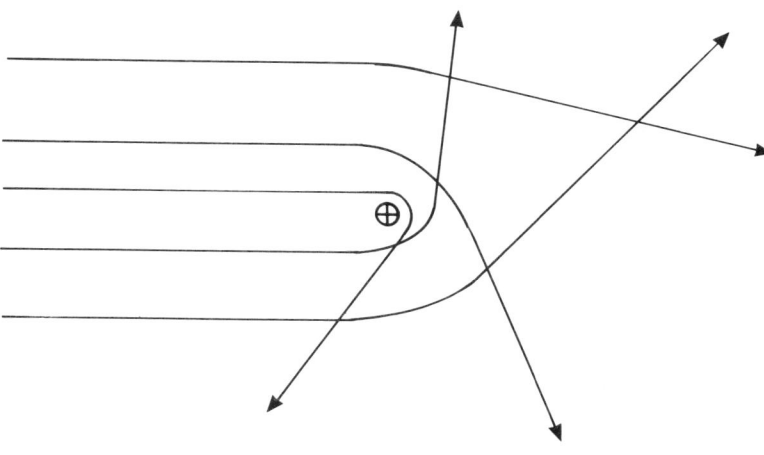

Fig. 11.3.

wavefunction: a uniform incident beam travelling steadily along the z-axis is given by the wavefunction

$$\psi_{in}(\mathbf{x},t) = \psi_{in}(x,y,z,t) = e^{i\omega z} \tag{11.8}$$

(we may ignore the time variable, since the beam is steady).

At large distances $r = \|\mathbf{x}\|$, the scattered beam will look like a wavefunction ψ_{sc} which is propagating *radially*. It will, moreover, be *axially symmetric*, so not dependent on the 'longitudinal' polar coordinate φ. Since the intensity (energy density) $|\psi_{sc}|^2$ of this radial wave decreases with $1/r^2$ at large r, we have, at large r,

$$\psi_{sc}(\mathbf{x},t) = \psi_{sc}(r,\theta,\varphi,t) = A(\theta)(1/r)e^{i\omega r} . \tag{11.9}$$

(Note that the scattered electrons again have momentum $\hbar\omega$. It is independent of the scattering angle θ because no momentum is transferred to the nucleus.) The complete wavefunction ψ of the electron beam must then look like $\psi_{in} + \psi_{sc}$ for large r.

The function $A(\theta)$ represents how the wave gets scattered by the obstacle; it is called the *scattering amplitude*. The intensity of the scattered beam is measured in mass (or energy) flow per unit solid angle, and is represented by $|A(\theta)|^2$.

To solve the scattering problem, then, we find the solution of Schrödinger's equation that satisfies these boundary conditions. The equation in this case includes a potential term:

$$i\hbar \frac{\partial \psi}{\partial t} = \frac{-\hbar^2}{2m} \nabla^2 \psi + \frac{Ze^2 \psi}{4\pi\varepsilon r} .$$

The solution involves a family of functions called the *confluent hypergeometric functions*: these are defined by the series formula

$$F_{a,c}(x) = 1 + \left(\frac{a}{c}\right)x + \left\{\frac{a(a+1)}{c(c+1)}\right\}\left(\frac{x^2}{2!}\right) + \ldots \tag{11.10}$$

In fact, it may be shown that

$$\psi = K e^{i\omega r} F_{1+\alpha,1}(-2i\omega r\sin^2\tfrac{1}{2}\theta) ,$$

where

$$\alpha = -imZe^2/4\pi\varepsilon\hbar^2\omega . \tag{11.11}$$

Now the behaviour of $F_{a,c}(x)$ for large x is known. It follows that

$$|A(\theta)|^2 = \alpha^2/(4\omega^2\sin^4\tfrac{1}{2}\theta) .$$

Since the kinetic energy of the incoming electrons is $E = p^2/2m = \hbar^2\omega^2/2m$, this reduces to

$$|A(\theta)|^2 = (Ze^2/16\pi E)^2 \text{cosec}^4\tfrac{1}{2}\theta . \tag{11.12}$$

Curiously, this is the same formula as is given by *classical* electrodynamics, based on electrons as point charges and hyperbolic orbits. This coincidence holds only for the very special case of a Coulomb potential (i.e. a potential term in the Schrödinger equation of the form k/r). It is reasonably accurate, according to experiment; though it can be improved by taking into account nuclear recoil, relativistic effects, and forces due to the particles, magnetic moments.

However, having included all these factors, there is still a discrepancy between theory and experiment. We find, in practice, a significantly *lower* fraction of the electron beam deflected through very large angles than (11.12) would indicate. That is, the observed scattering amplitude $\bar{A}(\theta)$ does not agree with the predicted function $A(\theta)$. What we are seeing is the quantum blurring of the nucleus.

In fact, we can use the results of scattering experiments to deduce the *actual* charge density profile $\rho(\mathbf{x})$ of the nucleus. We may assume that ρ is spherically symmetrical; so, then, is its (three-dimensional) Fourier transform $F_3\rho(\mathbf{s})$. We define the *form factor* of the nucleus as

$$F(\|\mathbf{q}\|^2) = F_3\rho((1/\hbar)\mathbf{q}) = \int e^{-i\mathbf{q}\cdot\mathbf{r}/\hbar} \rho(\mathbf{r})\, d\mathbf{r} . \tag{11.13}$$

Since ρ is symmetrical, $\rho(\mathbf{r}) = \rho(-\mathbf{r})$; it follows that $F(x)$ is real-valued (see section 3.10).

In this transform, $\mathbf{s} = \mathbf{q}/\hbar$, so $\mathbf{q} = \hbar\mathbf{s}$. We think of $\mathbf{q}$ as representing the electron's overall change of (vector) momentum. If it is scattered through angle θ, and has initial momentum p, then

$$\|\mathbf{q}\| = 2p(1-\cos\tfrac{1}{2}\theta) = 4p\sin^2\tfrac{1}{2}\theta . \tag{11.14}$$

Now it turns out that $F^2(\|\mathbf{q}\|^2)$ is the ratio between the *actual* scattering amplitude (from the spherically symmetrical, diffuse nucleus) and the *ideal* scattering amplitude (from a point nucleus, as calculated above) at angle θ. But this can be measured experimentally, for real nuclei. Take square roots, to get F; then an inverse Fourier transform yields ρ.

This has been done for a variety of atomic nuclei. Fig. 11.4 shows typical curves for the charge density profiles of hydrogen, oxygen, and gold nuclei. Heavy nuclei, like gold, contain many particles, and so are (for statistical reasons) well behaved: the charge density is almost uniform over a large central region, and falls off steadily

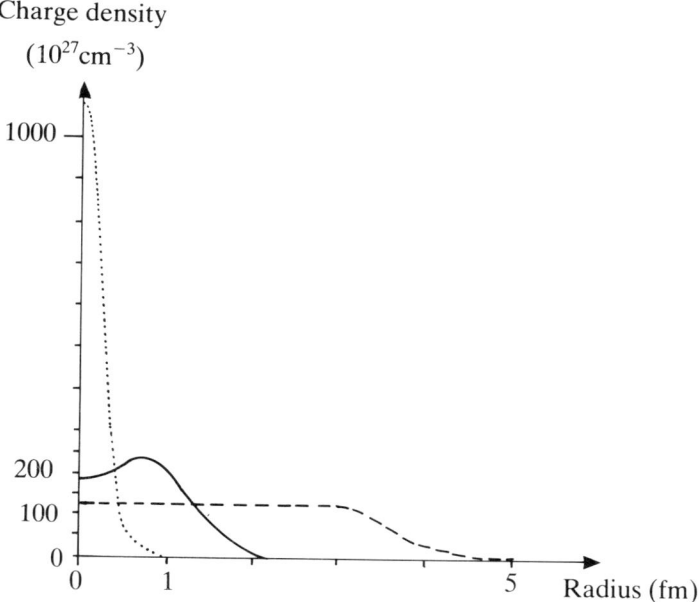

Fig. 11.4 — Charge density profiles of hydrogen (dotted), oxygen (solid), and gold (dashed) nuclei, as they might appear from scattering experiments. (See R. Hofstadter, Nuclear and nucleon scattering of high-energy electrons, *Ann. Rev. Nucl. Sci.* **7** (1957), p. 231.)

at the edges. By contrast, the hydrogen nucleus contains a single proton; its charge profile is much more peaked, and covers a much smaller volume. Curiously, the intermediate oxygen nucleus (16 nucleons) has a slightly 'hollow' charge profile. This can perhaps be interpreted as due to the mutual repulsion of the positively charge protons.

Note that the individual protons and neutrons cannot be discerned in these profiles. This is for four principal reasons. First, these particles show quantum 'blurring'. Secondly, the assumption of spherical symmetry in the nuclei 'averages' the various parts of the nucleus at the same distance from its centre. Thirdly, the scattering experiments do not just use a single target nucleus, but many thousands of independent nuclei. For all these reasons, that which is deduced as ρ is actually the *average* charge density profiles of such nuclei.

The final reason for the smoothness of the plots is the fact that the electron beam used to measure these form factors is itself of finite momentum. Thus, according to the uncertainty principle, the electrons are relatively 'large', and the spatial resolution of the scattering experiment is limited. Incident beams of higher momentum suffer less from this problem.

If the proton (the hydrogen nucleus) were a 'fundamental' particle, acting as a genuine *point* charge, then the experimentally deduced charge density profile ought to become sharper and more peaked as the electron beam energy rises. In fact this does not happen. Even at the highest energies, the proton retains a finite structure, with an approximately exponential profile, and mean radius of approximately

6×10^{-16} m. The interpretation of this is that the proton is a *composite* particle, and that what we are seeing is the blurred, average charge profile of its three constituent *quarks*.

11.4 Diffraction

The simplest sort of diffraction involves a plane wave, in two or three dimensions, encountering a semi-infinite obstruction (see Fig. 11.5). For example, we might have

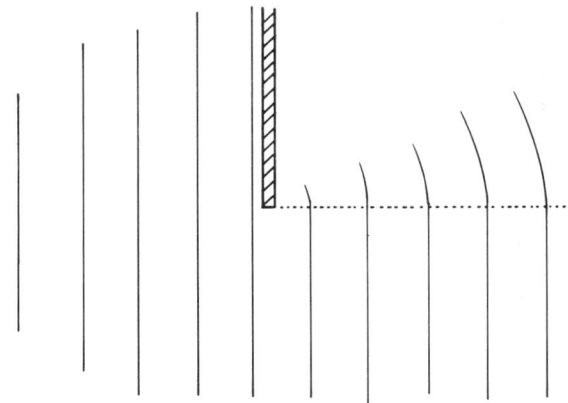

Fig. 11.5 — Diffraction of waves, travelling from left to right, past a semi-infinite obstruction.

a point light source at infinity shining at (and past) the edge of an object. Rather than casting a sharp shadow, the light 'spreads out' around the corner. This behaviour can be explained as the superposition of waves from an infinitude of 'virtual' point sources, one at each point of the plane where the light is *not* blocked, each generating its own small spherical wave. (This explanation is called *Huygens' principle*. It can, more formally, be deduced from quantum theory.)

Suppose, conversely, we have an opaque plane surface (in the $y - z$ plane, say) with a (small) cut-out shape R, which is illuminated with a uniform beam of monochromatic light. Let the light have spatial angular frequency ω and amplitude A units per square metre, and be travelling parallel to the x-axis. By Huygens' principle, the element $dydz$ at a typical point $(0,y,z)$ of R acts as a little light source, contributing a wavelet of the form

$$\frac{A\,dydz\,\exp(i\omega\|\mathbf{r} - (0,y,z)\|)}{\|\mathbf{r} - (0,y,z)\|} \tag{11.15}$$

to the light at point $\mathbf{r}$ in the 'downstream' region.

At large distances, the denominator is approximately $\|\mathbf{r}\|$. The exponent cannot be so easily approximated, however, because its phase is important. If we put $\mathbf{r} = (X,Y,Z)$, and set $r = \|\mathbf{r}\|$, then (provided that $y,z \ll Y,Z, \ll r$) we have

$$\|\mathbf{r} - (0,y,z)\| = \{X^2 + (Y-y)^2 + (Z-z)^2\}^{1/2}$$
$$= \{X^2 + Y^2 + Z^2 - 2Yy - 2Zz + y^2 + z^2\}^{1/2}$$
$$= r\{1 - 2(Y/r)(y/r) - 2(Z/r)(z/r) + (y^2 + z^2)/r^2\}^{1/2}$$
$$= r(1 - (Y/r)(y/r) - (Z/r)(z/r) + O(1/r^2))$$
$$= r - Y(y/r) - Z(z/r) + O(1/r) \ . \tag{11.16}$$

Now (11.15) becomes, to first order,

$$\sim \frac{A\,dy\,dz\exp[i\omega(r - (Yy/r) - (Zz/r))]}{r} \ .$$

The *total* radiation is then

$$\psi(X,Y,Z) \simeq \int_R \left(A\,dx\,dy\,e^{i\omega(r - (Yy/r) - (Zz/r))}/r \right)$$

$$= (Ae^{i\omega r}/r) \int \chi_R(y,z) e^{-i(\omega/r)(Yy + Zz)}\,dy\,dz \ , \tag{11.17}$$

the integral now being over the whole y–z plane.

More generally, suppose we have a *partially* transmitting screen. That is, suppose the point $(0,y,z)$ in R lets through a fraction $k(y,z)$ of the incident light amplitude. (We can make k complex-valued if we wish to allow the phase of the transmitted light to be altered.) In this case, the element $dy\,dz$ at $(0,y,z)$ acts as a source of intensity $A\,dy\,dz\,k(y,z)$, and the wave at (X,Y,Z) is

$$\psi(X,Y,Z) \simeq (Ae^{i\omega r}/r) \int k(y,z) e^{-i(\omega/r)(Yy + Zz)}\,dy\,dz$$

$$= (Ae^{i\omega r}/r) F_2 k(\omega Y/r, \omega Z/r) \ . \tag{11.18}$$

In other words, apart from the scaling factors and the phase term $e^{i\omega r}$, the result of diffraction is *the (two-dimensional) Fourier transform* of the transmissivity function. This unexpectedly simple formula, connecting diffraction with Fourier theory, is called the *Fraunhofer diffraction* model.

Note that the wave angular frequency ω is not a variable in this transformation. The x-domain is the y–z plane, while the ω-domain is the Y–Z plane, apart from the scaling.

Now the wavefunction ψ gives the amplitude and phase of the wave. The energy flux at a point is (as in quantum theory) given by $|\psi|^2$. If we record light patterns as an image on photographic film, or using a photomultiplier, or with the more modern charge-coupled device, what we are measuring is the energy flux. In any such measurement, we lose the phase information of ω. In particular, recording the light pattern ψ downstream of Fraunhofer diffraction gives us (via (11.18)), not the Fourier transform, but the *power spectrum* of $k(y,z)$.

This has practical consequences. One of the principal uses of diffraction is in the mapping of atoms in crystals. The diffracting object is now a *three*-dimensional lattice of atoms (or ions), so the analysis is more complicated, but the principle remains the same. By using appropriate combinations of distances r and angular frequencies ω, we can 'magnify' the transform to a readable size. It would be very handy if we could then simply perform an inverse transform, for the crystal structure would be immediately revealed. But since we have only the power spectrum, and no phase information, this is not possible. To determine this phase information, we must use indirect means, such as 'doping' the crystal with heavy atoms and noting the distortion in the resulting diffractogram. (See section 12.5 for more about diffraction crystallography.)

More recently, the low bandwidth of laser light has allowed a system (*holography*) to be developed in which the phase of a diffraction pattern can be recorded, directly, in a thin sheet of material. In one technique, the sheet records a diffraction pattern related to the transform of the original (three-dimensional) object. If we look at such a sheet, we see the diffraction pattern generated by this diffraction pattern. But the Fourier transform is (almost) self-inverse, so that what we see is a three-dimensional *image* of the original object, provided we look from far enough away. (If we look *very* closely at the recorded pattern we simply see the diffractogram. Under a microscope, the image disappears.) Unlike a photograph, the diffraction pattern is not a 'direct' representation of its object, so the object field can be much deeper than the physical carrier of the diffraction pattern.

Equation (11.18), then, deals with the diffraction of a *uniform* beam of light through a *complex* aperture. There is a similar effect in the diffraction of a *nonuniform* light source. Suppose the input beam has amplitude $A(y,z)$ at point (y,z) on the aperture plane. We retain, for comprehensiveness, the partial transmission $k(y,z)$. The element $dydz$ then produces, on Huygens' principle, a wave like

$$\frac{k(y,z)A(y,z)dydz\exp(i\omega\|\mathbf{r}-(0,y,z)\|)}{\|\mathbf{r}-(0,y,z)\|};$$

the total wave at a 'downstream' point (X,Y,Z) then has the asymptotic form

$$\psi(X,Y,Z) = (e^{i\omega r}/r)F_2(Ak)(\omega Y/r, \omega Z/r)$$
$$= (e^{i\omega r}/r)(1/2\pi)^2(F_2A *_2 F_2k)(\omega Y/r, \omega Z/r) \qquad (11.19)$$

from the converse of the vector convolution theorem.

There is a paradox here. If $k(y,z) = 1$ everywhere, then the y–z plane is empty, so there is no diffraction; yet (11.19) indicates that what passes through is the *transform* of the input A. The problem here is one of scale: in order that (11.19) hold, we need to look at distances r large in comparison with the extent of the transmission window to see the transform. If $k = 1$ *everywhere*, this window is infinite. If $k = 1$ only over a restricted region, then we are *not* looking through an empty y–z plane, but through a small aperture.

In fact, this is exactly how electron microscopy works. A tiny sample scatters (transmits) electron waves; a long way away these look like the transform of the emission pattern; an inverse transform recovers the image, for display.

Instead of using a single small hole, which transmits only a small amount of light, we could use a y–z plane screen with an *array* of holes. Such a screen is called a *diffraction grating*. Thus, for example, a screen with widely spaced small holes in a square array has a transmission function like $k(y,z) \simeq III_2(y/d,z/d)$. From equations (4.37) and (4.25), we have $F_2k(\alpha,\beta) = d^2 III_2(d\alpha/2\pi, d\beta/2\pi)$. Equation (11.19) gives the output of this 'aperture' as

$$F_2(kA)(\omega Y/r, \omega Z/r)$$
$$= (e^{i\omega r}/r)(d/2\pi)^2(III_2(d\alpha/2\pi, d\beta/2\pi) *_2 F_2A)(\omega Y/r, \omega Z/r)$$

$$= (e^{i\omega r}/r) \sum_{m,n=-\infty}^{\infty} F_2A((\omega Y/r) - (2\pi/md), (\omega Z/r) - (2\pi/nd)) \ .$$

Suppose the effective bandwidth of the function A is κ, so that $F_2A(\alpha,\beta)$ is only significant for $|\alpha|,|\beta| < \kappa$. If d is chosen to be small compared with $2\pi/\kappa$, the sum in (11.19) reduces to

$$F_2A(\omega Y/r, \omega Z/r) \qquad (11.20)$$

in the region $|Y|,|Z| < \kappa r/\omega$.

Again, though, having very small holes means that very little light gets through. Thus, most gratings have holes which are *not* small compared with their separation. The analysis of these gratings is slightly more complicated, but similar in kind; they yield a more complex convolution.

11.5 Telescopes

Consider the problem of seeing stars. They are large objects, but they are a long way away. Sirius, the brightest normal star in the night sky, has a radius of $\sim 10^9$ m, but is $\sim 10^{17}$ m from Earth. Seen from Earth, it has an angular diameter of $\sim 2 \times 10^{-8}$ rad, or 0.004 seconds of arc. This is far too small to show up through any (current) telescope as anything but a dot. The problem is not just one of magnification: there is a fundamental uncertainty-type relation between the *size* of the light collector and the (*angular*) *resolution* of the image.

Suppose we have a simple lens, focusing the light (of spatial angular frequency ω, say) from a distant *point* source. The light from such a source arrives at the lens as a plane wave. It passes through the lens and comes to a point at a distance d from the lens (the *focus*). Now, on Huygens' principle, each point on the back side of the lens acts as a new source. But the light passes more slowly through the lens material than through air, so that it is retarded at the (thick) centre of the lens compared to the edge. In this view, the focusing effect is nothing more than the fact that wavelets from the edge and wavelets from the centre arrive at the focus *in phase* (see Fig. 11.6). The wavelet from the centre travels a shorter distance, but suffers an exactly compensating phase delay in passing through the lens.

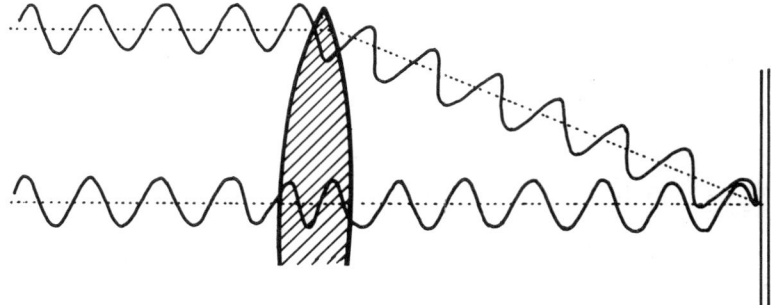

Fig. 11.6 — Operation of a lens. A wavelet travelling through the centre of the lens is delayed just enough to be in phase with wavelets from the periphery, at the distance of the focus.

Precisely, if the lens has radius r, the wavelet at the centre travels a distance d, corresponding to $d\omega/2\pi$ wavelengths of the light; while a wave at the periphery travels the greater distance $\sqrt{(d^2 + r^2)}$. When $r \ll d$, $\sqrt{(d^2 + r^2)} \simeq d + (r^2/2d)$, so the phase lag angle at the centre must therefore be $\omega r^2/2d$ to compensate.

But the light arriving at the focus does not yield a point image. Consider the situation a small distance $\varepsilon \ll r$ 'sideways' from the focus. The on-axis wave has still to travel a distance of approximately d (to first order in ε/d) to this point; but the wave from the edge on that side of the lens travels a slightly smaller distance than to the focus: $d + (r - \varepsilon)^2/2d$, or $(d + r^2/2d) - (r\varepsilon/d)$ to first order. This is a reduction of about $r\varepsilon/d$. The phase difference of the two waves at this point, then, is not zero, but $\omega r\varepsilon/d$. Other parts of the lens contribute wavelets whose phase at the focus differs by an amount between $\pm \omega r\varepsilon/d$.

When this range reaches π (at $\varepsilon = \pi d/\omega r$), the waves from centre and edges will interfere destructively. Near this point, the superposition of *all* the wavelets will have small amplitude, and we will have darkness. Thus the image of the point source is 'smeared' by the lens into an image of an *Airy disc*, of radius approximately $\pi d/\omega r$. Beyond this, there is (limited) constructive interference again, so that we see a series of faint halos around the principal disc.

A complex (nonpoint) source, provided it does not extend over too great an angle in the visual field, produces an image which is the *convolution* of the source (input) light pattern and the Airy disc pattern (cf. the analysis of frosted glass in section 5.13). Thus, the image resolution of the lens is about $\pi d/\omega r$; the *angular* resolution is this divided by d, or about $\pi/\omega r$. The larger the lens, the better the resolution; also, the larger the angular frequency (or the smaller the wavelength) of the light, the better the resolution. The same formula holds for telescopes that use mirrors instead of lenses. In practical terms, this means that an optical telescope of diameter 15 m (the largest currently being planned), working in visible light of wavelength $2\pi/\omega = 5 \times 10^{-7}$ m, can resolve images down to about 3×10^{-8} rad.

But we can use Fourier theory to help us overcome this limitation. On a space-based telescope, provided the engineering is adequate, the image will be a near-perfect convolution. (On Earth, atmospheric turbulence distorts the image. Indeed, this is a more important effect than the Airy disc 'smearing', and no Earth-bound telescope has a resolution of better than a few times 10^{-7} rad.) Given this, it is possible in principle to *deconvolve* the image with the Airy disc to obtain a much higher resolution than would be possible directly. However, the engineering requirements are stringent: even in a telescope with an effective focal length d of 1 km, the (deconvolved) image of Sirius' disc will have a diameter of just 20 μm. And there is the problem that the star is not transmitting monochromatic light, but emits across the spectrum; thus the image will be a superposition of various coloured Airy discs of different sizes.

11.6 Interferometry

Consider a 'part-telescope' that uses only two small areas from opposite sides of a lens, rather than the whole lens. We can achieve this by placing a mask in front of a lens. To collect as much light as possible, we might use slits rather than pinholes, and a cylindrical lens (Fig. 11.7). The distance between the slits (or pinholes) is the *baseline* of the device. This is the basic *interferometer*.

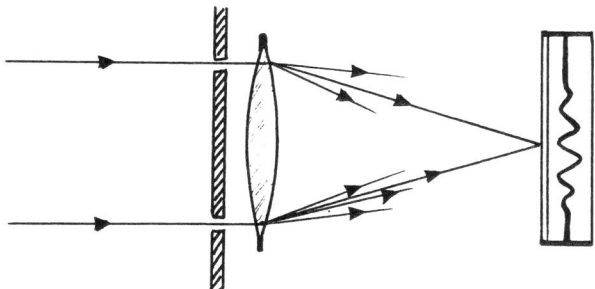

Fig. 11.7 — Basic interferometer.

A distant point source of (monochromatic) light now yields, not an Airy disc, but a uniform set of interference fringes. The fringe spacing is $2\pi d/\omega l$ (as before, d is the focal length of the lens and ω is the spatial angular frequency of the light, and l is the baseline).

As before, the image from a *distributed* source is a convolution. This means that the fringes generated by the various parts of the source interfere with each other, and begin to become less distinct. In particular, if we have a uniform (rectangular) source of small (but nonzero) angular width α, the fringes vanish when $\alpha = 2\pi n/\omega l$ for some n. We get, not the darkness of the Airy disc, but a uniform illumination. For a small *circular* source of angular diameter α, we have the same uniform illumination, when $\alpha \simeq 2.44\pi nd/\omega l$.

Given a distributed source, then, we can vary the baseline l of our interferometer from zero until the fringes first disappear. At this point, we have a measure of α, namely $\alpha = 2\pi/\omega l$.

Since the baseline is the same as the diameter of telescope that will actually provide an image of the source, it may not seem as if this has achieved a great deal. In fact, by using mirrors we can reduce the size of the focusing lens required, as in Fig. 11.8. This is *Michelson's stellar interferometer*. It was used to make the first direct measurement of a stellar diameter (of the supergiant Betelgeuse).

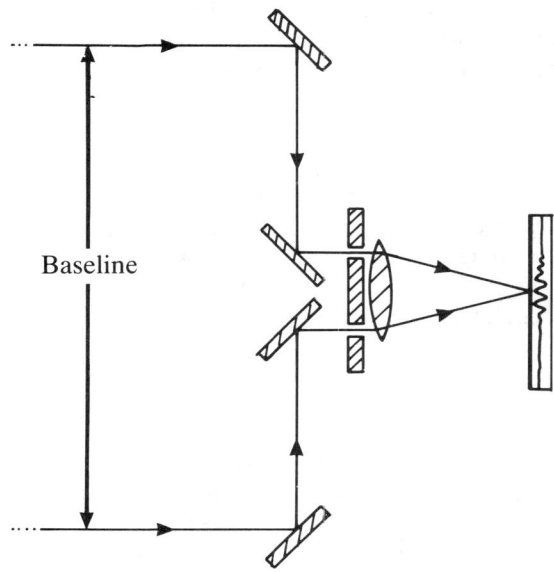

Fig. 11.8 — The Michelson stellar interferometer (schematic). The use of mirrors enables a longer baseline to be obtained using a small lens.

This principle of spatial interferometry has, more recently, led to the remarkable technique of *aperture synthesis*. In this technique, the outputs from an array of small telescopes are combined in such a way as to simulate a large telescope, of which they act as merely parts. The resolution of the resulting image is that of a telescope with the large 'effective aperture'. This technique is particularly useful in radio astronomy, where the long wavelengths mean that huge telescopes are needed for useful resolution. By this means, global networks of radio dishes have been linked, to simulate a telescope the size of the Earth; the images produced actually have a better resolution than any optical telescope has achieved. (However, although the *resolution* of the synthesized aperture matches that of a single large telescope, the *light-gathering* power is much smaller; therefore, aperture synthesis is only useful for relatively bright objects.) There is currently a project to construct a large aperture-synthesized optical telescope of effective diameter up to around 100 m. For technical reasons this is much more difficult than aperture-synthesis with radio waves.

A second type of interferometry is *spectral interferometry*. The equipment used in this kind of interferometry is shown in Fig. 11.9. An incoming beam of radiation is split by a semi-silvered mirror. Half of it is reflected back off a *fixed* mirror, half off a *movable* mirror. The two reflected beams are then allowed to interfere directly, and we measure the intensity of the resultant beam.

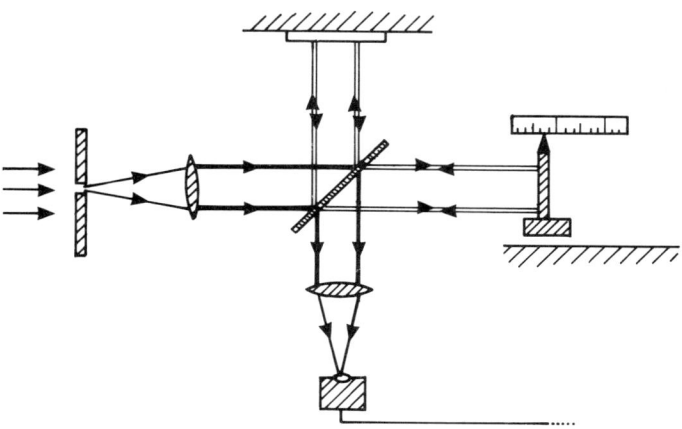

Fig. 11.9 — The Michelson spectral interferometer (schematic).

If the incoming radiation signal is $\psi(t)$, and the difference in time taken by the radiation in travelling along its two paths is τ, then the radiation arriving at the detector is the sum $\psi(t) + \psi(t+\tau)$. The intensity (energy density) of this signal is its absolute square; integrating, we define

$$L(\tau) = \int |\psi(t) + \psi(t+\tau)|^2 \, dt \ . \tag{11.21}$$

We can expand this:

$$L(\tau) = \int (\psi(t) + \psi(t+\tau))(\overline{\psi}(t) + \overline{\psi}(t+\tau)) \, dt$$

$$= \int (\psi(t)\overline{\psi}(t) + \psi(t)\overline{\psi}(t+\tau) + \psi(t+\tau)\overline{\psi}(t) + \psi(t+\tau)\overline{\psi}(t+\tau)) \, dt$$

$$= \int \{|\psi(t)|^2 + 2\operatorname{Re}(\psi(t)\overline{\psi}(t+\tau)) + |\psi(t+\tau)|^2\} \, dt$$

$$= 2\int |\psi(t)|^2 \, dt + 2\operatorname{Re}\int \psi(t)\overline{\psi}(t+\tau) \, dt \ .$$

Now $I = \int |\psi(t)|^2 dt$ is the total energy in the incoming signal ψ; while the function $\upsilon(\tau) = \int \psi(t)\overline{\psi}(t+\tau) \, dt$ is a complex analogue of the variance function, equation (7.10). We may write

$$L(\tau) = 2I + 2\operatorname{Re}\upsilon(\tau) \ . \tag{11.22}$$

If the *path differences* $c\tau$ between the two signals can be controlled, or at least

264   **Fourier applications**   [Pt. II

determined, then we have a system capabale of measuring $\text{Re}\upsilon[\psi](\tau)$. This is important in a system where ψ is oscillating at too high a frequency to be measured directly; visible light, for instance, has a frequency of approximately 10^{15} Hz.

Now let us take Fourier transforms. We obtain:

$$F\upsilon(\omega) = \int \left[\int \psi(t)\overline{\psi}(t+\tau)dt \right] e^{-i\omega\tau}d\tau$$

$$= \int \psi(t) \left[\int \overline{\psi}(t+\tau)e^{-i\omega\tau}d\tau \right] dt$$

(putting $z = t + \tau$)

$$= \int \psi(t)e^{i\omega t}\left[\int \overline{\psi}(z)e^{-i\omega z}dz \right] dt$$

$$= F\psi(-\omega)F\overline{\psi}(\omega) \ . \tag{11.23}$$

From equation (3.33), $F\overline{\psi}(\omega) = \overline{F\psi(-\omega)}$, so that

$$F\upsilon(\omega) = F\psi(-\omega)\overline{F\psi(-\omega)} = P\psi(-\omega) \tag{11.24}$$

(compare (7.18))

Again from (3.33), $F\overline{\upsilon}(\omega) = \overline{F\upsilon(-\omega)} = P\psi(\omega)$, since $P\psi$ is real. But $\text{Re}\upsilon = \frac{1}{2}(\upsilon + \overline{\upsilon})$. It follows that

$$F(\text{Re}\upsilon)(\omega) = \tfrac{1}{2}(F\upsilon(\omega) + F\overline{\upsilon}(\omega))$$

$$= \tfrac{1}{2}(P\psi(\omega) + P\psi(-\omega)) \ . \tag{11.25}$$

In a typical physical signal, the wave ψ will be travelling in one direction only, so that it has no components of negative angular frequency. In this case, $F(\text{Re}\upsilon)(\omega) = \frac{1}{2}P\psi(\omega)$ simply. Thus from the interferometer output $L(\tau)$ we can extract the power spectrum of the input signal ψ.

Spectral interferometry is of more importance in Earth-based sciences than in astronomy. It offers chemists, for example, a method of obtaining very precise information about the behaviour of molecules — see section 12.3.

11.7 Impedance

In section 6.8 we investigated, briefly, the properties of simple electrical circuits in the ω-domain. We saw how circuits composed of the basic linear devices — resistances, capacitances and inductances — could be used as filters on electrical signals. Since the resulting circuits are governed by linear equations, they act as linear and invariant filters; thus the output signal is the convolution of the input signal with the circuit's impulse response function. In the ω-domain, these circuits have a transfer function, which governs their frequency response.

Here, we develop this ω-domain analysis a little more. Suppose the voltage applied across a resistance R, a capacitance C, or an inductance L, is $V(t)$. If the current caused to flow through the device is $I(t)$, then we have the equations

$$V = RI$$

$$C\frac{dV}{dt} = I$$

and

$$V = L\frac{dI}{dt} \, . \tag{11.26}$$

Taking Fourier transforms of these and rearranging, we have the relations

$$FV/FI = R$$

$$FV/FI = 1/i\omega C$$

$$FV/FI = i\omega L \, . \tag{11.27}$$

(This is, properly, only possible if the functions V and I, and their derivatives, have Fourier transforms, but we ignore such considerations for the purposes of this discussion.)

What equations (11.27) say is that if $I(t)$ is the treated as the input to a device and $V(t)$ is the output, then the transfer functions of R, C, and L are (respectively) $RI(\omega)$, $1/i\omega C$, and $i\omega L$. These functions are the *impedances* of the devices. The symbol Z is usually used for a general impedance; thus, if an alternating current $I = \mathrm{Re}(I_0 e^{i\omega t})$ at angular frequency ω is passed through a device of impedance $Z = Z(\omega)$, the voltage across the device is $V = \mathrm{Re}(I_0 Z(\omega) e^{i\omega t})$. For fluency, we tend to omit the statement that the current is sinusoidal, and simple say: for a current I through Z, the voltage across it is ZI.

Any linear circuit element has an impedance. Those built up from the above three components have impedances that can be deduced from (11.27). Consider, for example, the circuit shown in Fig. 11.10(a): two impedances in series. The voltage

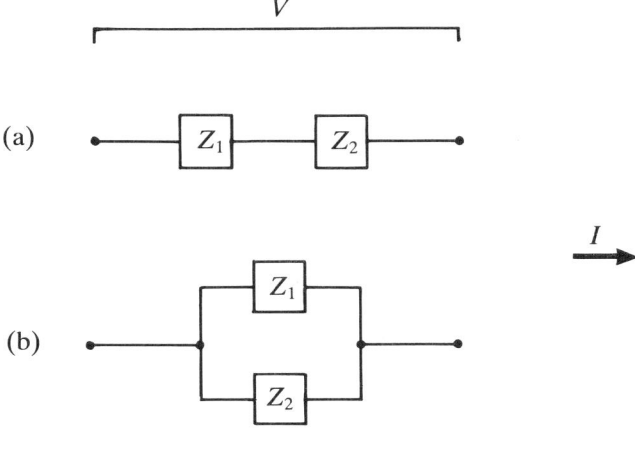

Fig. 11.10.

across Z_1 is $Z_1 I = W$ say, across Z_2 is $Z_2 I = V - W$. Hence $V = (Z_1 + Z_2)I$, and the impedance of the compound device is $Z = Z_1 + Z_2$. Similarly, the impedance of the element shown in Fig. 11.10(b) is calculable: since, in a parallel circuit, it is the currents that add, we have $I = V/Z_1 + V/Z_2$, so that $Z = (Z_1^{-1} + Z_2^{-1})^{-1}$.

One case where the idea of impedance is useful is in *transmission line* theory. A transmission line is a long conductive cable. Ideally, the voltage at the ends would always be identical; in practice, this is not so, for the line carries its current at a finite speed. To model this, we suppose that each metre of line has capacitance C and inductance L (we say that the line has a *distributed capacitance* of C, measured in Fm^{-1}, and a *distributed inductance* of L, in Hm^{-1}). Thus, the line acts as a long series of components like that shown in Fig. 11.11. A more complex model would

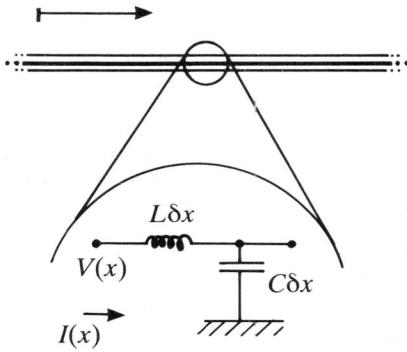

Fig. 11.11 — Model for a transmission line.

incorporate resistive terms as well; here, for simplicity, we consider the ideal, 'lossless', line.

Now let the voltage at a position x along the line be $V(x,t)$, the current $I(x,t)$. Then equations (11.26), applied to this typical element of line, yield

$$C\delta x \frac{\partial V(x+\delta x,t)}{\partial t} = I(x,t) - I(x+\delta x,t)$$

and

$$V(x+\delta x,t) - V(x,t) = -L\delta x \frac{\partial I(x,t)}{\partial t}$$

Dividing by δx and letting $\delta x \to 0$ gives the equations

$$C\frac{\partial V}{\partial t} = -\frac{\partial I}{\partial x}$$

and

$$\frac{\partial V}{\partial x} = -L\frac{\partial I}{\partial t}.$$

(11.28)

Combining these,

$$\frac{\partial^2 V}{\partial t^2} = \frac{-1}{C}\frac{\partial^2 I}{\partial x \partial t}$$

$$= \frac{1}{LC}\frac{\partial^2 V}{\partial x^2} . \tag{11.29}$$

But this is just the one-dimensional wave equation, equation (1.32′). The harmonic solutions are $V = V_0 e^{i\omega t} e^{i\Omega x}$, with $\Omega = \pm \omega\sqrt{(LC)}$; the sign here depends on the direction of the wave's propagation. Thus, voltage (and current) travel as waves along the line, with wave speed $c = |\omega/\Omega| = 1/\sqrt{(LC)}$.

The first of equations (11.28) then gives that $I = (C\omega/\Omega)V_0 e^{i\omega t} e^{i\Omega x}$. Therefore

$$V/I = \Omega/\omega C = \pm\sqrt{(L/C)} . \tag{11.30}$$

The quantity $Z_k = \sqrt{(L/C)}$ is called the *characteristic impedance* of the transmission line. Note that this is a constant; thus we have the curious fact that a lossless transmission line, incorporating no resistive terms, behaves — at least in this respect — like a simple resistance.

The way that an electrical load behaves in a circuit now depends, not only upon its own impedance, but also upon the characteristic impedance of the conductors connecting it to its power source. Consider, for example, the circuit shown in Fig. 11.12. An alternating source is connected, via a (lossless) transmission line of

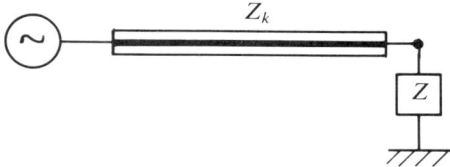

Fig. 11.12.

characteristic impedance Z_k, to an earthed impedance Z. A sinusoidal voltage wave of angular frequency ω passes along the line. At the junction with Z, some of the wave is reflected back. If position x along the line is measured backwards from the load, the voltage in the line has the form

$$V(x) = V_{in}\exp(i\omega(t + x/c)) + V_{ref}\exp(i\omega(t - x/c)),$$

where V_{in} and V_{ref} represent, respectively, the input and reflected wave amplitudes. (They may be complex-valued, if the phases are nonzero.) The current is $I(x) = (1/Z_k)\{V_{in}\exp(i\omega(t + x/c)) - V_{ref}\exp(i\omega(t - x/c))\}$, with the minus sign as in equation (11.30) for the 'backward-travelling' wave.

At the junction with the load, we have

$$V = V(0) = V_{in} + V_{ref} ,$$

while

$$I = I(0) = (1/Z_k)(V_{in} - V_{ref}) \ . \tag{11.31}$$

But the voltage across Z is V, so that $V = ZI$. Thus we have the relation

$$V_{in} + V_{ref} = (Z/Z_k)(V_{in} - V_{ref});$$

rearranging,

$$V_{ref} = \frac{(Z - Z_k)}{(Z + Z_k)} V_{in} \ ,$$

The *coefficient of reflection* at this junction is

$$\rho = \frac{V_{ref}}{V_{in}} = \frac{(Z - Z_k)}{Z + Z_k} \ . \tag{11.32}$$

Again, the electrical power in the input wave is $T_{in} = V_{in}I_{in} = V_{in}^2/Z_k$. The power transmitted through the load Z is

$$\begin{aligned} T_{tr} = VI &= (V_{in} + V_{ref})(I_{in} - I_{ref}) \\ &= V_{in}(1 + \rho)I_{in}(1 - \rho) \\ &= T_{in}(1 - \rho^2) \ . \end{aligned}$$

Thus the proportion of input power actually transmitted to the load is

$$\tau = (1 - \rho^2) = \frac{4ZZ_k}{(Z + Z_k)^2} \ .$$

When Z is very small compared with Z_k, the end of the line is effectively earthed; then $\rho \simeq -1$, and $\tau \simeq 0$. When Z is very large, the end of the line is effectively open; then $\rho \simeq 1$, and again $\tau \simeq 0$. When $Z = Z_k$, then $\rho = 0$ and $\tau = 1$. At this point, there is no reflected wave, and all the incident power is transmitted through Z. The impedances are said to be *matched*.

Matching impedances is important in the design of a great deal of circuitry, both in order to maximize the use of the available power and in order to minimize the (possibly disruptive) effects of the reflected wave. Impedance matching applies not only to discrete circuit elements at the end of transmission lines, but also to circuits composed of discrete elements, and to junctions and branching points in transmission lines.

The account we have given here of impedance and of transmission line theory is in terms of electrical signals and circuitry, for this is the longest-standing use of the technique. However, the concepts and calculations are more widely applicable. For instance, a standard analogy for the flow of electric current is the flow of fluids. Pipes correspond to conductors; pressure corresponds to voltage; volume flow corresponds to current. This analogy works because the equations that govern the fluidic system are identical to those governing the electric.

Circuit elements can be modelled too. A capacitor is analogous to an elastic-walled tank: the volume flow into the tank is proportional to the *rate of increase* of pressure at its mouth. Viscous forces are analogous to resistances. Inductance is related to the fluid mass: it takes a pressure difference across a fluid element to accelerate it.

So, just as a discrete electric circuit element has an impedance, so does a discrete fluidic component. Also, just as a long conductor has a characteristic impedance, so does a fluid conduit. The idea of characteristic impedance is then useful in understanding the behaviour of fluidic (liquid or gaseous) systems, particularly vibrating systems: in other words, in understanding the *acoustics* of fluids.

With physical systems in the solid phase, too, there is the concept of *mechanical impedance*. Thus, for a spring satisfying Hooke's law, the force F applied and the relative velocity v of its ends are related by the equation $dF/dt = kv$. For an ideal damper, $F = \eta v$; while for a mass, we have Newton' second law, $F = mdv/dt$. Taking Fourier transforms and rearranging, we have:

$$FF/Fv = k/i\omega$$

$$FF/Fv = \eta$$

and

$$FF/Fv = i\omega m \qquad (11.34)$$

(cf. equations (11.26) and (11.27)). These are the mechanical impedances of these components; thus, a spring is analogous to a capacitance, a damper to a resistance, and a mass to an inductance. (This idea is developed in the Project at the end of Chapter 6.)

12

Chemistry

12.1 Mass spectrometry

Mass spectrometry is a technique of separating a batch of ions, according to the ratio of *mass* to electric *charge*. By this means, the elemental content of a sample can be established; but the technique has more uses than that, as we shall see.

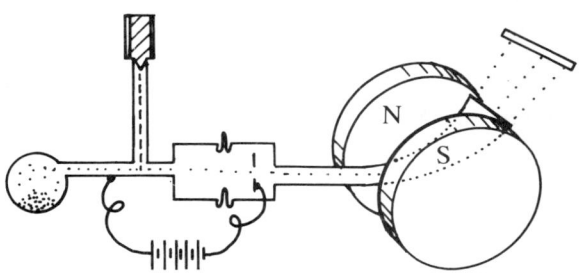

Fig. 12.1 — 'Conventional' mass spectrometer (schematic).

The basic mass spectrometer is shown in Fig. 12.1. A sample of the substance or mixture to be analysed is vaporized and bombarded with electrons. The molecules of the substance are thus broken up and the fragments ionized. Next, the ions so produced are accelerated by an applied electric field, provided by a charged plate. Charging the plate positively pulls out the negative ions for analysis, while charging it negatively extracts the positive ions. A hole in the plate lets through a beam of ions.

Suppose the potential of the plate above the source is V. An n-fold-charged ion will then acquire a (kinetic) energy neV, e being the charge on the electron. If its mass is m, then its velocity v after acceleration is given by the relation $\frac{1}{2}mv^2 = neV$, so that

$$v = \sqrt{(2neV/m)}. \tag{12.1}$$

Thus, among fragments of the same charge, *heavier ions travel more slowly.*

Now the beam of ions is passed through a (small) transverse magnetic field B, over a region of length l, say. An n-fold-charged ion experiences a lateral force, perpendicular to the field lines and to its direction of motion, of $Bnev$; by this force, it is accelerated laterally at the rate of $Bnev/m$. The time the ion spends in this region is l/v, so that it emerges from it with transverse velocity

$$u = (l/v)(Bnev/m). \qquad (12.2)$$

Since B is small, $u \ll v$, and the angle θ at which the particles emerge is

$$\theta \simeq \tan\theta = u/v$$
$$= lBne/mv.$$

Substituting for v from equation (12.1),

$$\theta \simeq \frac{lB\sqrt{e}}{\sqrt{(2V)}}\sqrt{\left(\frac{n}{m}\right)}. \qquad (12.3)$$

Thus, the angle at which the particles emerge depends on their *mass-to-charge* ratio, $\tilde{m} = m/n$.

Any given substance will, under ionization by electrons of a fixed energy, fragment and charge in a (statistically) fixed pattern. Thus, the proportion of ions of given mass-to-charge ratio yielded by a sample of a particular substance is characteristic of that substance (and of the ionizing radiation). Now, if we have an unknown substance, we might take a mass spectrogram; comparing the output with tables of spectra of known compounds, we may be able to identify the substance.

The presence of charge in equation (12.3) must not be overlooked. Although most of the ions will be singly charged — and for these, $n = 1$, so that $\tilde{m} = m$ — there will usually be some doubly (and multiply) charged ions. If the output is treated as a *mass* spectrum, then a doubly charged ion of mass m, with $\tilde{m} = m/2$, will masquerade as an ion of mass $m/2$; a triply charged ion, as an ion of mass $m/3$, and so on. On the whole, the lower the ionizing radiation energy, the smaller the proportion of multiply charged ions, and the more justification there is in treating the output as a mass spectrum. However, the lower the energy, the fewer ions of any kind there will be, so that the spectrum will take longer to build up; also, the fewer *kinds* of ion there will be, so that the resulting spectrum contains less information about the original sample.

An important special case of mass spectrometry is when the ionizing radiation is intense enough to split all the molecules of the sample into their component atoms — without, ideally, producing any multiply charged ions. In this case we can read off from the mass spectrum the proportion of atoms of each atomic mass in the substance, and therefore the (nonstructural) chemical formula of the substance to be identified.

It is true that a list of atomic *masses* does not quite give a list of atomic *numbers*, because of the isotopic nature of elements; in fact this is not a problem. The reason for this is the sensitivity of resolution in mass spectroscopy. Not only can it tell the difference between molecules of, say, carbon monoxide and nitric oxide (masses 28 and 30 respectively), it can distinguish between carbon monoxide (CO^+ has mass 27.9949) and diatomic nitrogen (N_2^+ has mass 28.0062). In order to achieve this, we need to be able to measure mass to an accuracy of around one part per million; with good engineering, this is quite possible. Thus, mass spectroscopy can reveal the elemental (indeed, the isotopic) composition of a substance.

A more recent development of the mass spectroscope is the *ion cyclotron resonance* (or *ICR*) spectroscope. It is with these systems that Fourier methods have, in the last decade or so, come into mass spectroscopy, in what is called FT-ICR spectroscopy; and it is with these that the impressive resolutions quoted above have been achieved. Fig. 12.2 shows a schema of the FT-ICR spectroscope. The two plates

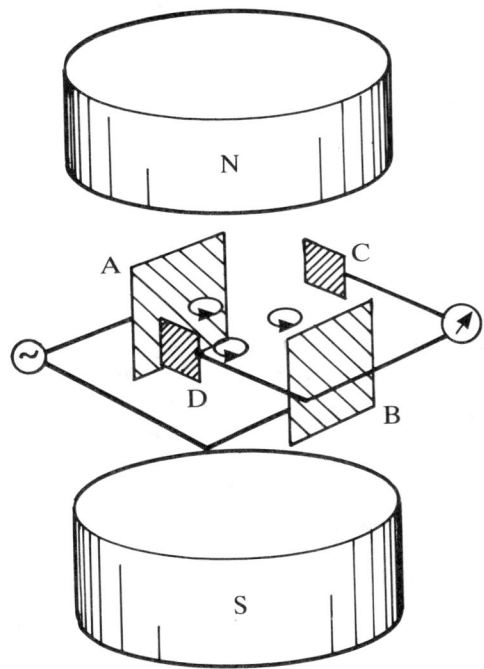

Fig. 12.2 — ICR mass spectrometer (schematic).

A and B carry an alternating electric potential, which accelerates the ions lying between them. A steady magnetic field, perpendicular to this, drives the ions into circular trajectories. The lighter ions are accelerated further, so they travel faster and in larger orbits than the heavier ions; but the *period* of each orbit is the same, for it is determined by the frequency of the alternating potential.

When the driving potential is switched off, we have, as before, a velocity spectrum mirroring the mass spectrum. Each ion is moving at its particular velocity, in a magnetic field; the force so generated makes the ion describe a circle. But now, the faster light ions experience *more* force than the slower heavy ions, so they travel in tighter circles. The periods of the orbits are no longer the same.

As they travel in their orbits, the ions are accelerating, and they therefore give off electromagnetic radiation. The frequency of the radiation is the frequency of the orbiting. This has two consequences: first, there is an *induced* potential variation in the receiver plates C and D; secondly, the ions gradually lose their kinetic energy, and slow down. Each ion thus induces a small, damped-harmonic potential variation in the receiver plates. With good design and a sample of reasonable size, the energy loss of the ions can be minimized, so that the damping is slight. In this case, each ion induces an *almost-harmonic* potential, of which the frequency is dependent on the ion's mass.

What is received by the plates is then a superposition of harmonic signals at fixed frequencies. The Fourier transfrom of this received signal will pick out these frequencies, and their relative strengths, which is the mass spectrogram of the sample.

One important advantage of FT-ICR spectroscopy over 'traditional' mass spectroscopy is its speed: a spectrum can be obtained in a fraction of a second. Multiple spectra give cleaner pictures, and therefore better resolution.

The mass resolution of FT-ICR is limited principally by the rapidity at which voltage measurements must be taken. This is a second advantage of the FT-ICR method: resolutions as high as one part in 10^8 have been obtained for low masses.

The third great advantage of FT-ICR is related to the second. The *absolute* resolution of any mass spectroscope decreases with increasing mass. Most 'conventional' spectroscopes, with resolutions of perhaps one part in a few thousand, are thus limited to molecular masses below about 1000; above this, the lines run together, and one cannot tell whether a given peak is at (say) 1728 or 1729. (In fact, even the *relative* resolution decreases with increasing ion mass, because of the form of equation (12.3).) The greater resolving power of Fourier transform machines pushes this mass limit up by a factor of a hundred or so, thus allowing the use of the technique on large organic compounds.

12.2 NMR spectroscopy

Nuclear magnetic resonance (or NMR) spectroscopy is, operationally, similar to the ICR method of mass spectroscopy. In NMR, however, it is the electromagnetic properties of atomic nuclei, rather than ions, that is investigated.

Many atomic nuclei possess an intrinsic *spin*; this turns them into tiny electromagnets, with a certain *magnetic moment*. These nuclear 'magnets' are not fixed in orientation, but are free to rotate. In particular, when they are placed in an external magnetic field, they will *precess* about the field lines, just as (in a gravitational analogy) a spinning top will precess about the vertical. The frequency of this precession depends on the magnetic moment and on the strength of the applied field.

In classical mechanics, a nucleus might precess at any angle to the external field. This is not true in quantum mechanics: only certain orientations are possible. Now a nucleus precessing at a high angle is in a high-energy state. It may, then, decay spontaneously to a lower-energy state, of precession at a lower angle. In doing so, it emits a photon whose energy corresponds to the energy difference between the two states. Because the orientation angle is quantized, these photons can only have certain values; but these values depend *continuously* on the applied field strength.

We can use this quantization diagnostically (Fig. 12.3). Suppose we have an unknown material. We put it into a magnetic field, and beam photons into it (using a high-frequency oscillator coil). When the frequency is exactly right, the photons will be strongly absorbed by low-energy states, turning them into high-energy states; these states will soon decay, releasing an identical photon. But the released photon will have random direction. If we put a receiver at right angles to the transmitter coil, we will pick up a proportion of the trapped and re-emitted photons, but no photons from the transmitter.

Because different nuclei have different magnetic moments, this technique is element-sensitive, indeed isotope-sensitive. It is also quantitative, for the more nuclei of a particular type are present, the more photons of the appropriate frequencies are absorbed and re-emitted. Thus to find out the elemental composition of our sample, we have only to measure the strength of its response at the appropriate frequencies. (Some nuclei have very small magnetic moments, which makes NMR on these isotopes difficult.)

That is not the end of the story, however, for an atom is surrounded by an electron cloud. This shields the nucleus to an extent from the applied magnetic field, and the precession frequency is altered accordingly. Now in chemical bonding, these electron clouds are distorted, and the shielding effect changes. Thus arises the *chemical effect*: the frequency of the photons absorbed by a nucleus depends not only on the nucleus, but also on what its atom is connected to. So it is not merely individual nuclei, but *compounds* which have individual NMR 'fingerprints'.

Mass spectroscopy and NMR are complementary techniques. Mass spectroscopy 'sees' the whole molecule and its large fragments; NMR 'sees' individual atoms and their local connections. Using the two together, the chemist can often deduce the molecular structure of an unknown substance.

This is the basis of the technique of NMR spectroscopy. As with mass spectroscopy, it is susceptible to considerable simplification by using Fourier techniques. Thus, instead of using an oscillator at a fixed (but variable) frequency, we might beam into the material a short, sharp pulse of white-noise radiation. This radiation has harmonic components (contains photons) of a wide range of energies, of uniform amplitudes (in equal numbers). Under the influence of this input, (some of) each type of nucleus in the sample will be excited into high-energy states. These decay and release their photons, which are detected by the receiver coil. The detected voltage pattern can be subjected to Fourier transformation; this distinguishes the harmonic components of the response, thus picking out the proportions of photons of each frequency.

As with FT-ICR, this technique (called *FT-NMR* spectroscopy) makes a big difference in the *speed*, in the *accuracy*, and (therefore) in the *scope*, of the method. For instance, the added accuracy means that NMR methods can be used on more dilute samples than was previously possible.

One consequence of the use of Fourier methods is that the input signal to the sample can be 'tailored'. Suppose we have a dilute solution in ethanol of higher alcohols which we wish to assay. The peaks in the output arising from the higher

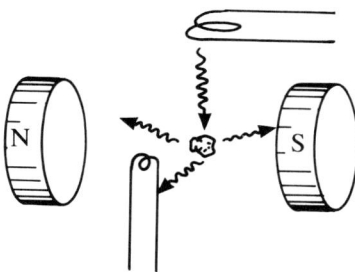

Fig. 12.3 — NMR spectrometer (schematic).

alcohols will be very close to much larger ethanol peaks. To make them more visible, we might filter the input pulse with a narrowband notch filter to remove some of the ethanol frequencies.

12.3 Infra-red spectroscopy

In molecules, atoms are connected by the net electromagnetic effect of their component nuclei and electron clouds. If a bond is stretched, there is a restoring force proportional (for small displacements) to the amount of stretch. As a result, molecular bonds act as tiny extension oscillators.

From quantum theory, we know that the oscillation of these bonds cannot have arbitrary amplitude. Rather, it (and therefore the vibrational *energy* in the bond) can only occur in certain discrete values. It follows that a change in vibrational energy involves the absorption or emission of photons of certain fixed energies. (The same is true for those vibrations in which the molecule *bends* or *rotates*. In each case, the oscillations can only have certain amplitudes, and the energies of photons that can be absorbed or emitted are quantized. The size of the energy steps depends on the nature of the bond.)

Each of these phenomena can be investigated in essentially the same way as NMR: we excite molecules in lower-energy states to higher-energy states, and either measure the absorption of input energy or detect the output energy. The energies that arise in these transitions are typically in the infra-red region of the spectrum, so the techniques used are those of *IR spectroscopy*. As with mass spectroscopy and NMR, the 'classical' method is to use a monochromatic source of variable frequency; the FT method inputs a burst of wideband noise, after which a Fourier transformation yields all the information much more quickly. The resulting IR spectrum yields information about how many bonds of each kind there are in the substance(s) of the sample.

The detailed mechanisms of the experimentation are not important for us here, but there is one serious problem that does not arise in either FT-ICR or FT-NMR, namely the problem of measuring the output. Both of the techniques discussed so far — FT-ICR and FT-NMR — involve output photons in the radio-frequency region, from a few kilohertz to a few hundred megahertz. Hence the potentials induced in the receiver plates vary in this frequency range. Modern electronics allows direct measurements to be taken at intervals as small as every 10^{-9} s, at a rate of 10^9 Hz, and retained for more leisurely calculation; thus the outputs for FT-ICR and FT-NMR can be recorded directly. But the transitions in bond-vibrations involve photons whose frequencies can be up to 10^{16} Hz (actually well into the ultra-violet, but nominally still 'IR spectroscopy'). It would seem as if the output signals cannot be measured quickly enough for an FT-IR system to work. The solution to this problem is, instead of direct sampling, to use a spectral interferometer (see section 11.6).

One interesting point about vibrational activity in particular concerns the nature of the bond 'spring'. It is not perfectly Hookean, so the vibrations are not perfectly harmonic. The discrepancy is called *anharmonicity*, and it has noticeable effects. The frequency of the vibration is no longer independent of its amplitude. Since the energy of a quantum is related to its frequency, this means that the transitions between states have slightly different energies. Also, the nonlinearity of the spring means that the 'harmonic modes' are no longer independent: they interact, and give rise to what is called *Fermi resonance*. Yet, although the system is not precisely linear, Fourier analysis remains a useful descriptive tool.

12.4 Visible light spectroscopy

The oldest of the spectroscopic techniques concerns (what we now understand as) the transition of electrons between energy states within their atoms. It works both in 'absorptive' and in 'emissive' modes. In the absorptive mode, an incoming beam of (white) light is absorbed, driving electrons into higher states; the output is the selectively depleted input beam. In the emissive mode, atoms are excited, often thermally, and emit photons as they decay to lower-energy states; these photons form the output. In either case, the output beam is spectrally analysed (e.g. by passing it through a prism or a diffraction grating). The presence of transitions of particular energies shows up as a series of lines — dark *Fraunhofer lines* in an absorption spectrum, bright lines in an emission spectrum.

This technique is of great importance for astronomers, for it means that they can distinguish the chemical content of stellar atmospheres, interstellar gas and dust, planetary atmospheres, and so on, without leaving the surface of the Earth. By this means, methane and ammonia were discovered in the atmospheres of the planets Jupiter, Saturn, Uranus and Neptune by spectroscopic means long ago; at the time of writing (1989), the first probe to sample the atmosphere of one of these planets, the Galileo probe to Jupiter, is still on the ground.

It is also of use to general chemists. A number of elements (caesium, rubidium, gallium, and so on) were actually discovered spectroscopically before there was any 'chemical' evidence for them. (Particularly spectacular was the spectroscopic discovery of helium — in the Sun!). Even the old-fashioned 'flame test' for metals depends on a rudimentary form of emission spectroscopy. Put a sodium salt into a flame, and it emits strongly at a characteristic wavelength in the yellow region; potassium salts give a red colour, and so on.

Astronomy is interested in the spectra of stars for other reasons than to establish their composition. At great distances, all the Fraunhofer lines in a stellar spectrum are displaced. It is as if the chemistry were different, with all the transitions having energies lower by a constant factor. Since the lines are displaced to lower frequencies, the 'red end' of the spectrum as it were, this phenomenon is called the *red shift*. The simplest interpretation of this startling feature is that it results from the *Doppler effect*: if the star is moving away from us at high speed, the light waves it emits get 'stretched out', so that they appear to have greater wavelengths and lower frequencies.

However, as we saw in section 11.2 on lasers, transitions between electron states do not occur with *precisely* fixed energies. Indeed, if they did, then all emission spectra would consist of bright lines, and they do not. (Household lighting, for instance — whether by incandescent bulb or fluorescent tube — is nearly continuous across the visible spectrum.) There are a number of 'line-broadening' mechanisms involved, among which we may mention three.

Firstly, molecules in a real substance are always moving and colliding, so the Doppler effect will affect all emissions to a small extent, depending on temperature. Since the atomic motions are pretty well random, the atoms in a sample will be affected by different amounts, and the (perceived) emission is over a range of wavelengths. Secondly, there is the 'natural' or 'lifetime' broadening that arises out of the uncertainty principle: a transient excited state implies an uncertain energy level, so the energy emitted is uncertain to that extent. And thirdly, the actual energy levels of an electron in an atom depend on what surrounds the atom. If there are other atoms close by, they will change the energies of the electron states, and therefore in the transitions. Thus, in anything other than a rarefied gas, there is a substantial amount of *pressure broadening*.

12.5 Crystallography

We saw in section 11.4 the close connection between the diffraction of waves and Fourier transformation: the diffraction pattern generated by a (small) aperture under uniform illumination is the (two-dimensional) transform of the function describing the shape. One of the most important uses of this fact is in the unravelling of crystal structure.

It is clear that the best waves to use are those whose wavelengths are comparable to the scale of the crystal structure: of the order of 10^{-10} m. For electromagnetic radiation, this represents a part of the X-ray region. (Other waves can be used — for example, neutron beams.) A beam of the chosen radiation — preferably monochromatic — illuminates a sample of the crystal to be analysed, and the diffracted radiation is captured on photographic film (Fig. 12.4).

If the sample is a single crystal, it turns out that the photograph (or *diffractogram*) consists of a pattern of spots. This indicates that the sample must have a *regular* structure. (This was the first direct evidence that was ever obtained of the regularity of crystal structure. It is also evidence of the wave nature of X-rays. For these two reasons, the first X-ray diffraction experiments won von Laue and the Braggs two separate Nobel prizes.)

Of course, crystals are *three*-dimensional structures, not two-dimensional apertures. Thus, in performing the diffraction, we have 'lost' a dimension. If we simply ignore the third dimension, we might use the diffractogram to deduce, not the full crystal structure, but at least the structure of a two-dimensional *projection*. Alterna-

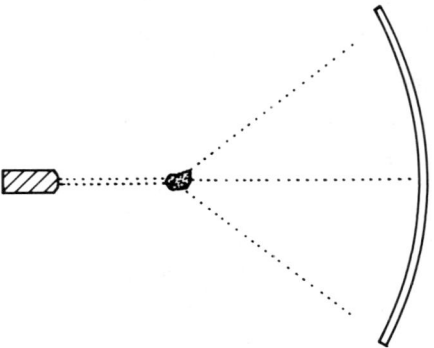

Fig. 12.4 — X-ray diffraction.

tively, if we take a range of diffractograms from different angles, we might build up a full three-dimensional picture, using the tomographic methods of the Radon transform (see section 8.8).

But as noted in section 11.4, there is a serious problem concerning the *phase* of the diffracted radiation. The problem is that the photographic film records only the power (or equivalently, the amplitude) of the scattered beam, and not the relative phases of the various spots. Thus we do not have the full Fourier transform of the crystal structure available; we only have its power spectrum. The power spectrum cannot be inverted, and so we cannot deduce the crystal structure.

A number of methods have been developed to overcome this problem. One is the *isomorphous crystal* method. In this, we take two closely related substances whose crystals have almost exactly the same structure (as judged from macroscopic evidence — the hope being that they are also very similar microscopically). Thus, if we are interested in the structure of crystals of phenyl chloride C_6H_5Cl, we might look to phenyl bromide C_6H_5Br as an analogue. Crystals of these two substances will yield similar, but slightly different, diffractograms. Since the structure is distorted a little in the isomorph, there is a slight readjustment in the *amplitudes* and *positions* of the spots on the diffractogram. Crucially, the change in amplitude is related to the *phase* of the component. For highly symmetric crystals, just two isomorphous substances will give a complete solution for the phases (and therefore, via a Fourier inversion, for the crystal structure); three isomorphs will 'solve' any crystal.

More recently, a second technique has been used. The atoms in a crystal lattice have electrons around them in low-energy states, but they can be excited by photons of appropriate energy to higher states. The effect is greatest when the incoming photons are very close in energy to the difference in electron energy levels. But for such electrons, some of the 'scattered' photons are, in fact, absorbed and re-emitted. This so-called *anomalous dispersion* effect means that the diffraction pattern of a crystal changes when the X-rays being used are near to certain frequencies. (X-rays of the right energy for producing diffractograms are absorbed only by the inner electrons of medium-to-heavy atoms, so the technique is not a universal one.) The effect of anomalous dispersion is to change the 'visibility' of some of the atoms in the sample, and this again enables the phase angles to be calculated.

It is the electron clouds about the atoms in a crystal that actually diffract the X-rays. Thus, when we have reconstructed the image of the crystal, what we have is an *electron-density map*. Heavy atoms hold many electrons tightly, and so produce local (roughly spherical) regions of high charge density; light atoms, by contrast, have few electrons, and may appear in the image only as a small 'bump' on the surface of a heavy neighbouring atom. Hydrogen atoms, indeed, with their single electron involved in bonding, will be almost invisible.

The basic level of crystal structure is the *unit cell*. The unit cell is repeated regularly on a *lattice* (see section 4.8). Thus, the electron density in a crystal is the *convolution* of the density function of the unit cell with the lattice function. Now the effect of diffraction is a Fourier transformation; from the convolution theorem, it follows that the diffractogram is the *product* of the transform of the lattice and the transform of the unit cell.

These two parts of the diffractogram can be distinguished. The Fourier transform of the lattice is again a lattice, with impulsive components of equal amplitude (equation (4.47)). The crystal lattice, then, determines the *position* of the spots on the diffractogram. Conversely, the unit cell is finite and nonperiodic; it has a 'proper' Fourier transform, and determines the *amplitude* (and phase) of the spots. Thus, the crystal lattice can be deduced without the problems of phase determination. The phase is needed for the mapping of the unit cell.

An alternative view is the following. A crystal is a regular, periodic structure; thus we can think of its having a (three-dimensional) Fourier *series*. In this view, the diffraction spots are in positions corresponding to the (vector) *frequencies* used in the series, and their amplitude and phase are the *coefficients* of the terms of the Fourier series. The unit cell electron map is the corresponding function on a finite 'interval'.

That deals with true crystals. Now it has been known for some time that there are *non-periodic* structures which nevertheless have certain regularities, such as the so-called 'Penrose tilings'. Recently, it has been discovered that there are real substances that adopt such structures. These are the *quasicrystals*. Among the interesting properties that such materials display is an indication of five-fold, or pentagonal, symmetry — something which is quite impossible for a true crystal.

The behaviour of these quasicrystals under X-ray diffraction is also interesting. They produce diffractograms that are neither the continuous blur of a glass nor the lattice of spots of a crystal. They have sharp spots, but in an infinite number: some bright, others progressively fainter and fainter.

Another interesting recent discovery comes from scattering experiments on heavy nuclei (see section 11.3). It seems that there is some discernible (fleeting?) 'crystalline' structure to such nuclei: the protons and neutrons pack in a lattice-like arrangement. These nuclear 'crystals' are very small, never more than a dozen nucleons on a side even in the largest nuclei, so that the resolution of such diffractograms is poor.

13

Life science

13.1 Vision

The eye is responsible for the most concentrated sensory information in the human body. The central part of the retina, the *fovea*, is only perhaps a millimetre across, but it contains some half-million sensory cells. Amost all of this prodigious analytical power is devoted to *spatial* discrimination: distinguishing objects in different physical positions.

It is true that we see not just position, but *colour* too. Yet the eye's colour analysis is extremely crude. There are just four different kinds of receptor cell — the 'rods' and three distinct sorts of 'cone' — each of which responds differently to different wavelengths of light. The rods are treated separately from the cones in the brain, and are not really used in colour vision. We might therefore say that our eyes perform (what is effectively) a three-point discrete Fourier analysis on the incoming light.

In addition to this (two-dimensional) spatial and (one-dimensional) spectral analysis, there is one further analysis performed on visual information: *temporal* analysis. This is what enable us to see motion, and is performed entirely within the brain. For each of these three areas, Fourier methods come in useful. (It is not clear to what extent Fourier methods are actually *used*, by nature, in the processing of visual signals in the brain. Nevertheless, they offer an illuminating *descriptive* tool.)

Thus, to investigate the *spatial discrimination* of the eye, we might show it grids or gratings, with black and white (or coloured) stripes of variable angular width. Making the stripes slightly different shades of grey, instead of black and white, tests the eye's *sensitivity* to spatial variation at a range of scales. (It turns out that, at least in conditions of bright light, the eye is most sensitive to variation over angles of around $\frac{1}{2} - 1°$.)

Again, the colour response of each of the four kinds of retinal receptor is known. Using these, we can predict the effect on the eye of light of any spectral characteristics. In particular, it is possible to predict whether two colours of paint or ink will appear the same colour, under given lighting, by using physical measurements only. (This is less obvious than it seems. The spectral characteristics of two colours might be very different, and yet the colours might be perceived identically, thanks to the crudeness of the eye's spectral analysis.)

Finally, *temporal* acuity can be investigated using a light whose brightness oscillates. A light that flickers dim and bright very fast cannot be followed by the eye; a light that dims and brightens very slowly will not seem to change. By varying the *amplitude* of the oscillations at fixed *frequency*, we might determine the sensitivity of the eye to brightness changes of given speed. (For sinusoidal intensity variations, the eye has its peak sensitivity at a rate of a few cycles per second.)

While we do not know the extent to which human visual processing uses Fourier techniques, it is clear that at least some of the processing is done by *non*-Fourier methods. For example, there are certain brain cells (called *simple cells*) that detect edges between dark and light areas. (The same sort of approach is used in some modern, computerized, 'visual processing' systems.) Using the output of the simple cells, there are *complex cells* which detect the orientation of lines, and so on.

But methods of linear processing, if not Fourier methods themselves, *are* used in the eye, and in a very direct way. The retina is a (spatial) convolution filter.

The rods and cones that receive light do not feed straight into the optic nerve. There are several layers of cells between the receptors and the brain. These layers have a variety of rôles; but the net effect is to collect together the inputs from the receptors and feed them the *ganglion cells* (see Fig. 13.1). Each ganglion cell takes its

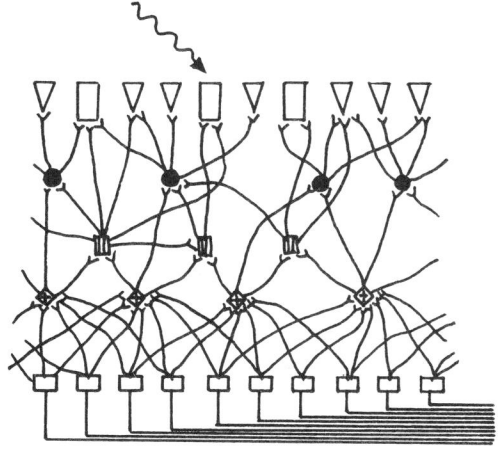

Fig. 13.1 — The structure of the retina. Light impinges on the receptor cells, the rods and cones (top), which 'fire'. These are connected, ultimately, to the ganglion cells (bottom), which feed along the optic nerve to the brain. (Actually the diagram is inverted: light entering the eye passes *through* the retina to reach the receptors.)

input from a *receptive field* covering a quarter of a degree or so (about the apparent size of the letter 'o' at normal reading distance). The receptive fields of the ganglions overlap considerably; indeed, there are about as many ganglion cells as there are receptors.

The inputs from the various parts of a ganglion cell's receptive field have different effects. A signal from receptors at the *centre* of the field will *stimulate* the ganglion, but it is *inhibited* by signals from receptors on the *periphery*. When *no* part of the receptive field is stimulated, the ganglion cell produces a small 'resting' output.

All ganglions behave in much the same way, so that the retinal processing is invariant across the visual field (or at least across the fovea). But a linear invariant filter operates by convolution; so that the connections in the retina act to *convolve* the incoming light pattern with the ganglion (spatial) response curve (Fig. 13.2). The form of the ganglion response curve means that this effect is a carefully controlled spatial high-pass filter.

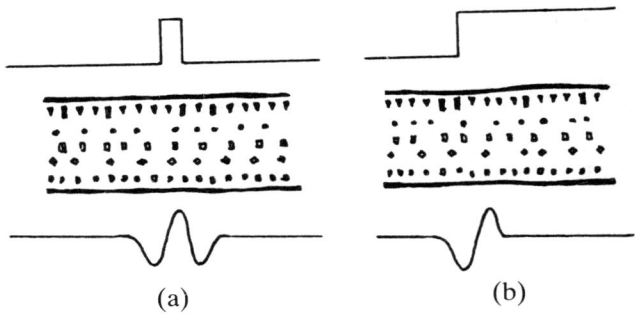

Fig. 13.2 — The effects of the retinal connections is that of a high-pass, spatial, convolution filter, as shown here on a point of light (a) and an edge (b).

This has a number of important consequences. First, it enhances the contrast at edges, at the cost of large-scale changes (Fig. 13.2(b)). Secondly, it begins to suggest an explanation of the phenomenon of *colour-invariance*, in which a red object on a green background (say) is perceived as such under widely different conditions of illumination: it is *local contrasts* — in this case contrasts of colour — that matter, and not simply the response of the receptors.

13.2 Hearing

The structure of the human ear is shown in Fig. 13.3. The outer ear consists of a 'sound funnel', the *pinna*; the middle ear is a series of small bones in a complicated linkage; and the inner ear contains a curled structure, the *cochlea* (as well as organs of balance, not shown, which are not involved in hearing). The middle ear is bounded on the outside by the eardrum and on the inside by a membrane called the *oval window*.

The middle ear is a gearbox. Its function is to 'gear down' the vibration of the eardrum in order to maximize the vibration energy transmitted to the inner ear. This is necessary because the impedance characteristics of air and of the fluid filling the cochlea are different; the effect of the gearing is to match the impedances. Thus, although the *amplitude* of vibration of the oval window is less than that of the eardrum, almost all the acoustic *energy* is transmitted.

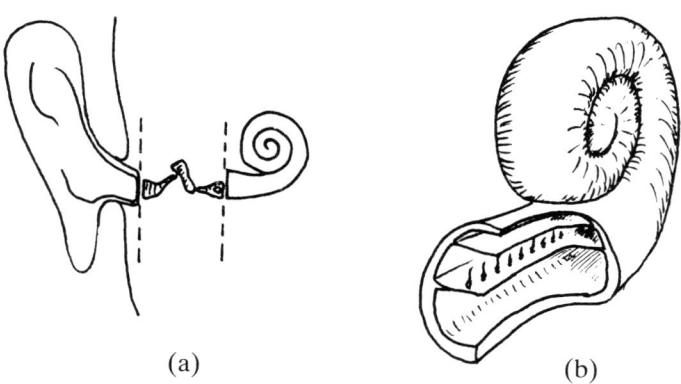

Fig. 13.3 — (a) The ear (schematic), showing outer, middle, and inner sections. (b) The cochlea of the inner ear in section. Note the hair cells running the length of the basilar membrane.

The outer and middle ears, then, are concerned with the *engineering* aspects of hearing, serving merely to transmit vibrations efficiently to the cochlea. It is within the cochlea that the *analysis* of this vibration begins. The mechanical vibrations are there converted into electrical nerve signals, which are sent along the auditory nerve to the brain.

Fig. 13.3(b) shows the cochlea in section. It is divided longitudinally, by two membranes, into three chambers, which are filled with body fluids. The middle chamber is therefore largely bounded by flexible membrane. It is to this chamber that the oval window connects.

One of the membranes bounding this middle chamber (the *basilar* membrane) bears a row of *hair cells*, running the length of the cochlea. These respond, electrically, to local distortions of the membrane and to movement in the cochlear fluid. Each hair cell is connected to its own neural fibre, feeding into the auditory nerve. Then, when sound vibrations — pressure waves — enter the cochlea, these waves move both membrane and hair cells, causing the hair cells to 'fire' in a way characteristic of the sound. The resulting electrical bursts are transmitted to the brain, to be perceived.

But the basilar membrane is not uniform along its length. It is thick and narrow near to the outside of the cochlea, becoming thinner and wider towards the inner end. Because of this, it is around a hundred times stiffer (as a membrane) at its outer end than at its inner. As a result, a harmonic input vibration of any given frequency will force, in response, a complex pattern of vibration along the length of the basilar membrane. Moreover, vibrations of different frequencies (which is to say sound of different pitches) will affect these various parts differently. More precisely, lower vibrational frequencies excite the 'floppy' inner section more than the 'stiff' outer section, while higher frequencies do the opposite; intermediate frequencies drive the membrane most at some intermediate position. (See Fig. 13.4.)

The effect of this tapering is thus that the basilar membrane becomes a sort of mechanical Fourier analyser. Given an input vibration, the electrical activity of a hair

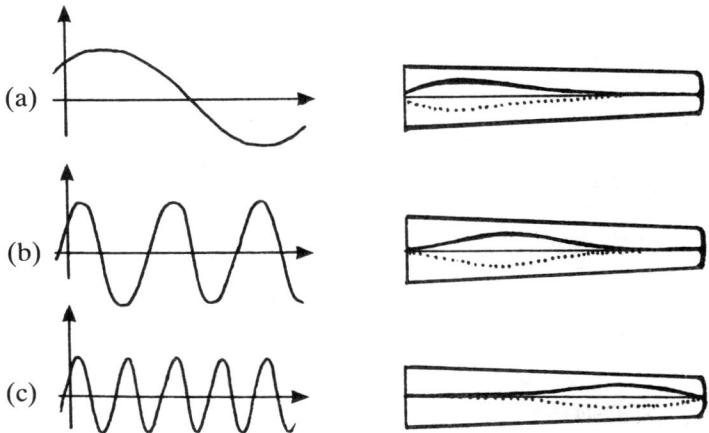

Fig. 13.4 — Vibration of the basilar membrane, under low-frequency (a), mid-range (b), and high-frequency (c) sound.

cell at any particular position along the membrane is proportional to the harmonic component of the input at its own particular characteristic frequency. Or, more correctly, since the hair responds to a (narrow) range of frequencies rather than to a single precise frequency, the neural output is a *smoothed* Fourier transform of the sound input.

But there is more subtlety to the engineering of the cochlea than just the tapering of the basilar membrane. Each individual hair cell is 'tuned' to vibration at a particular frequency. The mechanism of this tuning is not clear. It may be similar to the tuning of the membrane: the hairs at the inner end of the cochlea having a low natural frequency of vibration by being long and floppy, while the short stiff hairs towards the outside have a higher natural frequency. Whatever the mechanism, the bandpass effect of the hair cells compounds the bandpass effect of the local membrane. Each nerve output thus reflects the product of *two* filters in series, both of narrow bandpass type (see Chapter 6, Question 5).

The extent to which the neural output is smoothed from the true Fourier transform depends on the width of the pass band of this dual filter; that is, on the frequency-specificity of the hair cells at a particular point on the membrane. In man, this specificity is remarkably good. For instance, cells in the region designed to pick up sounds at 3.5 kHz (the frequency of the highest note, A^{iv}, on a piano) are 100 times as sensitive to this frequency as they are to 3.1 kHz (G^{iv}, one note lower). (At lower frequencies, the specificity is somewhat less.)

Thus, there is a curious difference between the ear and the eye. The ear has very little power of *positional* discrimination, but very good *spectral* discrimination. The eye, by contrast, has poor spectral properties, but excellent positional ones.

There is an auditory analogue to the visual 'edge-enhancement', convolution-filtering, mechanism. With hearing, this processing happens in the brain. In this part of the analysis, the outputs from neighbouring hair cells are compared and combined, in the same way as the ganglion cells combine rod and cone outputs. Thus the tuning of the hair cells is futher sharpened by a third, non-mechanical filter. In other words, the brain performs a *deconvolution* on the smoothed Fourier transform output by the cochlea. As a result, our ability to discriminate between sound pitches is even better than the discrimination of any individual hair cell.

One last piece of sophistication in auditory processing concerns the time of firing of the hair-cells. The hair-cells do not buzz continuously; rather, they produce sharp bursts or spikes of electrical activity when they reach a certain threshold distortion. After firing, a cell takes a finite time to recover: around a millisecond.

Now the vibration of the basilar membrane in the vicinity of any given hair cell is, thanks to the mechanical filtering process, almost sinusoidal. The hair-cells are thus bent backwards and forwards smoothly and regularly. At low frequencies, the hairs have time to recover between successive waves. At higher frequencies, the millisecond of 'dead time' becomes important. As the cell slowly recovers its firing potential, the hair may be bent through several cycles; it will then be 'tripped' preferentially when it is distorted its maximum amount. This means that the cell will tend to fire at a constant *phase* of this harmonic component (although not on every cycle). For frequencies above around 500 Hz, this becomes a noticeable feature.

This mechanism means that the cochlea yields not merely the *power* spectrum of the input, but some *phase* information as well, at least for higher frequencies. This is useful: by comparing the phase of a sound component in one's two ears, one has some information about the *direction* of the source of a sound. By comparing the phase for a range of components, one may be able to locate the source fairly precisely. (A pure tone, with a single component, yields only a single phase, which is not enough for precise location.)

Note, incidentally, that the wavelength of sound at 500 Hz is about 0.7 m. The distance between the ears — the width of the head — then corresponds to about a quarter of a wave at this frequency. A maximum phase difference of a quarter wave is about the minimum for useful directional information, which is a happy coincidence.

The evolution of the cochlea as an analytical structure is interesting. Fish and amphibians have a simple chamber structure containing hair cells. Reptiles and birds have elongated this to a cochlear sac; mammals have further extended and coiled the sac. With an extended cochlea, more hair cells can be packed into it. The more hair cells there are, the more points there are in the ω-domain of the Fourier analysis; thus the *range* of hearing, or the *discrimination* of the ear, or both, can be improved.

On range, fish, reptiles and birds can hear only up to about 10 kHz; but many mammals can hear up to 50 kHz. Indeed, those mammals (bats, dolphins) that use sonar systems for echo-location need to use yet shorter waves for precise direction-finding, and can hear sounds well over 100 kHz.

By contrast, the higher primates have lost their high-frequency sensitivity, and can hear only to around 20 kHz. This is because they have used their 'extended ω-domain' to improve their discrimination (perhaps because of their communication needs.) At around middle C, 250 Hz, man can discriminate tones separated by a fraction of a hertz. Thus he can tune musical instruments, by ear, to within 1/20 of a semitone. The macaque monkey (not a higher primate) has a discrimination ten times poorer; and the macaque does better than many nonmammals.

One final point concerns the perception of very *low*-frequency sound, from perhaps 50 Hz down to less than 1 Hz. These 'sounds' are too low-pitched to be heard, but they can be felt — particularly by the nerve endings in the abdomen (the *enteroceptors*) and in the muscles (the *proprioceptors*). High intensity sound at these frequencies can induce nausea and loss of consciousness, so it is particularly important to know where such sound is present. Motion sickness is, perhaps, a consequence of this. I know of no frequency-based investigation into motion sickness, but it seems unlikely that none exists.

13.3 Speech analysis

We have seen how the function of the ear may largely be described in the ω-domain, and how, without too much simplification, the cochlea can be called a natural (mechanical) Fourier analyser. It is not then surprising that the sounds we make — speech — should also benefit from being viewed in the ω-domain, i.e. via Fourier analysis.

The principal structures involved in the production of speech are (in order) the brain; the diaphragm and lungs; the larynx with its vocal cords; the oral, nasal, and pharyngeal cavities; and finally the tongue, soft palate, and lips. The brain is the controller; its rôle is far too complicated for discussion here. The diaphragm and lungs are the power supply, producing a ready source of pressurized air.

The larynx is the sound producer. If the muscles of the larynx are flexed, then the vocal cords are tautened, and close off the trachea. The pressure builds up behind them until it can force the cords apart; when it does so, it rushes past, relieving the pressure and allowing the cords to snap shut again. The pressure begins to build again, and so on. There is thus a nonharmonic, but periodic, vibration set up, and a nonharmonic, but periodic, wave of air pressure downstream of the larynx.

Fig. 13.5 shows an idealized time-domain record of vocal cord opening and of downstream pressure, together with the Fourier spectrum of the resulting sound. It is a line spectrum, since the pressure wave is periodic. Because of the sharp 'corners' in the pressure curve, its spectrum contains many harmonics of substantial amplitude. The period of the vocal cord oscillation, and therefore the frequency of the fundamental in the sound spectrum, depends both on the *size* of the vocal cords and on their *tension*. Vocal cord size varies from speaker to speaker: men tend to have larger cords than women, so their fundamental frequency is lower, which is why men have 'deeper' voices. Vocal cord tension, by contrast, is largely under voluntary control.

The rôle of the cavities (throat, mouth, and nose) above the larynx is to *filter* the buzzing sound generated by the vocal cords. The shape and size of these cavities determine their resonant frequencies (as air volumes); components near to those frequencies tend to be amplified, relative to components not near a resonant frequency. (Compare this with the behaviour of the forced extension oscillator, section 1.7.) But the geometry of the cavities is not fixed: it can be altered, voluntarily, for example by moving the tongue, soft palate, and lips.

Typically, the mouth–throat cavity has a length of about 200 m (from lips to larynx), and a width of about 80 mm (from cheek to cheek). The natural modes of such a cavity will have a whole number of half-waves in these dimensions, as in the vibrating string of section 1.8 and the rectangular membrane in the Project of Chapter 4. Thus the lowest-frequency modes of the cavity have wavelengths of about

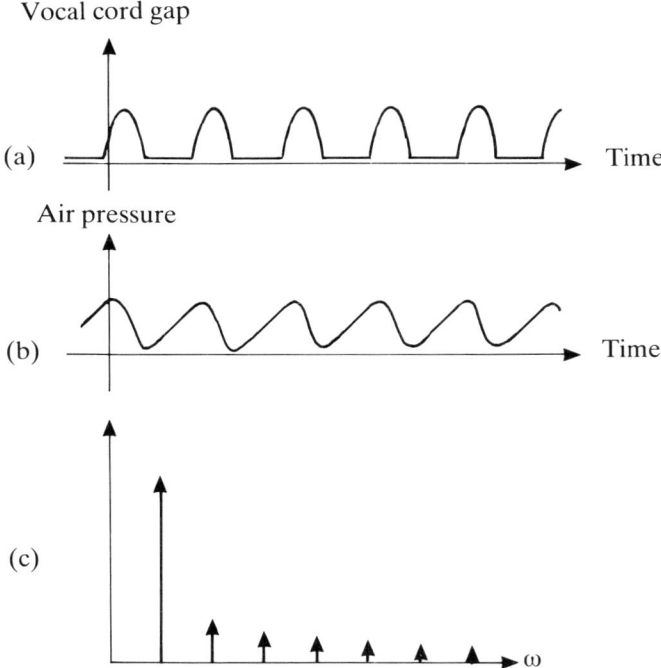

Fig. 13.5 — Voice. (a) The periodic opening of the vocal cords. (b) The consequent air pressure downstream of (above) the larynx. (c) The amplitude spectrum of this sound.

400 mm and 160 mm; since the speed of sound in air is around 320 ms^{-1}, the frequencies of these modes are about 800 Hz and 2 kHz.

Using a microphone and spectrum analyser, we can measure the air pressure variation in any given (vowel) sound. The pressure (power) spectrum of the vowel /ɑ/ 'aah' spoken by a man, for example, might be as shown in Fig. 13.6(a). A speaker

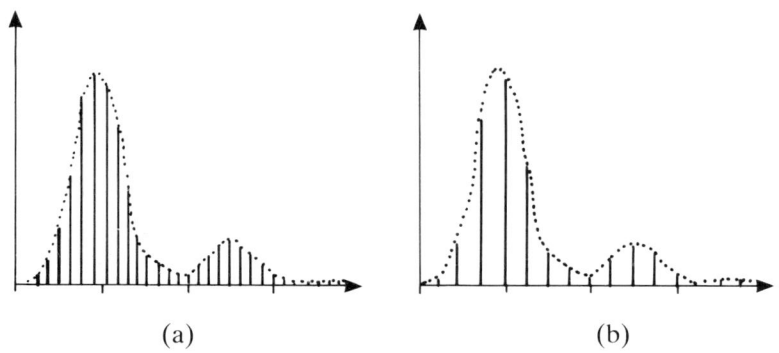

Fig. 13.6 — The filtering effect of the mouth cavity on vocal cord sound. The same vowel /ɑ/ 'aah' is shown, as spoken by someone with a deep (a) and a high (b) voice.

with a voice of higher pitch might produce something like Fig. 13.6(b) for the same vowel.

Interestingly, the relative phase of the components is not fixed for a given vowel. Thus the *waveform* varies enormously. Its power spectrum, however, always shows the same overall shape, or *envelope*. It is an indication of the Fourier-analytical properties of the ear that different waveforms with the same power spectra should be heard as the same sound.

For different vowels, the mouth cavity geometry is altered, most markedly by the change in tongue position. Indeed, the traditional phonetic (more properly, 'phonological') description of vowels characterizes them as 'front' or 'back', 'open' or 'closed', according to the position of the highest point of the tongue. Fig. 13.7 shows

Fig. 13.7 — The varying position of the tongue in the mouth during the articulation of four different vowels.

the position of the tongue for a range of 'pure' (static) vowels. /i/ denotes the vowel in 'mead', /æ/ in 'mad', /ʌ/ in 'mud', and /u/ in 'mood', in the English of southern Britain; Fig. 13.8 shows the power spectral envelopes for each of these four vowels. The effect of the altered geometry on the resonant frequencies is clear.

Consonants are more difficult to analyse. Many of them, including the 'stops' (such as /b/, /d/, /p/, and /k/) are identifiable only insofar as they affect adjacent sounds — usually the following vowel. Others, the 'fricatives' (such as /f/, /θ/ the initial sound of 'thin', and /χ/ the final sound of 'loch') involve a secondary 'buzzer': a constriction between tooth and lip, between tongue and lip, or between tongue and soft palate produces a sound pressure wave *independently* of the vocal cords. Finally, there are the 'nasals', 'semi-vowels' and 'glide vowels' (including /m/ and /n/; /r/ and /l/; /w/ and /j/, /j/ being the initial sound in 'you'). These, like the vowels, are formed by filtering the vocal cord sounds in a characteristically shaped resonant cavity.

In real speech, each sound occupies about 0.1 s. For a typical male voice, with

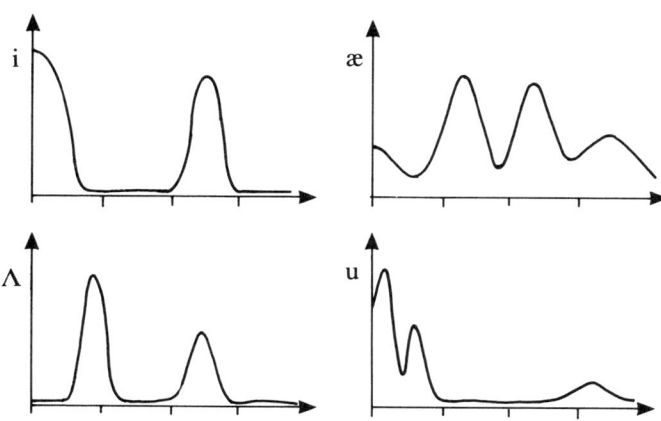

Fig. 13.8 — The spectral 'envelopes' for each of the four vowels depicted in Fig. 13.7.

vocal cord fundamental frequency around 150 Hz, the vocal cords open and close 15 times in this interval. This is long enough for the frequencies of the components to be identified approximately, but not long enough (thanks to the uncertainty theorem) for them to be measured precisely. It seems that an uncertainty in frequency of a few per cent is not enough to destroy the intelligibility of the speech.

13.4 Morphogenesis

One of the 'great questions' in biology is the question of how a single cell develops into a complex organism. This is a process not just of *growth* (the simple increase of mass by cell replication), but also of *morphogenesis*: the gradual assumption of a particular shape. Associated with morphogenesis is *cell differentiation*, the process by which initially identical cells become, after several divisions, quite different.

Some general mechanisms have been proposed for these phenomena, mainly qualitative rather than quantitative. Some involve the diffusion of proteinaceous 'growth factors', which appear to be nonuniformly distributed in an embryo. Others involve the weak electric field that the embryo generates, for this field affects many important molecules.

James Murray has looked at a relatively straightforward problem in morphogenesis: the formation of mammalian skin patterns. These patterns are caused by the presence of pigmented cells, in clusters over the animal's surface. The pigments involved are limited in number; indeed, the single pigment melanin accounts for most mammalian tones. Further, the patterns are generated at a fairly late stage of the animal's embryonic development, which minimizes the problem of the embryo's changing shape.

Simple observation shows that mammal skin patterns fall into a relatively small number of distinct types. Mice and elephants are uniformly coloured; cats are (sometimes partially) spotted or striped; giraffes are mottled. In addition to these general schemes, there are occasional curiosities, like the coloured face and genital patches of some monkeys.

What Murray has found is that all these patterns can be explained by a *reaction–diffusion* model, in which reversible reactions between chemical 'morphogens' set up patterns of (chemical) standing waves over the embryo's surface. Where the chemical wave 'motion' is large, pigmented cells develop; where it is small, there are no pigmented cells. Which modes of wave dominate depends on the size and shape of the embryo at the age when the pattern is set, as well as on the kinetics of the chemical reaction. The reaction kinetics imply a natural chemical 'wavelength'; the pattern that develops then depends on how many of these waves will fit across the animal's surface. (Think of the chemistry as being a 'forcing' vibration, the embryo as a filter, and the standing wave on its surface as the 'forced' vibration.)

Thus, for animals with small embryos (like mice), there is no room for any wave, and the colour is uniform. For slightly larger embryos, the dominant 'wave' mode is the fundamental, with one wave covering the entire organism; here, the surface becomes light on one half and dark on the other. Larger embryos still give patched or multipartite patterns, like black and white rats; larger still give stripes (for a long thin embryo) or spots (for a short fat one). For very large embryos, the spots blur into uniformity again. (See Fig. 13.9.)

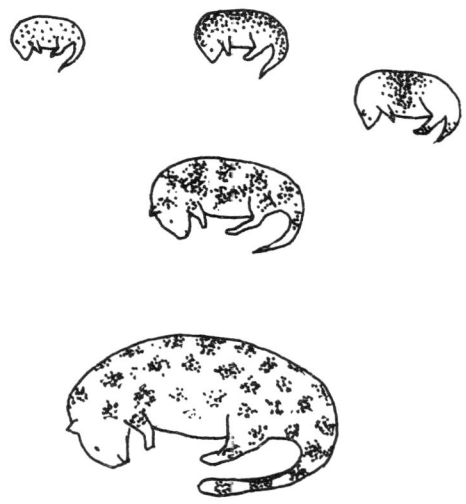

Fig. 13.9 — Size and shape, and pattern formation, in mammal embryos.

This model explains some interesting phenomena. For example, it is a fact that spotted animals often have a striped tail, but hardly ever vice versa. According to the model, this is because the tail is a 'long thin' part of the embryo. Again, there is the predominance of pale uniform bellies and dark patterned backs. We can explain this as being because the belly in the (curled-up) embryo is farther from the chemically-'waving' surface.

It would be interesting to know precisely how the shape determines the pattern (as far as I know, the calculations so far have all been numerical). Perhaps we might analyse the pattern of standing waves in terms of the 'natural modes' of the embryo's surface.

13.5 Medical diagnostics

Fourier methods and approaches find a wide variety of uses in medical circumstances. For instance, among the signals that may be recorded from a subject is the brain electrical activity (on an *electroencephalogram*, or *EEG*). This yields continuously varying functions of wavelike pattern, which change in subtle ways according to the 'state' of the subject.

EEGs are effectively functions of constant character. The power spectral density of an EEG typically shows substantial broad peaks in three ranges: 0.3–3 Hz, 8–12 Hz, and 20–30 Hz. Components in these ranges are called, respectively, δ *waves*, α *waves*, and β *waves*. The relative amplitudes of these peaks vary, most noticeably according to whether the subject is awake or asleep.

Other signals that may be treated in this way include measurements of heart activity (*electrocardiograms*, or *ECGs*). It is conceivable that there is a relationship between the 'cleanness' of a subject's ECG spectrum and his likelihood of heart problems, including 'mechanical' problems like *angina pectoris* and certain sorts of heart attack. However, the most spectacular use of Fourier methods in medicine is the application of the tomographic methods of Radon transforms (section 8.8) to *in vivo* medical diagnostics, in the class of techniques known as 'computer-assisted tomography' (or CAT).

Since the turn of the twentieth century X-radiography has been used as a diagnostic tool. But an X-ray is a two-dimensional picture of a three-dimensional object, and so a single X-ray picture is of limited use. For some purposes, one is adequate: it is possible to distinguish major bone fractures easily enough. For many other purposes, a second X-ray at a different angle will provide enough additional information: a small dense tumour may be precisely located with just two photographs from different angles.

But if the item of interest is large and of complex shape, or in the middle of other tissues, or of low visibility, it may not show up so clearly. The interpretation of vague smudges on X-rays is a precarious art. In such cases, the methods of tomographic reconstruction come into their own, for by using many projection images they can produce a full three-dimensional 'map' of the body's X-ray absorptivity. The necessary series of photographs is usually obtained by having the machine slowly revolve around the patient, wherefore such machines are often called *scanners*. Until relatively recently, such CAT methods were of no practical use, for technological reasons: there were simply no computers available that were capable of performing the enormous calculations involved. But by the mid-1970s, silicon technology had progressed to the point where CAT was a relatively routine technique. (CAT-scanners remain, however, very expensive.)

It is by no means only X-rays that can be used in CAT. Any beam or ray that is selectively attenuated by the body's tissues can be used. For instance, a second method that has become at least as widespread as X-ray CAT uses very high frequency sound waves. This is *ultrasound* tomography.

Typically, the frequencies used are of the order of a few megahertz, several hundred times higher than the limit of human hearing. Sound at these frequencies has a wavelength of a fraction of a millimetre. These frequencies are used for two reasons. The first is that the resolution of the image is connected to the wavelength of the 'probe' beam — compare the results of particle scattering experiments, section 11.3. The second reason is that ultrasound does not 'spread out' in the way that sound at lower frequencies does; rather, it travels in a well-defined beam. This is because the ultrasound source can be made large in comparison with the sound wavelength; Huygens' principle shows that this is necessary for the sound to form a beam.

More recently still, the (by now established) chemist's technique of NMR (see section 12.2) has been applied in medicine. This technique is particularly exciting, for it has a well-understood *chemical*, rather than merely physical, basis: a tomogram based on *magnetic resonance imaging* (or *MRI*) gives information about the local chemistry at each point of the body. MRI tomography uses essentially the same apparatus as chemical NMR, but a couple of sophistications are incorporated in order to discriminate the emissions from different parts of the body. First, the magnetic field is given a gradient across the sample. This means that the frequency of the emission excited in any given atom will depend on its position across the body. The resulting output is related to the (one-dimensional) Fourier transform of a (one-dimensional) *projection* of the body. If this is repeated with the magnetic field lines and field strength gradient in all possible directions, then a three-dimensional map of (say) hydrogen atom density can be built up, using a three-dimensional Radon transform.

A second sophistication uses a narrowband source of radio frequency excitation. The effect of this is that only atoms in a narrow 'slice' of the body are excited, where the magnetic field strength is such that this is the *resonant* frequency of the nuclei. All the atoms are excited to the same frequency. Then the radio excitation is switched off and the magnetic field gradient is changed to one at right angles. The decay of the excited atoms now releases photons whose energy is dependent on their position with respect to the *new* magnetic field gradient. this method thus yields the one-dimensional projection of hydrogen density from this slice only, and the analysis of thin-slice tomography described in section 8.8 can be followed.

Not all tomographic techniques are based on Fourier methods. With ultrasound, for example, there is a much simpler analysis that is more widely used. This is the *pulse–echo* method. Here, a burst of sound is emitted, which reflects off boundaries between tissue types. (This reflection is due to the mismatch of impedance characteristics; see section 11.7.) The echoes can be picked up and displayed, giving a crude but quick picture. The resolution and precision are significantly poorer than in the Fourier method, but the information is much easier to handle and process, so that the machinery can be much cheaper and faster. The speed of processing makes it possible for biological *activity* to be followed: a moving foetus, a beating heart, peristalsis in the gut can all be identified.

14
Miscellaneous

14.1 Water waves

Fourier theory is, of course, intimately connected with harmonic 'wave' motion. It is reasonable, then, to expect it to bear upon physical situations involving waves; and among these, the most familiar are water waves.

Newton's second law says that the force on a body is equal to its mass times its acceleration. In a non-viscous fluid, this becomes a connection between the pressure p, density ρ, and velocity $\mathbf{v}$, at a point; in the presence of gravity, the equation of motion is

$$\rho \left\{ \frac{\partial \mathbf{v}}{\partial t} + (\mathbf{v} \cdot \nabla)\mathbf{v} \right\} + \nabla p = (0, 0, -\rho g) \tag{14.1}$$

where g is the acceleration due to gravity, and the z-axis points vertically upwards. In addition, the conservation of mass leads to the *equation of continuity* for the fluid:

$$\frac{\partial \rho}{\partial t} + \mathbf{v} \cdot \nabla \rho + \rho \nabla \cdot \mathbf{v} = 0 \tag{14.2}$$

Equation (14.1) is not linear, thanks to the term $(\mathbf{v} \cdot \nabla)\mathbf{v}$. For convenience, suppose that the velocity $\mathbf{v}$ and its derivatives are everywhere 'small'; then this quadratic term is negligible, and equation (14.1) becomes

$$\rho \frac{\partial \mathbf{v}}{\partial t} + \nabla p = (0, 0, -\rho g) \ . \tag{14.3}$$

Again, for most purposes, a liquid (including water) may be taken to be incompressible. Then ρ is constant, and (14.2) reduces to

$$\nabla \cdot \mathbf{v} = 0. \tag{14.4}$$

This means (it may be shown) that $\mathbf{v} = \nabla \varphi$ for some function φ, the *velocity potential*; hence, from (14.3), we have that

$$p = p_0 - \rho \frac{\partial \varphi}{\partial t} - \rho g z. \tag{14.5}$$

Consider now the case of one-dimensional water waves in a body of water of constant depth (Fig. 14.1). If the waves are travelling in the x-direction, say, then

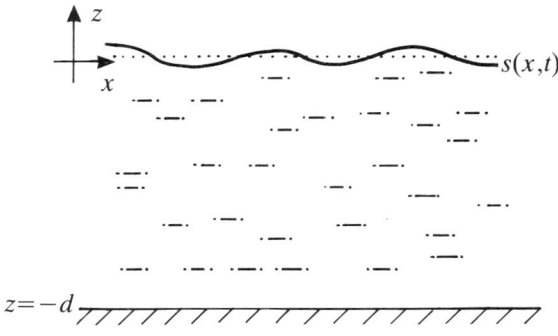

Fig. 14.1 — Water waves.

$\varphi = \varphi(x,z,t)$. (This is a useful approximation to waves in canals, for example, or in deep oceans). Suppose that the coordinate system is chosen so that the *mean free surface* is the x-axis, $z = 0$; then $p_0 = 0$. Let the actual surface of the fluid be given by $z = s(x,t)$. Ignoring surface tension, the pressure at the surface is zero (i.e. atmospheric), so one boundary condition is that

$$p(x, s(x,t), t) = 0. \tag{14.6}$$

The fluid at the surface is moving in such a way that the surface does indeed have the form $s(x,t)$. This means that a particle of water on the surface will remain on the surface; this may be shown to yield the condition

$$\frac{\partial \varphi}{\partial x}\frac{\partial s}{\partial x} - \frac{\partial \varphi}{\partial z} = -\frac{\partial s}{\partial t} \quad \text{at } (x,s,t) \tag{14.7}$$

Meanwhile, at the bed, at say $z = -d$, the fluid can have no vertical component of velocity. Since $\mathbf{v} = \nabla\varphi$, the vertical component of velocity is $\partial\varphi/\partial z$; hence the final boundary condition is that

$$\frac{\partial \varphi}{\partial z}(x, -d, t) = 0. \tag{14.8}$$

Now suppose that, not only are the particle velocities small, but the wave shape $s(x,t)$ and its derivatives are small too. Then the first term of (14.7) can be dropped, as a product of small quantities. Since $s \approx 0$, equation (14.7) becomes the approximation

$$\frac{\partial \varphi}{\partial z}(x,0,t) = \frac{\partial s}{\partial t} \tag{14.9}$$

Also, since s is small, (14.6) becomes $p(x,0,t) + s\partial p/\partial z = 0$. From (14.5), this means that $-\rho(\partial \varphi/\partial t) - s\rho(\partial^2 \varphi/\partial z \partial t) - s\rho g = 0$; but the central term in this is quadratic, and so negligible, and therefore

$$\frac{\partial \varphi}{\partial t} = -gs \quad \text{at } (x,0,t). \tag{14.10}$$

Finally, (14.4) becomes

$$\frac{\partial^2 \varphi}{\partial x^2} + \frac{\partial^2 \varphi}{\partial z^2} = 0. \tag{14.11}$$

We now have a system of linear equations in φ. Let us find the harmonic solutions of this system. A wave of (spatial) angular frequency ω and wavespeed c will have the form

$$s = A\sin(\omega(x - ct)); \tag{14.12}$$

hence, from (14.10),

$$\varphi(x,0,xt) = \frac{gA}{\omega c} \cos(\omega(x - ct)).$$

We guess that φ is separable, and has the form

$$\varphi = Z(z)\cos(\varphi(x - ct)); \tag{14.13}$$

then (14.11) becomes $(d^2Z/dz^2) - \omega^2 Z = 0$, so that $Z = a\cosh(\omega z + \psi)$ for some a and ψ. From (14.8), $\psi = -\omega d$, and then $gA/\omega c = a\cosh(\omega d)$, so that

$$\varphi = \frac{gA}{\omega c \cosh(\omega d)} \cosh(\omega(z - d)) \cos(\omega(x - ct)). \tag{14.14}$$

This gives the form of φ, and therefore of $\mathbf{v}$, for this harmonic solution. But (14.9) must be satisfied too; this becomes the relation

$$\frac{gA}{c\cosh(\omega d)} \sinh(-\omega d) = -A\omega c,$$

or just

$$c^2 = g\tanh(\omega d)/\omega. \tag{14.15}$$

Equation (14.15) is called the *dispersion relation* for the system; the graph of this function is plotted in Fig. 14.2.

Note that, just like the classical wave equations (equations (1.32') and (8.14)), the system supports harmonic waves of arbitrary frequency and amplitude (as long as the linearizing approximations are valid). Unlike the classical equations, however,

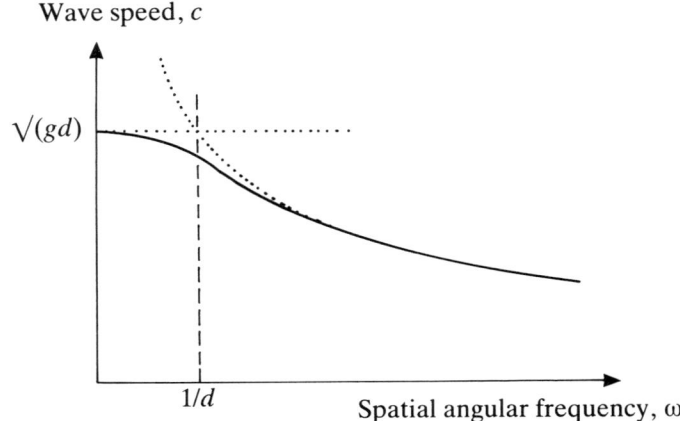

Fig. 14.2 — The speed of water waves.

the wave speed is *not* constant in this system; it varies according to the size of the wave. Where ω is small, so that the wavelength is large compared with the water depth d, we have $\tanh(\omega d) \simeq \omega d$, so that c is roughly constant at $\sqrt{(gd)}$. But when ω is large, and the wavelength is small, then $\tanh(\omega d) \simeq 1$ and $c \simeq \sqrt{(g/\omega)}$.

Since the controlling equations are linear, the principle of superposition still applies: waves of different frequency still travel independently of each other. The effect of this is that a travelling wave containing more than one harmonic component will gradually 'spread out', with the faster-moving long-wave components running ahead of the slower short waves (hence the term 'dispersion relation'). Thus the evolution of a water wave is intimately connected with its Fourier structure.

Incorporating surface tension into the analysis has an interesting consequence. If the water has constant surface tension τ, then the dispersion relation (14.15) becomes

$$c^2 = \left(\frac{g}{\omega} + \frac{\tau\omega}{\rho} \right) \tanh(\omega d). \qquad (14.15')$$

This implies that the wavespeed is least for waves with the critical value of $\omega = \omega_c = \sqrt{(\rho g/\tau)}$, and greater for larger or smaller ω. For water under normal circumstances, $\omega_c \simeq 360 \text{ m}^{-1}$, so that the slowest waves are those with wavelength $(2\pi/360)$ m, or about 17 mm. In water deeper than a few millimetres, these waves have speed $c \simeq 55 \text{ mm s}^{-1}$.

Now, suppose a stone is thrown into a pond or lake. Initially, the wave produced is compact and strongly nonharmonic; but as the faster waves disperse outwards, one is soon left with an almost-harmonic set of expanding ripples, of about this critical wavelength and travelling at about this minimal wave speed.

'Small' waves in other channels, and in two dimensions, succumb to a similar analysis. But many water waves are not 'small'. Wind-generated sea waves, for instance, are only 'small' on exceptionally calm days. Mostly, the sea is 'choppy': the waves adopt a pointed shape, with sharp peaks and flat troughs. This is not a peculiarity of Fourier synthesis, but a consequence of the importance of the terms neglected in the linearized analysis above. The presence of significant quadratic terms means that the principle of superposition fails; if you will, that the Fourier components of the waves no longer evolve independently. (This is the same process as the 'mode-coupling' mentioned in section 12.3.)

However, a 'statistical' description of a sea wave in the frequency domain is still possible. The mode-coupling process means that the power spectral density p_S of the wave surface is reasonably stable with time (for given stable conditions of wind, suspended matter, etc.). Fig. 14.3 shows the PSDF of a typical sea.

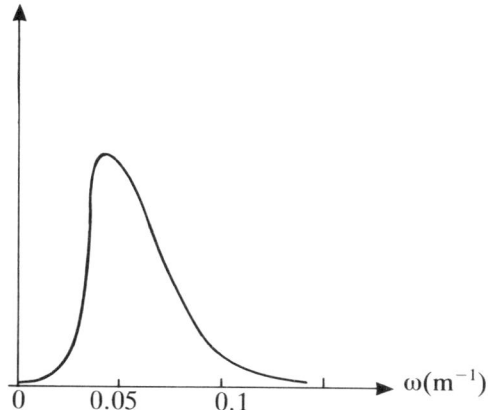

Fig. 14.3 — The PSDF of waves in a typical sea.

This Fourier structure is important for ships, oil rigs, sea walls, and other structures that are subjected to wave impacts. It is perhaps most important for the design of mechanisms to extract usable power from waves. A variety of schemes have been suggested. Many of these, including the famous Salter 'nodding duck', depend on a system of floats, whose motion under the waves drives a turbine. Any such system will have its own frequency-response characteristics, and it is important that the characteristics match the environment. In other words, if most of the wave energy expected over a period of time occurs around $\omega \simeq 0.5 \text{ s}^{-1}$, then the device should be able to respond efficiently to waves at this frequency.

14.2 Turbulence in fluids
In the absence of gravity, but allowing for viscosity, the equation of motion of a fluid changes from (14.1) to

$$\rho \left\{ \frac{\partial \mathbf{v}}{\partial t} + (\mathbf{v} \cdot \nabla) \mathbf{v} \right\} = - \nabla p + \mu \nabla^2 \mathbf{v}. \tag{14.16}$$

This is the *Navier–Stokes* equation; μ is the *(coefficient of) viscosity* of the fluid.

When **v** is small, the quadratic term disappears, the equation is linear, and the solutions are 'stable'. This is called *laminar flow*. When the velocities are large enough for the quadratic term to be significant, *turbulent* flow patterns arise. Because of the nonlinearity, these solutions cannot be obtained by Fourier methods; but they can be described in the ω-domain, like the sea waves of the previous section.

Suppose we have a turbulent flow pattern which, at fixed time t, has uniform properties across all of space (that is, the turbulence field is *isotropic* and *homogeneous*). Then we can describe it in terms of its (vector) PSDF; if $\mathbf{v} = (v_1, v_2, v_3)$, then define $\mathbf{G}(\omega, t) = (G_1, G_2, G_3)$, where

$$G_k(\omega, t) = \lim_{R \to \infty} \frac{1}{(2R)^3} \left| \int_{-R}^{R} \int_{-R}^{R} \int_{-R}^{R} v_k(\mathbf{x}, t) e^{i\mathbf{x}\cdot\boldsymbol{\omega}} \, d\mathbf{x} \right|^2. \tag{14.17}$$

(Compare equation (3.11).) By the three-dimensional Wiener–Khintchine theorem (see equation (7.18)),

$$G_k(\omega, t) = \mathbf{F}_3 u[v_k](\omega, t), \tag{14.18}$$

where $u[v_k](\mathbf{x}, t)$ is the *variance density function* or *VDF* of v_k:

$$u[v_k](\mathbf{x}, t) = \lim_{R \to \infty} (1/(2R)^3) \left\{ \int_{-R}^{R} \int_{-R}^{R} \int_{-R}^{R} v_k(\mathbf{y}, t) v_k(\mathbf{y} + \mathbf{x}, t) \, d\mathbf{y} \right\} \tag{14.19}$$

(cf. equation (7.10)). There are also corresponding *covariance density functions*, of v_j with v_k, but these turn out to be zero for isotropic and homogeneous turbulence.

Further, this isotropy and homogeneity means that the functions $G_k(\omega, t)$ and $u[v_k](\mathbf{x}, t)$ reduce to only two distinct functions: the *longitudinal VDF* $f(\|\mathbf{x}\|, t)$, the correlation of velocity components *parallel* to the displacement **x**; and the *transverse VDF* $g(\|\mathbf{x}\|, t)$, the correlation of velocity components *perpendicular* to the displacement. Thus, for instance,

$$f(r, t) = u[v_1](r, 0, 0, t)$$

and

$$g(r, t) = u[v_1](0, r, 0, t). \tag{14.20}$$

In the ω-domain, then, the three vector functions $G_k(\omega, t)$ reduce to two functions of the scalar $\omega = \|\boldsymbol{\omega}\|$, namely

$$F(\omega, t) = \mathbf{F}_3 f(\omega, t) \quad (= G_1(\omega, 0, 0, t), \text{ say})$$

and

$$G(\omega, t) = \mathbf{F}_3 g(\omega, t) \quad (= G_1(0, \omega, 0, t), \text{ say}). \tag{14.21}$$

Now we combine these into a single expression by putting

$$E(\omega, t) = (\rho \omega^2 / 4\pi^2)[F(\omega, t) + 2G(\omega, t)]. \tag{14.22}$$

This function represents the portion of the *kinetic energy density* of the fluid that lies between spatial angular frequencies ω and $\omega + d\omega$. To see this, note that the total kinetic energy in a volume V is

$$E = \int_V \tfrac{1}{2}\rho\|\mathbf{v}\|^2 \, d\mathbf{x} = \tfrac{1}{2}\rho \int_V (v_1(\mathbf{x})^2 + v_2(\mathbf{x})^2 + v_3(\mathbf{x})^2)d\mathbf{x}$$

$$= \frac{\rho}{16\pi^3}\int (\boldsymbol{p}_3 v_1(\boldsymbol{\omega}) + \boldsymbol{p}_3 v_2(\boldsymbol{\omega}) + \boldsymbol{p}_3 v_3(\boldsymbol{\omega}))d\boldsymbol{\omega}$$

by the three-dimensional Rayleigh–Plancherel theorem (cf. equation (5.23)). We write this integral in spherical polar coordinates ω, θ, φ. For fixed θ and φ, we take one Cartesian coordinate along the θ–φ direction and two perpendicular; then we have

$$E = \frac{\rho}{16\pi^3}\int\int\int \{\boldsymbol{p}_3 v_1(\omega,0,0) + \boldsymbol{p}_3 v_2(\omega,0,0) + \boldsymbol{p}_3 v_3(\omega,0,0)\}\omega^2 \sin\theta \, d\omega d\theta d\varphi$$

$$= \frac{\rho}{4\pi^2}\int \{\boldsymbol{p}_3 v_1(\omega,0,0) + 2\boldsymbol{p}_3 v_1(0,\omega,0)\}\omega^2 \, d\omega$$

$$= \int E(\omega,t) \, d\omega \; . \tag{14.23}$$

Now in $E(\omega,t)$ we have a (statistical) description of isotropic, homogeneous turbulence. But what is the form of this fuction? How might we use the Navier–Stokes equation (14.16) to determine $E(\omega,t)$?

As yet, there is no complete answer to this question. However, the properties of $E(\omega,t)$ are *qualitatively* well understood. Small-scale turbulence is most prone to the effects of viscous damping, which dissipates kinetic energy. Indeed, velocity fluctuations on a very small scale die away so quickly that they cannot realistically be called 'eddies'. Thus at large ω, $E(\omega,t)$ falls towards zero. This is the *dissipation range* of ω.

For slightly larger scales (smaller ω), viscosity is less important than the presence of nonlinear terms in the equations of motion. The effect of this is, as we have seen, that the harmonic modes of oscillation are no longer independent; this is the process of *mode-coupling*. Physically, we can understand this as turbulence at these intermediate wavelengths 'breaking up' into turbulence at smaller scales (thereafter to be dissipated by viscous loss). This is the *inertial range*.

The distribution of energy among the largest eddies depends on how the turbulence is created. In general, the larger eddies contain less and less energy, so that the graph of $E(\omega,t)$ falls to zero at $\omega = 0$. A typical curve of $E(\omega,t)$ is shown in Fig. 14.4.

More detailed and technical analysis suggests that $E(\omega,t) \simeq \omega^{-5/3}$ for medium-large ω; this is *Kolmogorov's law*. For very large ω, where the inertial term is of little importance, $E(\omega,t) \simeq \exp(-k\omega^2)$. These analyses have been checked experimentally, and found to be in good agreement for many situations. Indeed, the energy spectrum even of *ani*sotropic turbulence seems to behave isotropically at large ω.

14.3 Meteorology

A real circumstance where the above discussion has consequences of great importance is the practice of weather forecasting. The Earth's atmosphere is a vast, turbulent body of fluid. The largest single influence on the weather is, simply, the *kinetics* of the atmosphere, the way it moves as a fluid. Second to that is its *thermodynamics* — heat exchange both within the atmosphere, and with the land and oceans of the Earth's surface. The third major effect is the *water balance* of the

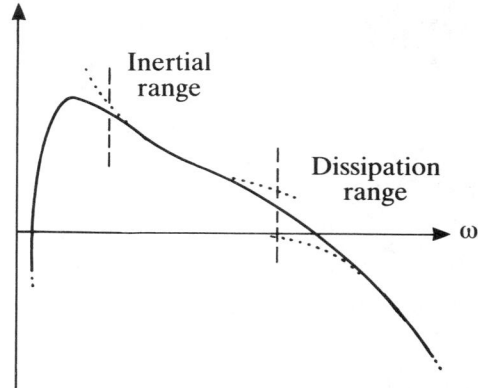

Fig. 14.4 — Energy spectrum of a typical turbulent flow field (the vertical axis is logarithmic).

atmosphere: evaporation, cloud formation, and precipitation. There are other 'local' factors, notably surface albedo (reflectiveness) and topography (hilliness).

The first detailed weather forecasting calculations were based on *spatial* approximations. The 'prognostic variables' (pressure, wind speed and direction, humidity and so on) are sampled in the spatial domain. The differential equations governing their behaviour are complicated, but comparatively well understood, and it is possible to perform numerical integration on any given sample data to deduce how the atmosphere should evolve with time. Of course, this makes the process sound much easier than it in fact is. Not least of the problems is that an accurate forecast requires a huge amount of accurate data, and collecting data at intervals precisely (say) ten miles apart in the middle of an ocean would not be easy. More seriously, the controlling hydrodynamical equations are, as we have seen, non-linear; one result of this is that small errors of measurement or approximation can grow exponentially. This puts a *theoretical* upper limit of around two weeks on accurate weather forecasting, no matter how much data is collected.

But the most obvious limitation on forecasting accuracy is simply the amount of computation required. A typical global circulation model might have a sample grid with a spacing of 2° in latitude and 3° in longitude — not spectacular spatial resolution, since this grid spacing is about the size of Switzerland, Bangladesh or Ohio. Vertically, perhaps ten sample points cover each geographical location. That implies 10^5 data points. The time step in the calculation might typically be about 10 minutes; so a 24-h forecast requires integration over about 150 time steps. At each point, for each time step, the value of each variable must be updated, according to its differential equation. If all this takes just a millisecond, the 24 h forecast will still take four hours to run. (Parallel computation represents a great boon in this area, for using it means many sample points can be updated simultaneously.)

Alternative to the spatial method, particularly for 'general circulation' models covering a large part of the Earth, are the *spectral* techniques. In this approach, the prognostic variables are sampled in the (*spatial*) *frequency* domain. Indeed, as the Earth's surface is finite, what this amounts to is approximating each variable by a truncated Fourier series. (This series is not the classical series, for the variables are functions on the surface of a sphere, not on an interval. Rather it is based on the *spherical harmonic* functions that we discussed briefly in section 11.1.) Comparative studies have been performed, which indicate that the spectral methods are considerably more efficient and accurate than spatial methods of comparable computational speed.

Some spectral models operate in the frequency domain only in respect of latitude and longtitude. Vertically, such models use the spatial domain. Other models operate in the frequency domain in all three directions. The reason for this variation is the difference in scale — horizontal distances are of the order of hundreds or thousands of kilometres, whereas the meteorologically significant part of the atmosphere is only a few tens of kilometres thick — which means that the vertical direction behaves quite differently from the horizontal.

As we are dealing with Fourier series rather than transforms, the frequency domain is naturally discrete. The computational errors introduced in *truncating* the spectra are therefore less problematical than those involved in *sampling* the variables in the spatial domain. Thus, there is a natural advantage in adopting the spectral approach.

But there are problems with the spectral approach too. The equations governing atmospheric behaviour are, as we said, nonlinear. Thus a (spherical) Fourier analysis of the state of the atmosphere *now* will not immediately yield its behaviour in the future, as it did in the (linear) case of the vibrating string (section 1.8). The numerical integration must still be performed along a sequence of time steps. By the converse of the convolution theorem, a quadratic product term in the spatial equations, like the $(\mathbf{v}\cdot\nabla)\mathbf{v}$ of equation (14.1), becomes the *convolution* of their transforms. Thus the energy in a mode at time step $n+1$ will depend on *every* mode at time step n.

We saw in Chapter 9 that calculation of convolutions is, in general, a long process. Owing to this complexity, the direct spectral approach (called the 'interaction coefficient' method, because in the calculation of the convolution any two modes interact by an amount given by this coefficient) takes a great deal of computation. Despite its attractiveness, therefore, it never has been widely used.

More recently, it occurred to some people that the calculations need not be performed *entirely* in the frequency domain. The model could be a spectral one, in the sense that the variables held in the computer represent the spectra at a particular time; while in the calculation for the next time step, where the nonlinear term arises, we use the convolution theorem. Thus, we transform back into the spatial domain, perform a simple multiplication, and transform forward once more. This spectral method is, confusingly, called the 'transform method', and the computation is much facilitated by it.

14.4 Glacier beds and 'roughness'

The previous three sections have all dealt with the advantages of the spectral viewpoint for problems involving fluids. But the Fourier method can be exploited in problems in the solid phase too. One very general such case is the physical interpretation of the subjective idea of the *roughness* of a surface. As an example of this, we look at the sliding of glaciers on their beds.

Glaciers are 'rivers of ice'. They form, typically, on mountains where a lot of snow falls; the snow compacts under its own weight into ice, and begins to move down the mountain. The terminus of a glacier may be a sea coast (where the ice breaks off into icebergs) or on land (if conditions are warm enough to melt the ice as fast as it moves down the mountain).

Under the conditions of high stress it experiences in a glacier, ice behaves as a ductile solid rather than as a brittle one. Because of this ductility, the ice *flows* down the mountain. Moreover, it does so almost exactly as a fluid would — though its effective viscosity is something like 10^{15} times that of liquid water. Such fluid behaviour may be modelled, using the Navier–Stokes equation (14.16) with a gravitational term. The viscosity term μ is large and the velocities are small, so that the quadratic term $(\mathbf{v}\cdot\nabla)\mathbf{v}$ can be neglected; for steady flow, $\partial \mathbf{v}/\partial t = 0$, and we have simply

$$\nabla p = (0,0, -\rho g) + \mu \nabla^2 \mathbf{v}, \tag{14.24}$$

where ρ is the density of the ice, g is the acceleration due to gravity, and the third coordinate is measured vertically upwards. In addition, we have the equation of continuity for an incompressible fluid, equation (14.4):

$$\nabla \cdot \mathbf{v} = 0.$$

These are both linear equations, so the system is soluble, given the boundary conditions. The upper boundary condition is that the pressure at the free surface is atmospheric. The lower boundary condition is trickier to formulate. If the ice is assumed to be bonded to the bed at all points, then it is that the ice velocity is zero at the bed.

Of course, in a real glacier, there are complicating factors. The ice of the glacier is not pure; it contains a lot of air bubbles near its free surface, and a lot of rock fragments near its base. The ice may be cracked, or have meltwater channels running through it. If it is flowing rapidly enough, it may not touch its bed at all points, being carried over the hollows before it has time to slump down into them. Finally, a combination of pressure, friction and the Earth's heat may melt the ice at some or all points on its base. Thus the above model is not adequate.

One improvement we can make is to free the ice from being bonded to its bed. Then it will move by a combination of a quasi-viscous bulk *flow*, and an independent *sliding* over its bed. The flow term is reasonably straightforward to estimate (as in section 14.1); the sliding term is much more complicated and variable.

Clearly, one of the features that will affect the sliding is the roughness of the glacier bed: the rougher the bed, the greater the resistance to sliding. We assume still that, although the ice is no longer *bonded* to its bed, it is nevertheless held *in contact* with it by the weight of the glacier. Since the governing equations are linear, we can now use Fourier analysis, and the principle of superposition, to reduce the problem to the case of a bed of sinusoidal shape.

Suppose, then, that the ice is flowing in the x-direction, over a bed which has small sinusoidal waves in the direction of its flow, say $A\cos(\omega x)$ with $\omega A \ll 1$ (Fig. 14.5). We ignore any transverse flow and any transverse unevenness in the bed, so that this becomes a problem in two dimensions. The velocity of the ice at a point (x,z) will be of the form

$$\mathbf{v} = (u,v) = (V + A(z)\cos(\omega x + \varphi), B(x,z)).$$

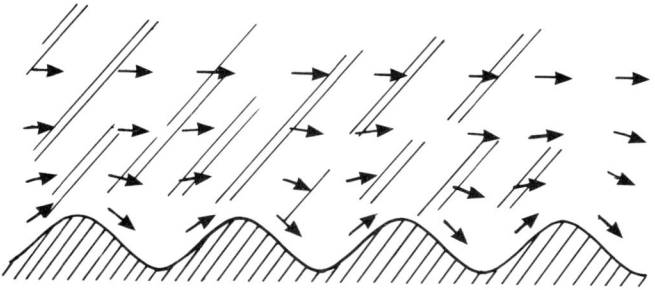

Fig. 14.5 — A glacier sliding over its bed by viscous deformation.

From equation (14.4), $\partial u/\partial x + \partial v/\partial z = 0$, or

$$-\omega A(z)\sin(\omega x + \varphi) = \frac{\partial B}{\partial z}.$$

If $a(x)$ is the indefinite integral of $A(x)$, then $A(z) = a'(z)$, and this says that $B(x,z) = -\omega a(z)\sin(\omega x + \varphi) + \beta(x)$, so that

$$u = V + a'(z)\cos(\omega x + \varphi)$$

and

$$v = -\omega a(z)\sin(\omega x + \varphi) + \beta(x). \tag{14.25}$$

Equation (14.24) now says that

$$\frac{\partial p}{\partial x} = -\mu\nabla^2 u = -\mu(-\omega^2 a' + a''')\cos(\omega x + \varphi)$$

and

$$\frac{\partial p}{\partial z} = -\rho g - \mu\nabla^2 v = -\rho g + \omega\mu(-\omega^2 a + a'')\sin(\omega x + \varphi) + \beta''(x).$$

Thus, evaluating $\partial^2 p/\partial x \partial z$ two different ways, we arrive at

$$-\mu(-\omega^2 a'' + a'''')\cos(\omega x + \varphi) = \omega^2\mu(-\omega^2 a + a'')\cos(\omega x + \varphi) + \beta''(x),$$

so we can take that $\beta'' = 0$ and

$$a'''' - 2\omega^2 a'' + \omega^4 a = 0.$$

The general solution to (14.26) is $a = (K + Lz)e^{-\omega z} + (M + Nz)e^{\omega z}$, but if the velocity high above the bed (large z) is to be approximately $(V, 0, 0)$, then we must have $M = N = 0$. Thus $a = (K + Lz)e^{-\omega z}$; equation (14.25) then become

$$u = V + (-\omega K - \omega L z + L)e^{-\omega z}\cos(\omega x + \varphi)$$

and

$$v = -\omega(K + Lz)e^{-\omega z}\sin(\omega x + \varphi). \qquad (14.26)$$

Now we incorporate the (lower) boundary condition. The ice is in contact with the bed; here, it is free to move only parallel to the bed. That is, at $z = A\cos(\omega x)$, we must have $\mathbf{v}\cdot(-\omega A\sin(\omega x), 1) = 0$, or

$$-A\sin(\omega x)\,(V + (-\omega K - \omega Lz + L)e^{-\omega z}\cos(\omega x + \varphi))$$
$$= (K + Lz)e^{-\omega z}\sin(\omega x + \varphi).$$

Since A is small, so is z here; so, therefore, are K and L. To first order, then, this equation is simply $-A\sin(\omega x)V = K\sin(\omega x + \varphi)$; thus $\varphi = 0$ and $K = -VA$. Since $e^{-\omega z} \simeq 1 - \omega z$, the second-order expansion is now

$$-A\sin(\omega x)(V + (-\omega K + L)\cos(\omega x)) = (K - K\omega z + Lz)\sin(\omega x);$$

with $z = A\cos(\omega x)$, this becomes

$$-VA - A(-\omega K + L)\cos(\omega x) = K + LA\cos(\omega x) - K\omega A\cos(\omega x)$$

so, equating the cosine parts, $L = \omega K = -\omega VA$. Thus, the velocity field is

$$u = V[1 + A(2\omega + \omega^2 z)e^{-\omega z}\cos(\omega x)]$$

and

$$v = VA[(\omega + \omega^2 z)e^{-\omega z}\sin(\omega x)].$$

We now have the velocity field. The pressure p is given in (14.24) by $\partial p/\partial x = -\mu\nabla^2 u = -2\mu VA\omega^3 e^{-\omega z}\cos(\omega x)$, so $p = -2\mu VA\omega^2 e^{-\omega z}\sin(\omega x) + c(z)$. On the bed $z = A\cos(\omega x) \simeq 0$, and $p = -2\mu VA\omega^2\sin(\omega x)$ to first order. This means that the pressure is higher on the upside of each wave than on the downside; this is the force resisting the sliding of the glacier. The net lateral force over a single wave is

$$\int_0^{2\pi/\omega} p\,\frac{d}{dx}(A\cos(\omega x))\,dx = 2\mu VA^2\omega^3 \int_0^{2\pi/\omega} \sin^2(\omega x)\,dx = 2\mu VA^2\omega^3\pi/\omega;$$

the *shear stress* is this force per unit glacier length, or

$$\tau = \frac{2\pi\mu VA^2\omega^2}{(2\pi/\omega)} = \mu VA^2\omega^3. \qquad (14.27)$$

That is, for the glacier to slide at velocity V over this bed, it must be pushed along with this lateral (shear) stress.

This analysis is for a bed of sinusoidal form. Because the governing equations are linear, a more general bed shape $Z(x)$ can be dealt with by Fourier analysis. A sinusoidal component of amplitude A and angular frequency ω contributes $\mu VA^2\omega^3$ to the shear; thus the total shear is

$$\tau = \mu V \int_0^\infty \omega^3 pZ(\omega)\,d\omega. \qquad (14.28)$$

In fact this analysis ignores many of the more complicated facts about glaciers that were mentioned above. We may particularly note the fact that the glacier can pass a wavy bed by additional mechanisms, like melting ahead of a wave and refreezing behind (the *regelation* mechanism), or by simply riding physically over the troughs (the *cavitation* mechanism). More involved analyses, such as that of Nye, suggest shear–velocity relationships of the form

$$\tau = \mu V \int_0^\infty W(\omega) pZW(\omega)\, d\omega. \tag{14.29}$$

where $W(\omega)$ is a spectral weighting (or 'weighing') function.

The presence of the monotonically increasing weighting function $W(\omega)$ in equations (14.28) and (14.29) is important. The 'roughness' of the glacier bed is, because of it, not simply the power spectral density pZ. Kamb defines a spectral roughness parameter $\xi(\omega) = \sqrt{(\omega^3 pZ(\omega))}$ to represent the drag effect of the component at angular frequency ω in the bed profile. Measurements of real glacier beds show that $pZ(\omega)$ drops sharply with increasing ω; however, $\xi(\omega)$ increases up to quite large values of ω, showing that small-scale bed roughness is as important in glacial drag as the more obvious large-scale unevenness (Fig. 14.6).

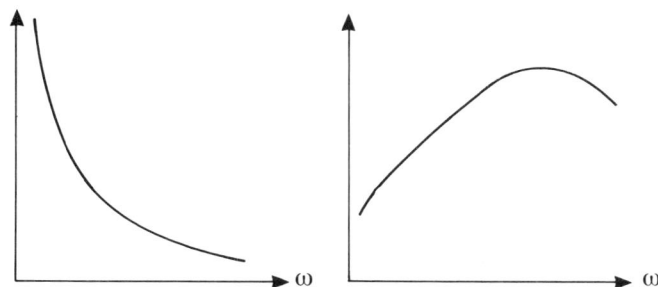

Fig. 14.6 — (a) The spectrum of a typical glacier bed profile. (b) The (spectrally weighted) Kamb roughness ξ for the same bed.

Actually, ice does not behave as an ideal fluid, with a viscosity μ independent of the stress upon it. Rather, it follows a *nonlinear* flow law, which of course compromises the whole basis of this analysis. In fact, empirical evidence suggests that V is more nearly proportional to τ^3 than to τ.

14.5 Seismology

A second, perhaps more obvious, geophysical application of Fourier methods is in the field of seismology. Seismology is more than merely the study of earthquakes; it includes the gathering and interpretation of data from all sources of terrestrial vibration, natural (as generated by earthquakes and volcanoes) or man-made (as from deliberate explosions).

The Earth is a very complicated medium for sound waves to travel in, for it is composed of a great variety of materials with different acoustic properties. These materials are, on the whole, elastic solids that behave more or less according to Hooke's law. Their governing equations of motion are (almost) linear; which is fortunate, for the principle of superposition then means that waves of different types can travel freely and independently through the Earth, without interfering with each other.

Because the Earth is largely solid, it admits of not one but two qualitatively distinct sorts of 'deep' or 'body' wave: *pressure* (or *P-*) *waves*, and *shear* (or *S-*) *waves*. In these waves, the material oscillates (respectively) parallel to, and perpendicular to, the direction of propagation of the wave. (Sound waves in air are P-type waves; waves on a string represent the S-waves.) P-waves travel faster than S-waves. In addition to these, various kinds of *surface waves* are distinguished, which travel yet more slowly than S-waves. These are modes of wave motion in which only the material near the Earth's surface moves significantly.

The different materials of which the Earth is composed transmit these waves at different speeds. At boundaries between materials, waves are partially reflected; where material properties change gradually, the waves travel in curves rather than straight lines. (For the reflection, compare section 11.7 on impedance-matching.) Thus, when a vibration event occurs somewhere on Earth, waves travelling out from it will arrive at a detector at any given point via a great variety of routes. Since the materials of the Earth are linear, the detection vibration is related to the input via a convolution; the convolving function is determined by the position of the source and detector on the Earth's surface. (Compare the acoustics of a theatre, section 5.1.)

The complexity of the Earth's structure means that this convolution is by no means easy to disentangle. However, by comparing the outputs of many detectors, for a range of vibration events, a fairly detailed model of the whole Earth can be built up. It is on this basis that geophysicists have deduced the 'onion-like' structure of the Earth: solid inner core, liquid outer core, solid lower mantle, and so on, arrayed as concentric spheres. Some of the physical properties of the materials comprising these various layers can be deduced — such as their density and compressibility, which determine (largely) the characteristic impedance of, and wave speed in, the layers.

Where earthquakes and volcanoes do not naturally occur, or where a more detailed picture is required than is provided by data from natural events, artificial events can be planned. Mineral (and in particular petroleum) exploration, for instance, often includes seismological methods. In these methods, an explosive charge is detonated in a chosen spot; seismic detectors placed in suitable nearby spots pick up the vibration. In this way, a three-dimensional 'map' of the survey site can be constructed.

This deduction process can work in reverse, too. From an (*a priori*) knowledge of the Earth's detailed physical structure, and the output of one or more seismometers, it is possible to deduce something about the location and the nature of the event detected. With three seismometers in different places, the site of the event can be pinpointed.

Such analysis has not only scientific, but even political, ramifications, for it can assist in the identification and characterization of underground (nuclear) explosions. Under treaty, explosions of greater than 150 kt yield are forbidden. But a 150 kt device puts out as much energy as an earthquake of magnitude around 6.5–7 on the

Richter scale, and this is easily detectable from anywhere in the world. By identifying the source of such events, a country's compliance with treaty can be monitored.

Neither is it possible for the culprit to dismiss the detected event as an earthquake, for earthquakes and nuclear detonations have different characteristics. For one thing, nuclear events are much shorter in duration and more localized than earthquakes; however, wave dispersion in the body of the Earth will blur this sharpness. But the *spectral* properties of the vibrations are also distinctive, and they do not suffer from this blurring (Fig. 14.7).

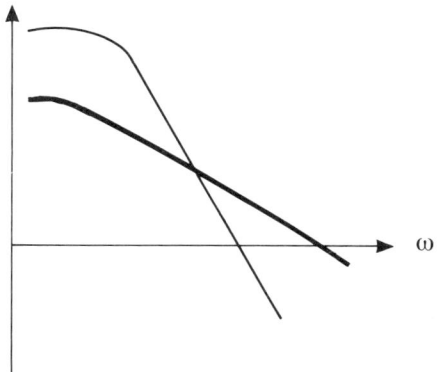

Fig. 14.7 — Power spectra for the seismic effects of an earthquake (light line) and of a nuclear detonation (heavy line). (The vertical scale is logarithmic.)

14.6 Vibration analysis

Engineering machinery often involves the transmission of substantial power through gears, rollers, belt drives, and other rotating components. Such components are never perfectly round or perfectly balanced. Now a slight eccentricity of the mounting, or localized wear, in a rotating component will generate a rotating centripetal force. Resolving this into its components, we can say that an imperfect rotator generates a *harmonic* force. Moreover, the angular frequency of the force is simply the angular velocity of the rotator. In a typical piece of machinery, there will be a number of rotating components, rotating at different angular velocities. Each of these generates an oscillating force at its own angular velocity.

A large, fast-spinning component can generate substantial forces. Such forces vibrate the entire machine — in extreme cases, to its eventual destruction. Even when the damage to vibrating machinery is not so drastic, it can result in loss of efficiency, loss of product quality, increased maintenance costs, reduced reliability, and simply a lot more noise in the vicinity. It is therefore desirable to keep the vibration in any machinery to a minimum, by ensuring that the components are in as good a condition as possible.

One approach would be to replace or overhaul every component at frequent intervals. But this is a waste of time and money: most of them will be in perfectly adequate condition. A better solution is to *monitor* the machinery, only replacing or overhauling flawed components. In order to discover which components need attention, we look for vibrations at their characteristic frequencies. All the while the vibration is small, no action is taken. When the vibration becomes unacceptable, the machine is stopped and the faulty component replaced or refurbished.

Fourier analysis is clearly a useful tool here. An accelerometer placed on the machine at a key point will record vibrations from all sources. If the output signal is subject to Fourier analysis, the spectrum will show the frequencies at which the machine is vibrating. The strength with which each component appears in the acceleration spectrum indicates the seriousness of that component of vibration. Hopefully, any important peaks in the spectrum will be at frequencies characteristic of a particular machine part (or small number of parts); in which case, that part has been identified as the one causing the trouble. With care, this diagnostic method can be made very precise, and even made to yield information about the nature of the flaw.

14.7 Economics

Economic data are a prolific source of time-dependent variables, many of which are said to exhibit 'cyclical' behaviour. Among these perhaps the most familiar are the stock market indices; other variables of this type include money supply figures, mean salaries, gross domestic product, retail price indices, and so on. To extract information about cyclical behaviour, we might consider using Fourier methods.

All these variables suffer (as far as Fourier analysis is concerned) from an overriding drawback: they have a *tendency to increase* with time. Thus they cannot be functions of constant character, and the idea of power spectral density is meaningless.

Neither will any finite sample have a sensible Fourier spectrum. Thanks to the trend and the 'wraparound' effect, there will be an effective discontinuity at the endpoints; the components this generates are likely to swamp any 'real' cyclical features of the curve. Fig. 14.8 compares the graph and amplitude spectrum of the FT 30-share index over the period 1974–84 with the sawtooth wave (equation (2.54)).

Thus, before we use Fourier methods, we must manipulate the data in such a way that we have (or suspect that we have) a function of constant character. In the case of a stock market index, we might argue that it is *percentage change* that controls it, leading over a long period to a curve that is roughly an increasing exponential curve. To correct for this effect, we might first take logarithms, and then correct (by linear regression or by some other method) for the trend (Fig. 14.9).

Other approaches are available to us. For example, if we are investing in stocks, we are really interested in (percentage) *changes* in the index. If $x(n)$ is the value of the index at the end of day n, then its change over the course of day $n + 1$ is represented by

$$y(n) = \log(x(n+1)/x(n)) = \log(x(n+1)) - \log(x(n))$$

(see Fig. 14.10(a)). To make a profit on our investments, we would like to be able to predict such changes; in order to do this, we could look at the (discrete) *autocorrela-*

Ch. 14] Miscellaneous 309

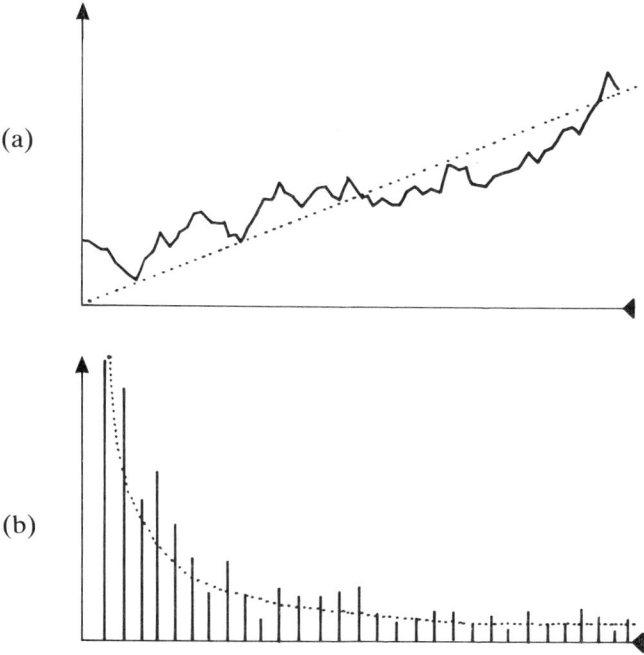

Fig. 14.8 — (a) The FT-30 share index, 1974–84. (b) The discrete amplitude spectrum of this curve (lines). Superimposed is the spectrum (dotted) of the sawtooth wave, shown dotted in (a).

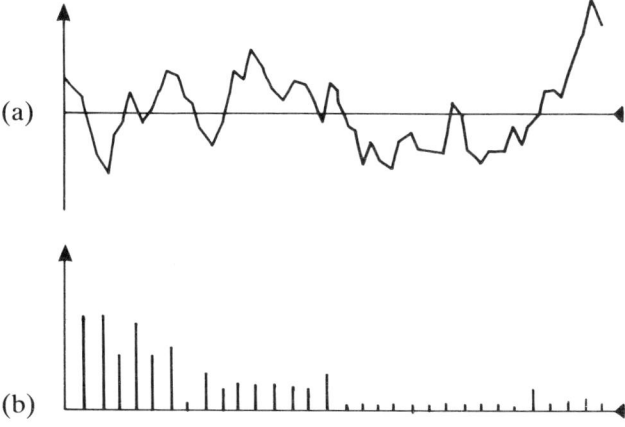

Fig. 14.9 — The effect of removing the trend from Fig. 14.8. (a) The index. (b) Its spectrum. The scale is the same as that in Fig. 14.8(b).

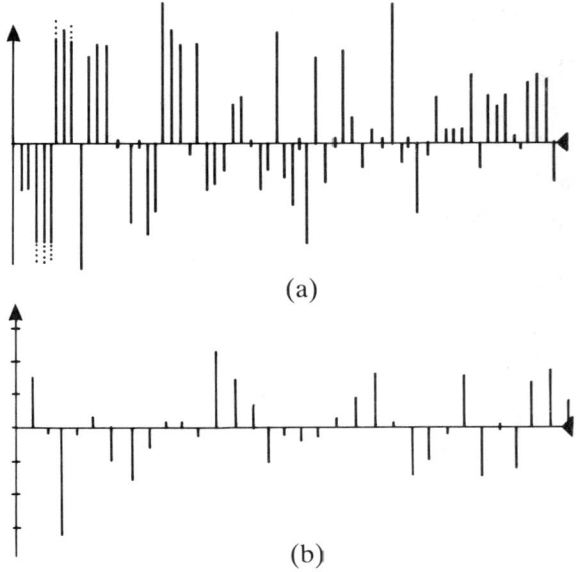

Fig. 14.10 — (a) Relative change in the FT-30 share index, 1974–84. (b) The discrete autocorrelation function of these data.

tion function of $y(n)$. This tells us the extent to which index changes in the near future are likely to resemble those of the recent past. The autocorrelation function of Fig. 14.10(a) is plotted in Fig. 14.10(b).

Unfortunately, every analytical approach seems to indicate that stock price indices have no regular Fourier structure. The index values are indistinguishable from the type of noise called *Brownian motion*. The autocorrelation function shown in Fig. 14.10(b), for instance, shows no (statistically) significant peaks, positive or negative.

Even discounting this complete dearth of results, there are two problems with this sort of analysis; one is insuperable, the other less so. The easier problem is the problem of aliasing. The index changes, not on a daily basis, but continuously, effectively every time a deal is made. Hence there will be a great deal of harmonic power at frequencies far above the Nyquist folding frequency (0.5 day^{-1}, if the index is sampled daily). These components confuse the results, and are partly to blame for the 'noise-like' spectrum of the output. This problem could, in principle anyway, be solved by low-pass filtering the index as it is generated, before sampling; though for this we would need a large computer and real-time access to all stock prices.

The more difficult problem concerns the nature of the stock pricing mechanism, the 'system' under study, as it were. There is absolutely no indication that this is governed by linear equations. As a result, Fourier analytical techniques are not likely to yield financially useful information. The best we are likely to get out of them is a statistical description of stock price movement, like the energy distribution of turbulence among eddies of various scales (Fig. 14.4).

Needless to say, neither of these problems have prevented the development of spectral 'betting systems' for equity investment.

Solutions to selected exercises

Chapter 1

2. $C = \sqrt{(A^2 + B^2)}$; $\varphi = \cos^{-1}(A/C)$.

4. $g/4\pi^2$, or about 0.25 m.

6. (a) $\sqrt{2/5}$ rad, or about 16°.
   (b) $\sqrt{2/50}$ rad, or about 1.6°.

10. (a) $A\exp(-\omega_0 t)\cos(4\varepsilon\omega_0 t + \varphi)$.
    (b) $A\exp(-(1 + 4\varepsilon)\omega_0 t) + B\exp(-(1 - 4\varepsilon)\omega_0 t)$.
    (Here $\omega_0 = \sqrt{(k/m)}$.)

11. $X(\omega) = \sqrt{\left\{\dfrac{1 - \dfrac{\omega^2}{\omega_0^2 Q^2}}{\left(1 - \dfrac{\omega^2}{\omega_0^2}\right)^2 + \dfrac{\omega^2}{\omega_0^2 Q^2}}\right\}}$

    (i) 1.   (ii) $\sqrt{(Q^2 + 1)}$.   (iii) 0.

Chapter 2

2. $(x-1)^2 \to \displaystyle\sum_{n\ \text{odd}} (4/n\pi)\sin(\tfrac{1}{2}n\pi x)$.

4. (a) $K = A_n = 0$; $B_n = (-1)^{n+1}(2/n\pi)$.

7. $\displaystyle\int_0^x (1 - \mathrm{III}(\tfrac{1}{2}t - \tfrac{1}{2}))\,dt = \mathrm{Saw}(x)$. By (2.19), $\mathrm{III}(t) \to \displaystyle\sum_n 2\cos(2n\pi t)$; by (2.22),

   $1 - \mathrm{III}(\tfrac{1}{2}t - \tfrac{1}{2}) \to -\displaystyle\sum_n 2\cos(n\pi t - n\pi) = \sum_n (-1)^{n+1} 2\cos(n\pi t)$; by (2.26), the

Solutions to selected exercises   313

integral should generate the series $\text{Saw}(x) \to \sum_n (-1)^{n+1}(2/n\pi)\sin(n\pi x)$, which indeed it does (see solution to Question 4 above).

10. $f(x) = 2/\pi + \sum_n [-4/\pi(4n^2-1)]\cos(2n\pi t)$.

12. If $f(x) = \sum_n Y_n \exp(2in\pi x/c)$ and $g(x) = \sum_n Z_n \exp(2in\pi x/c)$, then

$$\int_{-\infty}^{\infty} f(x)\overline{g}(x)\,dx = c\sum_n Y_n \overline{Z}_n.$$

15. $f_N(x) \to \sum_n (N\sin(2\pi n/N)/n\pi)\cos(2n\pi x)$.
When $n/N \ll 1$, $\sin(2\pi n/N) \approx 2\pi n/N$, and the coefficient is ~ 2. As $N \to \infty$, each coefficient tends to 2. This makes sense, because $f_N \to \text{III}$.

16. $\int_0^l y^2 \, dx = (l/2) \sum_n B_n^2$.

Chapter 3

1. $Ff(\omega) = 1/(\lambda + i\omega)$; Fig. S.1 shows the amplitude and phase spectra.

6. $x(t) = (-A/\omega_0)\sin(\omega_0 t)H(t)$ (where $\omega_0 = \sqrt{(k/m)}$).

7. $x(t) = a\alpha \int_{-\infty}^{\infty} \{e^{i\omega t}/((\gamma+i\omega)^2 + \alpha^2)(1-(\omega^2/\omega_0^2))\}\,d\omega$ (where $\omega_0 = \sqrt{(k/m)}$).

11. $F\tilde{\delta}_n(\omega) = (n/\sqrt{(2\pi)})(1/n)\sqrt{(2\pi)}N(\omega/n) = \exp(-\tfrac{1}{2}\omega^2/n^2)$.
$F\tilde{\delta}_n(\omega) \to 1(\omega)$ as $n \to \infty$.

14. $A\chi_{(a,b)}(\omega) = 2\sin[\tfrac{1}{2}(b-a)\omega]/\omega$.

16. $f(x)$ is odd; $Ff(\omega) = -i\omega\sqrt{(2\pi)}\exp(-\tfrac{1}{2}\omega^2) = -i\sqrt{(2\pi)}f(\omega)$.

18. $\theta(x,t) = F[F^{-1}A\chi_{(-1,1)}(\omega)\exp(-\omega^2 t/\kappa)](x)$
$= (A/2\pi)F[\sin(\omega)\exp(-\omega^2 t/\kappa)](x)$.
When t is large, $\exp(-\omega^2 t/\kappa)$ falls to zero rapidly away from $\omega = 0$. Here $\sin(\omega)/\omega \approx 1$; so $\theta(x,t) \approx (A/2\pi)F[\exp(-\omega^2 t/\kappa)](x) = (A/2)\sqrt{(\kappa/2\pi t)}\exp(-\kappa x^2/4t)$.

Chapter 4

1. $F_2 f(\alpha,\beta) = \sin(\alpha)\sin(\beta)/\alpha\beta$.

Solutions to selected exercises

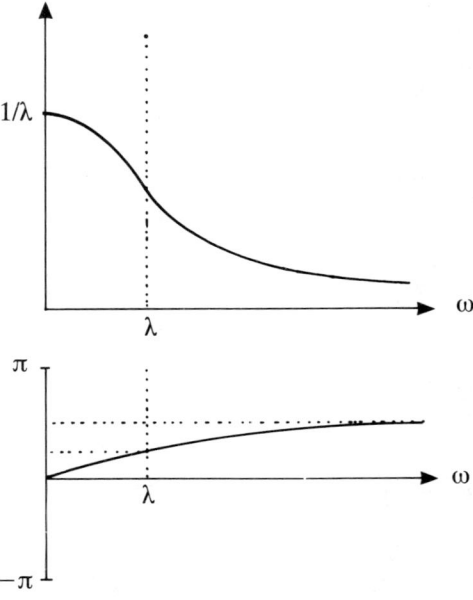

Fig. S.1

2. (i) $Sq(x+y) \to \sum_{n \text{ odd}} (4/n\pi)\cos(n\pi x)\sin(n\pi y) + \sum_{n \text{ odd}} (4/n\pi)\sin(n\pi x)\cos(n\pi y)$

(ii) $Sq(x+y) \to \sum_{n \text{ odd}} (-2i/n\pi)\exp(n\pi ix)\exp(n\pi iy)$.

6. $[iF_n f(\omega)]\omega = (F_n \dfrac{\partial f}{\partial x_1}(\omega), \ldots, F_n \dfrac{\partial f}{\partial x_n}(\omega))$ (or simply $F_n \nabla f(\omega)$).

8. The lattice generated by $(\pi 2/a, 0)$ and $(0, \pi 2/b)$.

9. The lattice generated by $(2\pi)(3, -1, -1)$, $(2\pi)(-1, 3, -1)$, and $(2\pi)(-1, -1, 3)$.

Chapter 5

1. $\sqrt{\left(\dfrac{\pi}{(a+b)}\right)} \exp\left\{\dfrac{-x^2}{(1/a)+(1/b)}\right\}$.

2. (i) $\left(\int f\right) I(\omega)$.

   (ii) $f(\omega)$.

3. $\sin((\alpha - \omega)b) + \sin((\alpha + \omega)b)$.

Solutions to selected exercises

5. $(f*f)(x) = 2\pi/(4+x^2) = \frac{1}{2}\pi f(\frac{1}{2}x)$.
7. $F\Lambda(\omega) = 4\sin^2(\frac{1}{2}\omega)/\omega^2$.
8. $(f*\Pi)(x) = 2(\sin\alpha/\alpha)\cos(\alpha x)$.
9. $(f*N)(x) = \sqrt{(2\pi)}N(\alpha)\cos(\alpha x)$.
10. $\int(\sin u/u)^2 \, du = \int|\frac{1}{2}F\Pi(u)|^2 \, du = \frac{1}{2}\pi\int\Pi(x)^2 \, dx = \pi$.
12. $W(\exp(-\alpha x^2)) = 1/(2\sqrt{\alpha})$; $W(F[\exp(-\alpha x^2)]) = \sqrt{\alpha}$; so the product is *equal to* $\frac{1}{2}$.

Chapter 6

3. It is a low-pass filter. $T(\omega) = 1/(1 - \omega^2 LC)$.
4. $T(\omega) = 1/(1 + i\omega RC)^2$, so $|T(\omega)| = 1/(1 + \omega^2 R^2 C^2)$.
   Hence the 3-dB point is $\omega_c = \sqrt{(\sqrt{2}-1)}/RC \approx 0.644/RC$.
5. $T(\omega) = 1/(1 + 3i\omega RC - \omega^2 R^2 C^2)$; so $|T(\omega)| = 1/\sqrt{(1 + 7\omega^2 R^2 C^2 + \omega^4 R^4 C^4)}$.
   Hence the 3-dB point is $\omega_c \approx 0.374/RC$.

Chapter 7

1. $\text{cov}(H,I) = 60$; $r(H,I) = 3/\sqrt{340} \approx 0.165$.
2. $\rho(f,g) = 1/\sqrt{3} \approx 0.577$.
4. $S[f](x) = \frac{1}{4}(x^4 - 2x^2 + 3)\exp(-\frac{1}{2}x^2)$.
5. It is harmonic (see equation (7.19)).
7. $S[f](x) = \exp(-\frac{1}{2}x^2)\left\{\left(\dfrac{\cos(\omega x) + \exp(-\frac{1}{2}x^2)}{(1 + \exp(-\frac{1}{2}x^2)}\right)\right\}$.

Chapter 8

4. $_2LH(s) = 1/s$.
6. $RN_2(r,0) = RN_2(r,0)$ (by circular symmetry)
$$= \int_{-\infty}^{\infty} \exp(-\frac{1}{2}r^2)\exp(-\frac{1}{2}s^2) \, ds$$
$$= \sqrt{(2\pi)}N(r).$$

Chapter 9

1. (i) $Z_{n,\tau}1(z) = (1 - z^{-n})/(1-z)$.
   (ii) $Z_{n,\tau}[e^{-ax}](z) = (1 - (ze^{-a\tau})^{-n})/(1 - (ze^{-a\tau}))$.
   (iii) $Z_{n,\tau}\delta(z)$ does not exist.

2. $D_{n,\tau}1(2\pi k/n\tau) = \begin{cases} 1 & \text{if } k = 0 \\ 0 & \text{if } k \neq 0. \end{cases}$
   Let the 'discrete delta function' $\delta_D(x)$ be

Solutions to selected exercises

$$\delta_D(x) = \begin{cases} 1 & \text{if } x = 0 \\ 0 & \text{if } x \neq 0 \end{cases}.$$

This function satisfies the relations

$$\sum_{r=0}^{n-1} \delta_D(r\tau) = 1$$

and

$$\sum_{r=0}^{n-1} f(r\tau)\delta_D(r\tau) = f(0) \qquad \text{for any function } f$$

(compare equation (2.17)). Its discrete transform is

$$D_{n,\tau}\delta_D(2\pi k/n\tau) = 1/n \text{ for all } k.$$

9. $|T(2\pi k/n\tau)| = |\tfrac{1}{2} + \tfrac{3}{4}\cos(2\pi k/n) + \tfrac{1}{4}\cos(6\pi k/n)|$ (see Fig. S.2).

It is a low-pass filter.

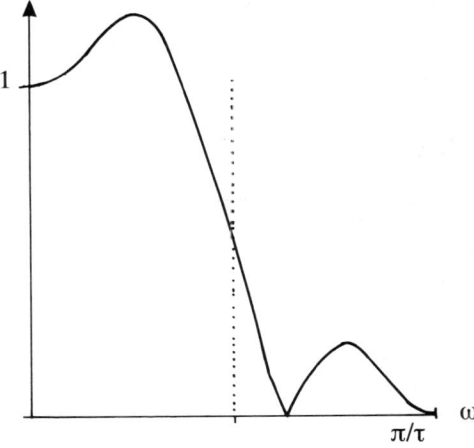

Fig. S.2

Bibliography

Fourier series
Lanczos, C., *Discourse on Fourier Series*, Oliver & Boyd, Edinburgh, 1966.

Transforms
Bracewell, Ronald N., *The Fourier Transform and Its Applications* 2nd edition, McGraw-Hill, New York, 1986.
Deans, Stanley R., *The Radon Transform and Some of its Applications*, Wiley, New York, 1983.
Körner, T. W., *Fourier Analysis*, Cambridge University Press, Cambridge, 1988.
Lighthill, M. J., *Introduction to Fourier Analysis and Generalised Functions*, Cambridge University Press, Cambridge, 1958.
Watson, A. J., *Laplace Transforms*, van Nostrand-Reinhold, New York, 1981.

Signal processing and filters
Lynn, Paul A., *The Analysis and Processing of Signals* 2nd edition, Macmillan, London, 1982.
Thompson, William, J., *Computing in Applied Science*, John Wiley, New York, 1984.

Mathematical applications
Churchill, Ruel V. and Brown, James Ward, *Fourier Series and Boundary Value Problems* 4th edition, McGraw-Hill, New York, 1987.
Davies, B. *Integral Transforms and Their Applications*, Springer–Verlag, New York, 1978.
Dym, H. and McKean, H. P., *Fourier Series and Integrals*, Academic Press, New York, 1972.

Bibliography

Quantum mechanics

Bowler, M. G., *Nuclear Physics*, Pergamon Press, Oxford, 1973.

Mathews, P. M. and Venkatesan, K., *A Textbook of Quantum Mechanics*, Tata–McGraw-Hill, New Delhi, 1977.

Pearson, D. B., *Quantum Scattering and Spectral Theory*, Academic Press, London, 1980.

Rae, Alastair, M., *Quantum Mechanics* 2nd edition, Adam Hilger, Bristol, 1986.

Optics

Brown, J. G., *X-rays and their Applications*, Plenum/Rosetta, New York, 1966.

Chamberlain, G. J. and Chamberlain, D. G., *Colour: its measurement, computation, and application*, Heyden, London, 1980.

Hutley, M. C., *Diffraction Gratings*, Academic Press, London, 1982.

Steel, W. H., *Interferometry* 2nd edition, Cambridge University Press, Cambridge, 1983.

Steward, E. G., *Fourier Optics: an introduction* 2nd edition, Ellis Horwood, Chichester, 1987.

Spectroscopic methods

Akitt, J. W., *NMR and Chemistry* 2nd edition, Chapman and Hall, London, 1983.

Blackburn, James A. (ed.), *Spectral Analysis: methods and techniques*, Marcel Dekker Inc., New York, 1970.

Demey, Ronald C. and Sinclair, Roy, *Visible and Ultra-violet Spectroscopy* (Analytical Chemistry by Open Learning), John Wiley & Sons, Chichester, 1987.

Hill, H. C., *Introduction to Mass Spectroscopy*, Heyden & Son, London, 1971.

Howarth, Oliver, *Theory of Spectroscopy*, Thomas Nelson & Sons Ltd., London, 1973.

Williams, Dudley, H. and Fleming, Ian, *Spectroscopic Methods in Organic Chemistry* 3rd edition, McGraw-Hill, London, 1980.

Crystallography

Ladd, M. F. C. and Palmer, R. A., *Theory and Practice of Direct Methods in Crystallography*, Plenum Press, New York, 1980.

Ramachandran, G. N. and Srinivasan, R., *Fourier Methods in Crystallography*, Wiley Interscience, New York, 1970.

Sensory analysis

Barlow, H. B. and Mollon, J. D., *The Senses* (Cambridge Texts in Physiological Sciences), Cambridge University Press, Cambridge, 1982.

Brindley, G. S., *Physiology of the Retina and Visual Pathway* 2nd edition (Monographs of the Physiological Society), Edwards Arnold, London, 1970.

Stein, J. F., *An Introduction to Neurophysiology*, Blackwell Scientific Publications, Oxford, 1982.

Morphogenesis

Murray, J. D., 'How the Leopard gets its spots', in *Scientific American*, vol. 258 no. 3, March 1988.

Medical imaging

Kreel, Lewis and Steiner, Robert E., *Medical Imaging: a basic course*, HM + M, Aylesbury (Bucks), 1979.

Natterer, F., *The Mathematics of Computerised Tomography*, B. G. Teubner, (Stuttgart) and John Wiley & Sons, Chichester, 1986.

Water

Crapper, G. D., *Introduction to Water Waves*, Ellis Horwood, Chichester, 1984.

Leslie, D. C., *Developments in the Theory of Turbulence*, Oxford University Press, London, 1973.

Schowalter, William R., *Mechanics of Non-Newtonian Fluids*, Pergamon, Oxford 1978.

Shaw, Ronald, *Wave Energy: a design challenge*, Ellis Horwood, Chichester, 1982.

Earth science

Bullen, K. E. & Bolt, Bruce, A., *An Introduction to the Theory of Seismology* 4th edition, Cambridge University Press, Cambridge, 1985.

Chang, J. (ed.), *General Circulation Models of the Atmosphere* (Methods in Computational Physics), Academic Press, New York, 1977.

Leggett, Jeremy, 'The noise of the Bomb', in *New Scientist*, no. 1482, 14 Nov. 1985.

Paterson, W. S. B., *The Physics of Glaciers* 2nd edition, Pergamon, Oxford, 1981.

Oscillating systems

Baldock, T. R. and Bridgeman, T., *The Mathematical Theory of Wave Motion*, Ellis Horwood, Chichester, 1987.

Meirovitch, Leonard, *Elements of Vibration Analysis* 2nd edition, McGraw-Hill, New York, 1986.

Economics

Cooter, Paul H., *The Random Character of Stock Market Prices*, MIT Press, Cambridge (Mass), 1964.

Index

absolute coherence, *see* coherence
acoustics, 128–129, 268–269
additive noise, 97, 151
Airy disc (in telescope image), 260–261
aliasing, 205
almost-periodic functions, 176, 178
almost-periodic noise, 152
AMPLITUDE, 21
  demodulation, 162–163
  modulation (AM), 137–138
  -phase expression of Fourier series, 52, 66–67
  response function, 34, 145, 165, 216–218
  spectrum, 48, 52, 80, 202
analogue filters, 163–166
angular frequency, 22, 98
  critical (=cut-off), 159, 165
  vector, 113
  see also frequency
antiphase, 27
aperture synthesis, 262
approximation of a function, 67–68, 74, 183–184
approximation, harmonic, 22–24
area, greatest, in curve of unit perimeter, 225–229
autocorrelation, 175–177, 178, 308–310
average, running, as low-pass filter, 215–216
averaging, signal
  x-domain, 153–155
  ω-domain, 155–157

band-pass and band-stop filters, 160, 165, 215, 284–285
bandwidth, 143–144, 151
baseline (of an interferometer), 261–262
'bed of nails' function, 119
bell curve, 80–81
  n-dimensional, 117–118
Bessel function, 121–122, 184–187, 195–196
  see also Fourier–Bessel series, Hankel transform
boundary conditions, 37
broadcasting, 143–144, 149, 160
  see also amplitude modulation, frequency modulation

broadening of spectral lines, 251–252, 277
Butterworth-type filter, 217–218

central limit theorem, 232–235
characteristic function:
  of a random variable, 232
  of an interval, 81
characteristic impedance, 267
charge density profile (of an atomic nucleus), 254–256
chemical effect (NMR spectroscopy), 274
circularly symmetric waves, 187–191
classical Fourier series, 44
cochlea, *see* ear
coefficients, Fourier, calculation of, 44–46, 51–52, 112, 114–115
coherence, 179–180
  of sampled functions, 214
comb filters, 161
complex formulation of Fourier series, 66–67
computers, and Fourier methods, 201, 208–210
conjugate variables (and uncertainty), 248
consensus cross-spectral power (CCSP), 179–180
consensus power spectrum (CPS), 156, 180
constant character, functions of, 80, 151, 174–177
continuity
  equation of (for fluid motion), 293
  of Fourier transforms, 93–94
convergence of Fourier series, 55, 70–73
CONVOLUTION, 128–149, 177–178, 258–259, 260, 261, 279, 306
  and amplitude modulation, 137–138
  and truncation, 135–137
  as 'smoothing', 133–135, 145, 148, 149
  by the eye, 281–282
  discrete, 204, 210–212
  distribution of a sum of random variables, 232–233
  echoes, 128–129
  filters, digital, 215–217
  in optics, 258–259, 260, 261, 279
  theorem, 132, 145–146, 233, 244
    discrete, 212
    n-dimensional, 145–146

vector, 145–146
CORRELATION
   auto-, 175–177, 178, 308–310
   coefficient
      of two random variables, 169
      of two functions, 169
   cross-, 172, 174–175, 178, 180
   delayed, 171
cos–sin expression of Fourier series, 52
cosine transform, 77
   one-sided, 195
covariance, 168
   density function, 298
   function, 174–175, 177–178, 180
critical (angular) frequency (of a filter), 159, 165
cross-correlation function, 172, 174–175, 178, 180
cross-spectral power, 140
   consensus, (CCSP), 179–180
crystallography, 258, 277–279
   see also lattice
cut-off (angular) frequency (of a filter), 159, 165

damped oscillator, 30–32, 42
damping, 28–32
   critical, 32
   ideal, 29
damping constant, 29
decay of Fourier coefficients, at large n, 64–65, 71
decibels, 153
deconvolution, 145, 261, 285
degrees of freedom (of vibrating system), 25, 35
delayed correlation, 171
delta 'function', 55, 85, 105, 131
   discrete, 219, 315–316
   n-dimensional, 118
demodulation, 162–163
density
   charge density profile of a nucleus, 254–256
   energy (flux) density, 63, 253, 257, 263, 298–299
   power spectral density function (PSDF), 80, 151–153, 304–305
   (co)variance density function, 298
derivative theorem
   for Fourier series, 62
   for Fourier transforms, 91–93, 98, 229
differential equations, transform methods for solving, 94–97, 229–231
diffraction, 247, 256–259, 277
   Fraunhofer, as Fourier analyser, 257
   grating, 259
digital computers, and Fourier methods, 201, 208–210
digital filters, 214–219
Dirac delta 'function', *see* delta 'function'
Dirichlet kernel, 68, 206
DISCRETE convolution, 204, 210—212
   theorem, 212
discrete delta function, 219, 315—316
discrete Fourier transform (DFT), 202
   of a sinsoid, 206–208, 220
discrete inversion theorem, 202–203

discrete transfer function, 212, 216–218
dispersion relation, for water waves, 295–296
distribution
   of a sum of variables, 232–233
   of the mean of a sample, 233–235
domain
   finite, 201–202
   frequency, 96, 300–301
   x- and ω-, 96, 153–157, 285
double oscillator, 24–28
double-pass filter, 166
drumskin, vibrations of a, 187–191

ear, 282–286
   discrimination and sensitivity of the, 284–285
   inner, as Fourier analyser, 283–284
echoes, 128–129
economic data, Fourier analysis of, 308–311
effective width (of a function), 141
electrical circuits, 163–166, 264–268
electron cloud in an atom, 248–250
electron density map of a crystal, 279
electron microscope, 259
electron scattering, *see* scattering
energy (density) of a wave, 63, 253, 257, 263, 298–299
EQUATION
   of continuity (for a fluid), 293
   of motion, 21, 23, 25–26, 30, 32, 37
   harmonic, 19–20
   heat (=diffusion), 94
   linear system of, 28
   Navier–Stokes, 297, 301
   Parseval's, 63–64, 67, 74
   Schrödinger's, 241, 147–248, 253
   transform method for solving differential, 94–97, 229–231
   wave
      one-dimensional, 38
      two-dimensional, 127, 188–189
even functions, 53, 91
expansion, Fourier, 44, 46, 113
extension oscillator, 20–22
eye
   convolution by, 281–282
   spatial, temporal, and spectral properties of, 280–282

fast Fourier transform (FFT), 209–210
   programme, 211
Fejér kernel, 75–76
FILTER
   accidental filtering, 158, 166
   analogue, 163–166, 264–268
   band-pass and band-stop, 160, 165, 215, 284–285
   Butterworth-type, 217–218
   comb, 161
   convolution, 129–130, 215–217, 281–282
   critical (=cut-off) frequency of, 159, 165
   digital, 214–219
   double-pass, 166

Index

dynamic, 158
high-pass, 159, 165, 215
instability, 218
limiting, 157–158
linear invariant, 129–130, 158
low-pass, 159, 163–165, 205, 215–218
(frequency-)mask, 158–162, 218
mechanical, 166–167
mouth cavity as (in speech), 286
'operator-convenient', 157–158
phase response of, 218–219
physical and non-physical, 130, 157, 166
recursive and nonrecursive, 217
running average, 215–216
finite domains, 201–202
finite orthogonality conditions, 203
finite Rayleigh–Plancherel theorem, 203–204, 208
folding frequency, 205, 310
forced oscillation, 32–35, 58–59, 290
forecasting, weather, 299–301
FOURIER SERIES, 44–76, 223—229, 231, 252
as a limit of transforms, 99–101
calculation of coefficients, 44–46, 51–52
classical, 44
complex formulation of, 65–67
generalized, 182–184, 186–191, 199–200, 250, 300–301
n-dimensional, 109–113, 114–114
truncated, 67–70, 75–76, 300–301
FOURIER TRANSFORM, 77–105, 229, 232–234, 244–245, 248, 254, 257–279, 264, 265, 273, 274, 275, 278, 287, 296–297, 304–305, 308
discrete, (DFT), 202–203
fast (FFT), 209–210
inversion of, 78, 99, 102–103
n-dimensional, 108–109
table of, 90
vector, 113–114
Fourier–Bessel series, 186–191
Fourier cosine transform, 77
one-sided, 195
Fourier frequency transform, 98, 133
Fourier–Legendre series, 183, 199–200
Fourier sine transform, 77
one-sided, 195
Fraunhofer diffraction, 257
Fraunhofer lines, in optical spectra, 276
FREQUENCY, 22
domain, 96, 300–301
modulation (FM), 138–139
critical/cutoff (of a filter), 159, 165
(Nyquist) folding, and aliasing, 295, 310
natural, 27
see also angular frequency
frosted glass, 146–148
FUNCTION
amplitude response, 34, 145, 165, 216–218
approximation of, 67–68, 74, 183–184
autocorrelation, 175–177, 178, 308–310
'bed of nails', 119
bell curve, 80–81

n-dimensional, 117–118
Besel, 121–122, 184–187, 195–196
characteristic
of an interval, 81
of a random variable, 232
coherence of sampled, 214
correlation of two, 169
cross-correlation, 172, 174–175, 178, 180
covariance (density), 174–175, 177–178, 180, 298
(Dirac) delta, 55, 85, 105, 131
discrete, 219, 315–316
n-dimensional, 118
effective width of, 141
even, 53, 91
families of orthogonal, 183–184, 186–187
'grating', 120–121
Heaviside, 84, 104
impulse response, 128, 131, 158, 212
impulsive, 55–59, 85, 118–122
lattice, 123
odd, 53, 91
of constant character, 80, 151, 174–177
periodic, 49, 85–89, 109–113, 122–123, 286
power spectral density (PSDF), 80, 151–153, 304–305
'row of spikes', 120–121
rectangle (=pi), 105, 148
sampled, 201–202
sampling, 57–59, 88–89
n-dimensional, 119–120
smooth, 70, 78, 99, 109, 113
n-fold, 149
step, 81–82
transfer, 144–146, 212, 216–218
triangle (=lambda), 105, 148
unity, 53, 104, 119
variance (density), 175, 177, 178, 263, 298
window, 212–213
see also polynomials
fundamental (mode of vibration), 39

generalized Fourier series, see Fourier series
Gibb's phenomenon, 68–70
glacier beds, roughness of, 301–303
grating diffraction, 259
grating function, 120–121
guitar strings, vibration of, 46–49, 73

Hankel transforms, 195–197
harmonic approximation, 22–24
harmonic equation, 19–20
harmonic motion, 21
harmonic solutions, see separation of variables
harmonics
of fundamental frequency, 39
spherical, 250, 300–301
Hartley transform, 195
hearing, see ear
heat conduction, in a wire, 94–96, 104–105
Heaviside function, 84, 104
Hermite polynomials, 184

high-pass filter 159, 165, 215
holography, 258
Hooke's law, 20
Huygens' principle, 256, 258, 259, 292

image reconstruction, 147
impedance
    characteristic (of a transmission line), 267
    mechanical, 268–269
    of electrical circuit elements, 264–265
impedance matching, 268, 282
impulse function, *see* delta function
impulsive functions, 55–59, 85, 118–122
impulse response function, 128, 131, 158, 212
infra-red (IR) spectroscopy, 275–276
initial conditions, 22
instability (of a digital filter), 218
integral theorem (for Fourier series), 62–63
interferometry, 261–264, 276
inversion
    of discrete Fourier transforms, 202–203
    of Fourier transforms, 78, 99, 102–103, 108–109, 114
    of Laplace transforms, 193
isoperimetric problem, 225–229

kernel
    Dirichlet, 68, 206
    Fejér, 75–76
    of a transform, 195

lambeda (=triangle) function, 105, 148
Laplace transforms, 191–193, 199
    solution of differential equations by, 230–231
    table of, 194
lasers, 251–252
lattice, 123–126, 278–279
    dual, 125
    see also grating
lattice function, 123
Legendre polynomials, 182–183, 198
length of a sample, and resolution of components, 205–206, 207–208
lens, focusing action of a, 259–260
limiting, filtering by, 157–158
line broadening, spectral, 251–252, 277
linear invariant filter, 129–130, 158
linear operators, 89–90, 98, 115, 130, 202
linear phase response (of a digital filter), 219
linear system of equations, 28
low-pass filter, 159, 163–165, 205, 215–218

mask filter, 158–162, 218
mass spectrometry, 270—273
matching of impedances, 268, 282
mean of a sample, distribution of a, 233–235
mechanical filters, 166–167
mechanical impedance, 268–269
medicine
    ECGs and EEGs, 291
    tomography, 197–198, 291–292
Mellin transform, 192–194

membrane, 283
    circular, 187–191
    rectangular, 126–127
meteorology
    Greenwich temperatures, 96–97, 161–162
    spectral methods in forecasting, 299–301
mode, natural, of oscillation, 27, 39, 251
    see also vibration
mode-coupling, in nonlinear systems, 276, 297, 299
modulation
    amplitude, 137–138
    frequency, 138–139
morphogenesis, as putative wave phenomenon, 289–291
motion sickness, 285–286

n-DIMENSIONAL
    bell curve, 117–118
    convolution, 145
    delta function, 118
    Fourier series, 109–113, 114–115
    Fourier transforms, 108–109
    impulsive functions, 118–122
    see also vector
n-fold
    -periodic function, 109–113
    -smooth function, 65, 93, 149
natural (angular) frequency, 27
natural mode of oscillation, 27, 39, 251
    see also vibration
Navier–Stokes equation, and turbulence, 297–301
NOISE
    additive, 97, 151
    almost-periodic, 152
    narrowband, 152, 166
    (pure) pink, 151–152
    (pure) white, 151
    wideband, 152, 310
see also filters
nonrecursive filters, 217
normal distribution, 80, 235
nuclear magnetic resonance (NMR) spectroscopy, 273–275
nuclear weapons testing, and seismometry, 306–307
nuclei, atomic, 254–256, 279
Nyquist folding frequency, 205, 310

odd functions, 53, 91
operators, 77, 129
    see also transforms
orthogonal functions, 182–184, 186–187
orthogonality relations, 25, 183–184, 186–187, 203
OSCILLATOR
    damped, 30—32, 42
    double, 24–28
    extension, 20–22
    forced, 32–35, 58–59, 290
    see also vibration

Index

packing theorem, 204
Parseval's equation, 63–64
pass band (of a filter), 158
peaks in a spectrum, interpretation of, 96–97, 206–208
pendulum, 22–23, 41
period
  of harmonic motion, 22
  of a periodic function, 49
  vector, 123, 126
  x_r-, 109, 114–115
periodic component, 96–97, 161–162
periodic function, 49, 85–89, 122–123, 286
phase, 21, 278, 285
phase response (of a digital filter), 218–219
pi (=rectangle) function, 105, 148
polynomials
  Hermite, 184
  Legendre, 182–183, 198
  see also function
POWER
  cross-spectral, 140
    consensus (CCSP), 179–180
  wave power, extraction of, 297
power spectral density function (PSDF), 80, 151–153, 304–305
power spectrum, 80, 178, 202, 278, 285
  consensus, (CPS), 156, 180
power transmission, through an electrical load, 268
power theorem, 93, 140–141
prime number theorem, 235–246

Q-value (selectivity), 31
QUANTUM THEORY
  electrons in atoms, 248—250
  lasers, 250–252
  scattering, 252–256
  Schrödinger equation, 241, 247–248, 253
  uncertainty principle, 248, 251–252, 255
quarks, scattering evidence for, 255–256
quasicrystals, 279

Radon transform, 197–198, 277–278
random variables, 169, 232–235
Rayleigh–Plancherel theorem, 141, 203–204, 208, 299
rectangular membrane, vibration of a, 126–127
recursive filters, 217–218
resolution
  angular, of a telescope, 259–261
  frequency, and sample length, 205–206, 207–208
  time, of a band-limited signal, 143–144
Riemann–Lebesgue lemma
  for Fourier series, 64–65, 71
  for Fourier transforms, 93, 245
  roughness, 301, 305
'row of spikes' function, 120–121
running average, as a low pass filter, 215–216

sample
  coherence of a sample of functions, 214
  distribution of a sample mean, 233–235
  length and frequency resolution, 205–206
  of a function, 201–202
sampling function, 57–59, 88–89
  n-dimensional, 119–120
sawtooth wave, 73
scattering
  of particles, 252–256
  of X-rays, and crystallography, 258, 277–279
Schrödinger equation, 231, 247–248, 253
Schwarz' inequality, 142
seismology, 305–307
selectivity (Q-value), 31
separation of variables (in solution of partial differential equations), 37, 94–95, 189–190, 249–250, 295
series
  Fourier, *see* Fourier series
  sum of a, by Fourier methods, 223–225
  Taylor, and Fourier–Legendre series, 183
share prices, 308–311
shift theorem, 202
signal, 150
  averaging, 153–157
signal-to-noise ratio, 152–153
similarity theorem, 202
sine transform, 77
  one-sided, 195
sinusoid, discrete Fourier transform of a, 206–208, 220
smooth function, 70, 78, 99, 109, 113
  n-fold-, 65, 93, 149
smoothing (effect of convolution), 133–135, 145, 148, 149
spatial analysis, and image reconstruction, 147
spatial analysis in the eye, 290–282
spatial domain, 96, 106
spatial variable, 106
spectroscopy
  infra-red, 275–276
  mass, 270–273
  NMR, 273–275
  visible light, 276–277
SPECTRUM
  amplitude, 48, 52, 80, 202
  interpretation of peaks in a, 96–97, 206–208
  phase, 52, 80, 202
  power, 80, 202, 264
    consensus, (CPS), 156, 180
speech analysis, 286–289
spherical harmonics, 250, 300–301
spring constant, 20
square wave, 50, 54–55, 68–70, 88–89
STATISTICS
  central limit theorem (mean of a sample), 232–235
  coherence, 179–180, 214
  correlation coefficient, 169
  covariance, 168
  distribution of a sum of variables, 232–234
  variance, 168

step function, 81–82
stop band (of a filter), 158–159
stretch theorem, 204
string, vibrating, 35–41, 46–49, 73
sum
 of a series, by Fourier methods, 223–225
 of random variables, distribution of, 232–234
superposition, principle of, 28

Taylor series, 183
telescope, 259–261
THEOREM
 central limit, 232–235
 convolution, 132, 145–146, 233, 244
 derivative
  for Fourier series, 62
  for Fourier transforms, 91–93, 98, 229
 integral, for Fourier series, 62–63
 inversion
  discrete, 202–203
  Fourier transform, 78, 99, 102–103
  Laplace, 193
  vector transform, 108–109, 114
 packing, 204
 power, 93, 140–141
 prime number, 235–246
 Rayleigh–Plancherel, 141, 203–204, 208, 299
 Riemann–Lebesgue lemma, 64–65, 71, 93, 245
 shift, 202
 similarity, 202
 stretch, 204
 uncertainty, 141–143, 288–289
 Wiener–Khintchine, 178, 298
3-dB point, 165
tomography, 197, 198, 291–292
transfer function (of a filter), 144–145, 212, 216–218
TRANSFORM
 cosine, 77, 195
 Fourier, see Fourier transform
 Hankel, 185–197
 Hartley, 195
 kernel of a, 194–195
 impulsive, 122–123
 Laplace, 191–193, 199, 230–231
 Mellin, 193–194
 Radon, 197–198, 277–278
 self- and conjugate-inverse, 195
 sine, 77, 195
 transform pair, 79
 z-, 201–202
transient, 34
transmissibility, 34
transmission line, 266–268
triangle wave, 53–54, 87, 89
truncation
 of a function, 81–82, 135–137, 148
 of Fourier series, 67–70, 75–76, 300–301
turbulence in a fluid, Fourier description of, 297–299

uncertainty principle, 248, 251–252, 255
uncertainty theorem, 141–143, 288–289
unit cell of a crystal, 278
unity function, 53, 104
 n-dimensional, 119

variance, 168
 density function, 298
 function, 175–177, 178, 263
vector
 Fourier vector transform, 113–114
 angular frequency, 113
 convolution, 145–146
 period, 123–126
 see also n-dimensional
VIBRATION
 of chemical bonds, 275
 of circular membrane, 187–191
 of double oscillator, 24–28
 of extension oscillator, 20–22
 of machinery, monitoring and diagnosis, 307–308
 of rectangular membrane, 126–127
 of string, 35–41, 46–49, 73
 of vocal cords, 286
 see also oscillator
viscosity, 297–305
vision, see eye
vowels, harmonic structure of, 287–288

water waves, 293–297
WAVE
 fully rectified, 74
 half rectified, 74
 wave equation:
  one-dimensional, 38
  two-dimensional, 127, 188–189
 wave packet, 247
 wave particle, duality, 247
 wave power, extraction of, 297
 wave speed, 78, 267, 295–296
weather forecasting, 299–301
white noise, 151
Wiener–Khintchine theorem, 178, 298
window function, 212–213
wraparound, 202, 215–216, 308
 and discrete convolution, 211–212

z-transform, 201–202

3-dB point, 165

Mathematics and its Applications

Series Editor: G. M. BELL, Professor of Mathematics, King's College London, University of London

Gardiner, C.F.	Algebraic Structures
Gasson, P.C.	Geometry of Spatial Forms
Goodbody, A.M.	Cartesian Tensors
Goult, R.J.	Applied Linear Algebra
Graham, A.	Kronecker Products and Matrix Calculus: with Applications
Graham, A.	Matrix Theory and Applications for Engineers and Mathematicians
Graham, A.	Nonnegative Matrices and Applicable Topics in Linear Algebra
Griffel, D.H.	Applied Functional Analysis
Griffel, D.H.	Linear Algebra and its Applications: Vol. 1, A First Course; Vol. 2, More Advanced
Guest, P. B.	The Laplace Transform and Applications
Hanyga, A.	Mathematical Theory of Non-linear Elasticity
Harris, D.J.	Mathematics for Business, Management and Economics
Hart, D. & Croft, A.	Modelling with Projectiles
Hoskins, R.F.	Generalised Functions
Hoskins, R.F.	Standard and Nonstandard Analysis
Hunter, S.C.	Mechanics of Continuous Media, 2nd (Revised) Edition
Huntley, I. & Johnson, R.M.	Linear and Nonlinear Differential Equations
Irons, B. M. & Shrive, N. G.	Numerical Methods in Engineering and Applied Science
Ivanov, L. L.	Algebraic Recursion Theory
Johnson, R.M.	Theory and Applications of Linear Differential and Difference Equations
Johnson, R.M.	Calculus: Theory and Applications in Technology and the Physical and Life Sciences
Jones, R.H. & Steele, N.C.	Mathematics in Communication Theory
Jordan, D.	Geometric Topology
Kelly, J.C.	Abstract Algebra
Kim, K.H. & Roush, F.W.	Applied Abstract Algebra
Kim, K.H. & Roush, F.W.	Team Theory
Kosinski, W.	Field Singularities and Wave Analysis in Continuum Mechanics
Krishnamurthy, V.	Combinatorics: Theory and Applications
Lindfield, G. & Penny, J.E.T.	Microcomputers in Numerical Analysis
Livesley, K.	Mathematical Methods for Engineers
Lord, E.A. & Wilson, C.B.	The Mathematical Description of Shape and Form
Malik, M., Riznichenko, G.Y. & Rubin, A.B.	Biological Electron Transport Processes and their Computer Simulation
Massey, B.S.	Measures in Science and Engineering
Meek, B.L. & Fairthorne, S.	Using Computers
Menell, A. & Bazin, M.	Mathematics for the Biosciences
Mikolas, M.	Real Functions and Orthogonal Series
Moore, R.	Computational Functional Analysis
Moshier, S.L.B.	Methods and Programs for Mathematical Functions
Murphy, J.A., Ridout, D. & McShane, B.	Numerical Analysis, Algorithms and Computation
Nonweiler, T.R.F.	Computational Mathematics: An Introduction to Numerical Approximation
Norcliffe, A. & Slater, G.	Mathematics of Software Construction
Ogden, R.W.	Non-linear Elastic Deformations
Oldknow, A.	Microcomputers in Geometry
Oldknow, A. & Smith, D.	Learning Mathematics with Micros
O'Neill, M.E. & Chorlton, F.	Ideal and Incompressible Fluid Dynamics
O'Neill, M.E. & Chorlton, F.	Viscous and Compressible Fluid Dynamics
Page, S. G.	Mathematics: A Second Start
Prior, D. & Moscardini, A.O.	Model Formulation Analysis
Rankin, R.A.	Modular Forms
Scorer, R.S.	Environmental Aerodynamics
Shivamoggi, B.K.	Stability of Parallel Gas Flows
Smith, D.K.	Network Optimisation Practice: A Computational Guide
Srivastava, H.M. & Manocha, L.	A Treatise on Generating Functions
Stirling, D.S.G.	Mathematical Analysis
Sweet, M.V.	Algebra, Geometry and Trigonometry in Science, Engineering and Mathematics
Temperley, H.N.V.	Graph Theory and Applications
Temperley, H.N.V.	Liquids and Their Properties
Thom, R.	Mathematical Models of Morphogenesis
Toth, G.	Harmonic and Minimal Maps and Applications in Geometry and Physics
Townend, M. S.	Mathematics in Sport
Townend, M.S. & Pountney, D.C.	Computer-aided Engineering Mathematics
Trinajstic, N.	Mathematical and Computational Concepts in Chemistry
Twizell, E.H.	Computational Methods for Partial Differential Equations
Twizell, E.H.	Numerical Methods, with Applications in the Biomedical Sciences
Vince, A. and Morris, C.	Mathematics for Computing and Information Technology
Walton, K., Marshall, J., Gorecki, H. & Korytowski, A.	Control Theory for Time Delay Systems
Warren, M.D.	Flow Modelling in Industrial Processes
Wheeler, R.F.	Rethinking Mathematical Concepts
Willmore, T.J.	Total Curvature in Riemannian Geometry
Willmore, T.J. & Hitchin, N.	Global Riemannian Geometry

Statistics, Operational Research and Computational Mathematics
Editor: B. W. CONOLLY, Emeritus Professor of Mathematics (Operational Research), Queen Mary College, University of London

Abaffy, J. & Spedicato, E.	ABS Projection Algorithms: Mathematical Techniques for Linear and Nonlinear Equations
Beaumont, G.P.	**Introductory Applied Probability**
Beaumont, G.P.	Probability and Random Variables
Conolly, B.W.	Techniques in Operational Research: Vol. 1, Queueing Systems
Conolly, B.W.	Techniques in Operational Research: Vol. 2, Models, Search, Randomization
Conolly, B.W.	Lecture Notes in Queueing Systems
Conolly, B.W. & Pierce, J.G.	Information Mechanics: Transformation of Information in Management, Command, Control and Communication
French, S.	Sequencing and Scheduling: Mathematics of the Job Shop
French, S.	Decision Theory: An Introduction to the Mathematics of Rationality
Griffiths, P. & Hill, I.D.	Applied Statistics Algorithms
Hartley, R.	Linear and Non-linear Programming
Jolliffe, F.R.	Survey Design and Analysis
Jones, A.J.	Game Theory
Kapadia, R. & Andersson, G.	Statistics Explained: Basic Concepts and Methods
Lootsma, F.	Operational Research in Long Term Planning
Moscardini, A.O. & Robson, E.H.	Mathematical Modelling for Information Technology
Moshier, S.L.B.	Mathematical Functions for Computers
Oliveira-Pinto, F.	Simulation Concepts in Mathematical Modelling
Ratschek, J. & Rokne, J.	New Computer Methods for Global Optimization
Schendel, U.	Introduction to Numerical Methods for Parallel Computers
Schendel, U.	Sparse Matrices
Schmi, N.S.	Large Order Structural Eigenanalysis Techniques: Algorithms for Finite Element Systems
Späth, H.	Mathematical Software for Linear Regression
Stoodley, K.D.C.	Applied and Computational Statistics: A First Course
Stoodley, K.D.C., Lewis, T. & Stainton, C.L.S.	Applied Statistical Techniques
Thomas, L.C.	Games, Theory and Applications
Whitehead, J.R.	The Design and Analysis of Sequential Clinical Trials